IN CHINA'S BACKYARD

The **ISEAS – Yusof Ishak Institute** (formerly Institute of Southeast Asian Studies) is an autonomous organization established in 1968. It is a regional centre dedicated to the study of socio-political, security, and economic trends and developments in Southeast Asia and its wider geostrategic and economic environment. The Institute's research programmes are grouped under Regional Economic Studies (RES), Regional Strategic and Political Studies (RSPS), and Regional Social and Cultural Studies (RSCS). The Institute is also home to the ASEAN Studies Centre (ASC), the Nalanda-Sriwijaya Centre (NSC), and the Singapore APEC Study Centre.

ISEAS Publishing, an established academic press, has issued more than 2,000 books and journals. It is the largest scholarly publisher of research about Southeast Asia from within the region. ISEAS Publishing works with many other academic and trade publishers and distributors to disseminate important research and analyses from and about Southeast Asia to the rest of the world.

IN CHINA'S BACKYARD

Policies and Politics of
Chinese Resource Investments in Southeast Asia

Edited by Jason Morris-Jung

ISEAS YUSOF ISHAK
INSTITUTE

First published in Singapore in 2018 by
ISEAS Publishing
30 Heng Mui Keng Terrace
Singapore 119614

E-mail: publish@iseas.edu.sg
Website: bookshop.iseas.edu.sg

*The responsibility for facts and opinions in this publication rests exclusively with the
authors and their interpretations do not necessarily reflect the views or the policy of the
publisher or its supporters.*

ISEAS Library Cataloguing-in-Publication Data

In China's Backyard : Policies and Politics of Chinese Resource Investments in
    Southeast Asia / edited by Jason Morris-Jung.
    1. Energy industries—Southeast Asia.
    2. Power resources—Southeast Asia.
    3. Investments, Chinese—Southeast Asia.
    4. Investments, Foreign—Southeast Asia.
    5. Southeast Asia—Foreign economic relations—China.
    6. China—Foreign economic relations—Southeast Asia.
    I. Morris-Jung, Jason.
HD9502 A9I35   2018

ISBN 978-981-4786-09-6 (soft cover)
ISBN 978-981-4786-10-2 (E-book PDF)

Typeset by International Typesetters Pte Ltd
Printed in Singapore by Markono Print Media Pte Ltd

CONTENTS

Foreword by Ho-fung Hung, Johns Hopkins University vii

Acknowledgements ix

About the Contributors xi

1. Introduction 1
   Jason Morris-Jung

2. Mixed Motivations, Mixed Blessings: Strategies and 27
   Motivations for Chinese Energy and Mineral Investments
   in Southeast Asia
   Philip Andrews-Speed, Mingda Qiu and Christopher Len

3. Mineral Resources in China's "Periphery" Diplomacy 57
   Yu Hongyuan

4. Energy Entanglement: New Directions for the 79
   China–Indonesia Coal Relationship
   Cecilia Han Springer

5. Indonesia–China Energy and Mineral Ties: The Rise and 104
   Fall of Resource Nationalism?
   Zhao Hong and Maxensius Tri Sambodo

6. The Direction, Patterns, and Practices of Chinese Investments 129
   in Philippine Mining
   Alvin A. Camba

7. Development Cooperation with Chinese Characteristics: 154
   Opium Replacement and Chinese Rubber Investments in
   Northern Laos
   Juliet Lu

8. The High Cost of Effective Sovereignty: Chinese Resource   182
   Access in Cambodia
   Siem Pichnorak

9. Complex Contestation of Chinese Energy and Resource   204
   Investments in Myanmar
   Diane Tang-Lee

10. Anti-Chinese Protest in Vietnam: Complex Conjunctures   229
    of Resource Governance, Geopolitics and State–Society
    Deadlock
    Jason Morris-Jung and Pham Van Min

11. Complexities of Chinese Involvement in Mining in the   256
    Philippines
    Menandro S. Abanes

12. Conclusion   277
    Tai Wei Lim

Bibliography   289

Index   321

FOREWORD

Over the last three decades, one of the most significant changes in the context of development in the Global South is the transformation of China from a capital deficient developing country to a rising capital exporter in the world economy. For a long time, the main form of China's capital export is its massive purchase of US Treasury bonds. But since the Hu Jintao era (2002–12), the Chinese state has diverted ever larger part of its foreign reserves to outgoing direct investment in infrastructures, mines, and other assets in the developing world. China's official financial institutions also started offering concession loans to other developing countries.

Africa has been the most prominent recipient of Chinese investment, partly because of the PRC's longstanding presence in the continent since the height of Cold War and partly because of Africa's abundance in fossil fuel and mineral resources that China needs desperately for its roaring developmental machine. "China in Africa" has been an established field in development studies. Numerous papers and books debated whether China's increasing presence in Africa would elevate the region's growth prospect or whether it represents little more than a new form of extractive colonialism.

China's ambitions in becoming a major capital exporter to the developing world is surely not restricted to Africa. Chinese investments in Latin America, the Middle East, and Southeast Asia has been increasing rapidly. Among all these regions, Southeast Asia, which is right at the doorstep of China, is the most interesting one, and it is poised to becoming the most important arena for China's overseas projection of political-economic influences.

China's presence in Southeast Asia dates back to premodern times, when many states in the region paid tribute to the dynastic state of imperial China and conducted trade with China via Chinese diasporic traders in the region. Time and again, some states in the region attempted to challenge Chinese dominance and sever its tributary ties to China. But at least in the perception of many officials and intellectuals in China, these challenges did not alter the reality that China had been hegemonic over Southeast Asia all along until Western imperialism shattered this Sinocentric other in the mid-nineteenth century. China's link to the region receded in the mid-twentieth century because of Cold War division, only to be restored after China's reintegration into the capitalist world in the 1980s. In the early twenty-first century, restoring China's historical influence in Southeast Asia became a crucial part of Beijing's project for the "Great Revival of China". China's recent One Belt One Road Initiative exactly manifests this ambition.

But Southeast Asia today is not Southeast Asia in the premodern times. Southeast Asian states are all fully sovereign entities now. The region has been receiving substantial investment from the United States, Japan, and many other developed countries besides China. No matter whether restoring Chinese hegemony in the region is the real purpose of China's rising investment in the region, such capital export will certainly create new tensions, anxiety, and conflicts on top of the new opportunities it offers. Given Southeast Asia's large population, economic size and economic dynamism, China's capitalist inroad into the region and the resulting contradictions will generate large impact that ripples through the global economy. Such important topic, unfortunately, has been so far understudied. Academic writings on this issues have been scant.

In China's Backyard: Policies and Politics of Chinese Resource Investments in Southeast Asia, edited by Jason Morris-Jung is one of the very few pathbreaking efforts in analyzing the impact of China's capital export on Southeast Asia. With experts specializing in different countries and sectors in the region, this book shows readers both the forests and the trees of the issue, illustrating in an accessible way the variegated and complicated dynamics and impacts of Chinese investment in Southeast Asia's resources sector. It is a book that will open up new debates and new directions of research. It is a volume that all students of development, global political economy, and Southeast Asia should not miss.

Ho-fung Hung
Johns Hopkins University
September 2017

ACKNOWLEDGEMENTS

Like many a book before it, this book took root in a seminar organized at the ISEAS – Yusof Ishak Institute, Singapore in May 2014. In the context of China's growing political and, especially, economic influence across Southeast Asia, our interest as contributing authors was to examine the importance of "resource deals" to the evolving situation, as had been written about extensively in the African and Latin American cases but very little in the Southeast Asian one. We wondered what advantages (or disadvantages) Southeast Asia might offer to China's "resource quest" with its geographic proximity, shared histories, and socio-cultural similarities, as well as what the typically contentious politics of resource development might illuminate about the bigger picture of China's return to Southeast Asia. This seminar enabled us to bring together a diverse group of authors to share experiences on how our own research and field sites in Southeast Asia had been unignorably affected by the "China factor". By bringing together authors from different disciplines and types of institutions, and cultivating dialogue between scholars from both inside and outside the region, we hope to have fashioned not any single one but many answers to these and other questions.

For making that seminar and this book possible, we owe our first debt of gratitude to the Konrad Adenauer Stiftung (KAS) Foundation and ISEAS – Yusof Ishak Institute. Without their generous support, our project would surely have been defeated by far-flung geographies, institutional silos, and our own disparate feelings about the meaning of China's rise for Southeast Asia. We also wish to thank the administrative staff at ISEAS, whose preparation and organization of the seminar was seamless.

For the preparation of this book, we thank Ho-Fung Hung, Jonathan Rigg, and Erik Harms for their kind words and intellectual support for this project. We also thank Terence Chong, Oh Su Ann and again Jonathan Rigg for their rich commentary during the seminar. As editor, I thank Mike Dwyer and Lim Tai Wei for their substantial editorial support on specific chapters. I also thank Zhao Hong for his support in organizing the ISEAS seminar and helping to move this project forward in its early phases.

Not least of all, we thank ISEAS Publishing for their excellent support in bringing material to this project, and especially Sheryl Sin Bing Peng for her diligent editing and compilation of the manuscript.

As always, we heartily acknowledge your vital contributions to this book, but we claim all responsibility for errors and oversights as entirely our own.

Jason Morris-Jung
Singapore 2017

ABOUT THE CONTRIBUTORS

Menandro S. ABANES is an Associate Professor of Social Sciences at Ateneo de Naga University and a Lecturer at the Safety and Security Management Studies (SSMS), The Hague University of Applied Sciences. He has earned fellowships from the Ford Foundation for Graduate Studies in International Development at the Institute of Philippine Culture (IPC), Ateneo de Manila University, and the Nippon Foundation in International Peace Studies at the United Nations-mandated University for Peace in Costa Rica. In 2010, he joined a research project financed by the Netherlands Organization for Scientific Research (NWO) on ethno-religious conflicts in Indonesia and the Philippines (ERCIP). He has more than sixteen years of research and work experience in the Philippines.

Philip ANDREWS-SPEED is a Senior Principal Fellow at the Energy Studies Institute of the National University of Singapore. He has thirty-five years of experience in the field of energy and resources, starting his career as a mineral and oil exploration geologist before moving into the field of energy and resource governance. Until 2010 he was Professor of Energy Policy at the University of Dundee and Director of the Centre of Energy, Petroleum and Mineral Law and Policy. During the academic year 2011/12 he was a Fellow of the Transatlantic Academy at the German Marshall Fund of the United States in Washington DC. Recent books he has written include *China, Oil and Global Politics* (2011, with Roland Dannreuther), *The Governance of Energy in China: Transition to a Low-Carbon Economy* (2012), and *Want, Waste or War? The Global Resource Nexus and the Struggle for Land, Energy, Food, Water and Minerals* (2015, with five co-authors).

Alvin A. CAMBA is a doctoral candidate in Sociology at Johns Hopkins University. His research focuses on Chinese investments in Southeast Asia, the political economy of natural resource extraction, and contentious politics. He combines detailed ethnographic fieldwork, quantitative methods, and comparative-historical methods to analyse historical mechanisms that embed states in and toward particular pathways within the global circuits of capital. His doctoral dissertation seeks to situate Chinese investments in Southeast Asia within the context of contentious politics and global elite competition. His research articles have received the Terence K. Hopkins Best Graduate Student Paper Award (honourable mention) from the American Sociological Association's (ASA) PEWS section and the Postdoctoral and Graduate Student Publication Research Award (honourable mention) from the Critical Realism Research Network. In addition, he has received fellowships from and collaborated with the Southeast Asian Research Group (SEAREG), the Middle East Institute (MEI), and the Asian Research Institute (ARI). His works have appeared in major journals and book publications, such as *Extractive Industries and Society*, *The Everyday Political Economy of Southeast Asia* and *New Directions in the Study of Africa and China*.

Christopher LEN is a Senior Research Fellow at the Energy Studies Institute, National University of Singapore. He obtained his PhD from the Centre for Energy, Petroleum and Mineral Law and Policy (CEPMLP) at the University of Dundee in Scotland. Before that, he studied at the University of Edinburgh, Scotland (Philosophy and Politics) and the Department of Peace and Conflict Research, Uppsala University, Sweden. He was previously a Research Fellow at the Stockholm-based Institute for Security and Development Policy (ISDP) where he was responsible for the Energy and Security in Asia Project. Len was also a Visiting Associate under the Energy Programme of the Institute of Southeast Asian Studies (ISEAS) in Singapore between 2006–12. His research interests include Asian energy and maritime security, Chinese foreign policy, Arctic energy security and sustainable development, and the growing political and economic linkages between the various Asian subregions.

LIM Tai Wei is a Senior Lecturer at SIM University and a Research Fellow adjunct at National University of Singapore East Asian Institute. He is a historian and an East Asian area studies specialist.

Juliet LU is a doctoral candidate at UC Berkeley's Department of Environmental Science, Policy and Management with a focus on political ecology. She worked from 2009–11 at the World Agroforestry Centre in Yunnan, China and from 2012–13 with the Centre for Development and Environment in Vientiane, Laos. Her most recent research examines the political economy of Chinese agribusiness companies' investments in Laos, primarily in rubber plantations. Her broader research interests involve the impacts of China's growing demand for raw materials on land and natural resource management in Southeast Asia, and the political economy of Chinese development cooperation initiatives.

Jason MORRIS-JUNG is a Senior Lecturer in Social Research at the Singapore University of Social Sciences (SUSS). He earned his PhD in Environmental Sciences, Policy and Management (ESPM) at the University of California – Berkeley. His research focuses on intersections of politics, society and environment in Vietnam and East Asia. He has published several journal articles and book chapters on environmentally-related regime contestation in Vietnam, online activism, and the Vietnamese bauxite mining controversy. Prior to his academic career, he worked for many years with non-governmental and community organizations on issues of rural poverty, ethnic minority rights, women's empowerment, upland development and biodiversity conservation in Vietnam and Southeast Asia.

PHAM Van Min is currently teaching at the University of South Australia. He formerly taught in the Department of International Studies, Vietnam National University, Hanoi. Pham Van Min earned his MA in International Studies from the University of Oregon and PhD in International Relations from the University of South Australia. He has published several peered-reviewed articles on Vietnam as well as international journals and book chapters. Pham Van Min has recently participated in a national research project on International Relations which is funded by the Vietnamese government. His research interests include international relations theories, the rise of China, and energy security in East Asia.

Mingda QIU is a Research Associate with the Freeman Chair in China Studies at the Center for Strategic and International Studies (CSIS) in Washington DC. His research interests involve China's political economy, energy strategy, innovation policy, and cross-Strait relations. Mr Qiu earned a Master in International Affairs, with concentrations in International Politics and Economy, from the School of Global Policy and Strategy at the University of California, San Diego. He also received a Bachelor of Social Sciences (Honours) in Political Science from the National University of Singapore.

Maxensius Tri SAMBODO is a Researcher at the Indonesian Institute of Sciences (LIPI) –Economic Research Center. He is also a Visiting Fellow alumni from the ISEAS – Yusof Ishak Institute, Singapore. His research interests are on economic development, energy, environment, and natural resources. He obtained a bachelor's degree in Economics from Pajdadjaran University, Indonesia; a master's degree in International and Development Economics from the Australian National University, Australia; and a PhD in Public Policy from the National Graduate Institute for Policy Studies (GRIPS), Japan. His latest publications include several journal articles and books, such as *From Darkness to Light: Energy Security Assessment in Indonesia's Power Sector* (2016) and "Developing a 'Green Path' Power Expansion Plan in Indonesia by Applying a Multiobjective Optimization Modeling Technique", *Journal of Energy Engineering* 143, no. 3 (June 2017) (co-authored with Hozumi Morohosi and Tatsuo Oyama).

SIEM Pichnorak is a graduate student in International Areas Studies at Seoul National University, Republic of Korea.

Cecilia Han SPRINGER is doctoral student in the Energy and Resources Group at the University of California, Berkeley. She is interested in interdisciplinary methods merging quantitative economics and qualitative fieldwork, and the application of these methods to evaluating energy policy issues. Her topical focus is on production and trade of energy-intensive commodities within and between countries in Asia. Previously, she worked at Climate Advisers, a policy consulting firm, and conducted research as a Fulbright scholar in Tianjin, China. Cecilia received her B.S. in Environmental Science from Brown University.

Diane TANG-LEE completed her PhD at the University of Manchester in 2016. Her thesis examines how Chinese state and SOE actors are socialized to adopt a public participation norm. It focuses on Myanmar civil society actors as they engage in argumentative action with Chinese interlocutors, and captures a shift in the public engagement approach of Chinese overseas investments. She conducted semi-structured in-depth interviews with non-governmental organizations (NGOs) and activists in Yangon, villagers at Letpadaung and Pyin Oo Lwin, as well as environmental NGOs/think-tanks, scholars and Ministry of Commerce of the People's Republic of China (MOFCOM) officials in Kunming and Beijing.

YU Hongyuan is a Professor and Director of the Institute for Comparative Politics and Public Policy at the Shanghai Institutes for International Studies. In 2010, he was recognized with an award for the Shanghai Social Science Newcomer. He is also an Honorary Fellow of the Center for International Energy Strategy Studies, Renmin University of China, and a Visiting Fellow at the Sustainable Developmental Research Center of China Academy for Social Sciences. He has also authored numerous articles and books in Chinese, including *Strategic Resource Politics and China's Strategies* (2016), *Challenges and Innovation in the Low-carbon Economy* (2015), *Environmental Change and Power Transfer: System, Game and Response* (2011), and *Global Warming and China's Environmental Diplomacy* (2008).

ZHAO Hong is a Visiting Senior Fellow of ISEAS – Yusof Ishak Institute in Singapore. He obtained his PhD in Economics from Xiamen University, where he formerly taught for eighteen years. Before joining ISEAS, he was a Senior Research Fellow at the East Asian Institute, National University of Singapore. His research interests mainly cover China–ASEAN economic and political relations, and East Asian economic integration. Currently he is working on a book entitled *The Rise of China and the ASEAN Community: Energy Security, Cooperation and Competition*.

1

INTRODUCTION

Jason Morris-Jung

Introduction

Across Southeast Asia, the "rise of China" has inspired both anticipation
and anxiety. A common aspiration throughout the region is that a
strong China means economic prosperity for all. The prevailing wisdom
is that the sheer size of the world's second largest economy and its
demands for trade will translate into jobs, business opportunities, and
economic growth for neighbouring countries. These aspirations have
been bolstered by China's leadership in establishing the US$100 billion
Asian Infrastructure Investment Bank (AIIB), as well as its ambitions
to retrofit ancient trade routes with twenty-first century infrastructure
as part of its "One Belt, One Road" initiative. Now with hints of
American influence in decline, many eyes are looking to China as the
region's new benefactor.

   If anticipation has been fuelled by visions of economic splendour,
the anxieties have been more polysemic. Influxes of Chinese investment
have been accompanied with waves of Chinese companies, workers
and migrants, whose intermixing with local populations has generated
both enthusiasm and distrust. The Chinese knack for keeping costs

low has come with familiar concerns about social and environmental management practices, exploitation of local labour, and networks of chronyistic corruption. China's territorial claims and expanding military installations in the South China Sea have only helped to heighten geopolitical tensions both within and among individual Southeast Asian nations. Hence, while Chinese economic influence has been welcome, a stronger political, military and socio-demographic presence has been met with more equivocation, if not outright consternation.

Amid these tensions, Chinese resource sector investments have emerged as flashpoints of protest and controversy. From government orders to suspend the Chinese-backed Myitsone dam and Letpadaung copper mine in Myanmar to national protests over Chinese resource sector investments in Vietnam, the Philippines and elsewhere, resource projects have come to be a "focal point"[1] for tensions and anxieties surrounding China's rise in Southeast Asia. Indeed, resource development is an interesting angle from which to examine China's future roles in Southeast Asia precisely because it reflects the contested nature of development.

Resource development is contested development. From "conflict minerals" to "resource curses", political economies of resource development have been known to incite inter- and intra-state conflict, hinder socio-economic development, and bind already poor countries and communities into dependent relations with other more powerful nations. Yet for China and all Southeast Asian countries, perhaps only with the exception of Singapore, natural resources have been and, for many, continue to be vital to national development. Not only has resource development helped to boost economic growth and state revenue, it has also supplied raw materials and energy to domestic industries (sometimes at subsidized rates), contributed to infrastructure development (especially in difficult to access remote regions), and provided much needed foreign currency (especially in the least industrialized countries). Indeed, China's own development success — which Beijing has been keen to share with developing nations — highlights many of the advantages of resource development.

Yet these tensions between development and immiseration, opportunity and exploitation are not unique to resource development.

Their extreme forms help to make more visible the conflicts and tensions that underlie development of all kinds. Highlighting both the tremendous potential and grave concerns at stake, this volume proposes to examine China's rise in Southeast Asia from a resource lens. *In China's Backyard* raises a question of what difference it makes being located next door to a rising super power for the developing regions of Southeast Asia. It draws attention to the geographical proximity of Southeast Asia to China, their shared histories, and many of the economic, social and cultural characteristics that bind them, if unevenly, together. Yet in resource debates the backyard can also have negative connotation as a nearby region to source raw materials while exporting the very literally — and sometimes figuratively — dirty work of resource production. Which of these scenarios will describe Southeast Asian relations with China going forward is a recurring theme in this book.

While outgoing Chinese investments and development cooperation now commands a sizeable literature for Africa and, increasingly, Latin America, China's new wave of investments in Southeast Asia and their implications have only recently begun to draw significant scholarly attention. While centuries of Chinese trade, migration and diplomacy in the region may make studying China in Southeast Asia less spectacular and, historically, more mundane, it is nonetheless ironic that the region that would appear to have the most lessons to share about growing with China has so far said the least. Through multi-disciplinary, multi-level and multi-sited analyses, featuring regional views, Chinese policy views, and detailed cases studies from six different Southeast Asian countries, this book aims to address this gap.

We use the term "investment" loosely in this volume. It is not meant to be limited strictly to a monetary investment, but rather encompasses all kinds of investments in time, energy, people, ideas and, hopefully, goodwill. It also focuses on the different levels of governments, businesses, publics and local communities invested in transnational resource cooperation with China. The volume makes no attempt to provide a comprehensive overview of China's expansive resource-based investment portfolio in the region, or each of the Southeast Asian countries those investments touch upon. Rather, *In China's Backyard* aims to provide multiple lines of sight into Chinese investments in Southeast Asia and present a diversity of perspectives on complex and varied phenomena that can never be fully understood from any single one.

The "Chinese Model" through a Resource Lens

Over the past few decades, China has become a major investor of international development cooperation around the world. A sizable literature now exists on a so-called "Chinese model" for development. The label is not just academic. Chinese officials and their development partners use the label widely and felicitously. The Chinese model is meant to contrast with conventional development as a neo-colonial Euro-American project forged in the post-war era, which enables the countries of the Global North to impose its political ideologies upon while continuing to exploit the Global South.[2] In contrast, the Chinese model emphasizes South–South cooperation, mutual benefit (between donor and recipient countries), non-interference in domestic affairs, and unconditional aid — at least in terms of governance requirements, as conditions related to the use of Chinese equipment, technology and labour are common. The model reflects China's own historical experience of development, along with its emphasizes on top-down planning and heavy investment in infrastructure and industrialization, and it has been embraced by many developing country leaders in Africa, Latin America and Asia.[3]

One of the more controversial aspects of the Chinese model, however, has been precisely its alleged ties to natural resources.[4] As Mel Gurtov has suggested, for China, the "name of the game is all-out economic diplomacy to harness energy and industrial resources from developing countries". This is the notion that has inspired the titles of such books as *Winner Take All: China's Race for Resources and What It Means for Us* by development economist Dambisa Moyo and *By All Means Necessary: How China's Resource Quest is Changing the World* by Elizabeth Economy and Michael Levi.[5] Nearly 80 per cent of China's imports from Africa are minerals, where China has opened up twenty-one mining bureaus across the continent.[6] In Latin America, 24 per cent of Chinese FDI is concentrated in the extractive sector.[7] Springer, in this volume, notes that, according to the Chinese Ministry of Commerce, approximately 23 per cent of outbound Chinese investment has been for mining alone. China's large State-Owned Enterprises (SOEs) for mineral and especially petroleum development are typically seen as primary vehicles for advancing Beijing's political will.[8]

These views, however, have not been without their detractors. Deborah Brautigam, one of the most widely respected scholars on China's outgoing investment, has described them as *"at best partial and misleading"*.[9] Brautigam rejects suggestions of *quid pro quo* exchanges of Chinese overseas assistance for resources (or that China gives more aid to countries with more resources) and challenges myths about the exploitative nature of Chinese "resource-secured" loans by comparing them with those of other countries.[10] Others, who recognize the preponderance of Chinese investments in resource production, suggest that it may characterize an early phase of development cooperation, where resource deals are "easy deals".[11] Later, these resource investments can act as springboards for economic diversification and investment in higher value sectors.[12] Sceptics further note that even where SOEs are leading the way, they are typically followed in droves by Chinese private companies.[13]

What these debates perhaps reflect most is one of the most salient messages to emerge from the scholarship on China abroad: China is not any monolithic entity. States consist of many diverse parts that are often fragmented and compete with one another.[14] For a state as large and diverse as China, the challenges of coordinating all of the actors involved with investing abroad would be, as Armony and Strauss suggest, "rife with principal-agent problems".[15] Brautigam and Ekman have suggested that rather than pursuing policy mandates from Beijing, Chinese corporations are "often rather reluctantly, investigating investment opportunities that are being eagerly put before them by African governments and potential private sector partners".[16] Even SOEs can have very different policy and business agendas, depending on whether they are owned at central or provincial levels, which sector they are investing in, and where. The oil industry, for example, where connections between government and China's largest and most successful SOEs would appear strongest, has developed "with little coordinated state control".[17]

Emphasizing complexity and diversity, however, should not be a reason for diffusing or obfuscating networks of accountability. Struggles to control its many parts should not provide Beijing or any other state a *carte blanche* to do as it wishes, much less to excuse dismissive comments of being no worse than the rest.

Rather, the purpose of emphasizing diversity and complexity is to understand with more clarity and precision the networks of wealth and power that come together around development cooperation. Ching Kwan Lee — whose own work has been exemplary in this regard — argues for more relational understandings of Chinese resource investments that examine how specific instances of Chinese investment combine in contingent ways with local labour relations, domestic politics and regional history.[18] In this regard, such an analysis is necessarily place-based, attentive to the nodes of transnational political and economic power that form in a particular place.

A resource lens can be helpful here in at least three ways. The first is in specifying the role of the state. Resource policy discussions tend towards state centricity. From dependency theory to the "resource course", discussions on transnational resource cooperation typically define those engagements in terms of one state (or set of states) against another. Unsurprisingly, then, government responses to these dilemmas often come in one form or another of resource nationalism. Yet, whether trying to protect one's own resources or using statecraft to acquire them from other nations, resource nationalism depends upon a state's ability to exert exclusive sovereign control over those resources.[19]

This view of sovereign control over a fixed territorial boundary, however, is somewhat at odds with the global production networks and fluid boundaries over which resources are actually produced today. The production of resources today typically depends upon production chains that traverse multiple state boundaries and involve an array of actors from different states, including producers, financers and regulators, especially in developing countries that lack necessary capital or technical know-how. The so-called resource-producing state depends upon these global networks to produce its "own" resources but has limited control over them.[20] Hence, the idea that a state can exert an absolute or even an end control over resources located within its sovereign territory reflects what political geographers refer to as a "territorial trap".[21]

Critical perspectives of resource production urge for a more relational understanding of states and the networks of actors connected to resource production. The state is a vital node within those networks, but one whose ultimate authority or "sovereign control" over resources are made possible by its relations with various extra-national actors. Hence, understanding the role of the Chinese or other states in Southeast Asian resource production demands tracing out networks of state relations with a wide range of local, national and transnational actors. In some places, these partnerships reinforce existing relations of power, in others they may create new opportunities for resistance and opposition, and in others still they may lead to surprising or uncertain outcomes.

A second area where a resource lens can be helpful is in understanding territory not simply as bounded geographical space encompassing people and resources, but rather as a set of social relations expressed through strategies of territorialization. As classically defined by Robert Sack, territoriality is an attempt to control people or things by "delimiting and asserting control over a geographic area".[22] Resource projects are territorializing for the simple reason that they take up space, or even just aspire to take up space.[23] Their efforts to impose a new order on space — sometimes a radically new order — reshapes existing socio-ecological relations between land, resources and people.[24] Hence, resource struggles should not be seen simply as struggles over areas of land that hold resources, but rather as wider struggles over territorializing strategies and their attempts to impose a particular order on existing social relations.

A third area, closely related to the second, is that resource conflict is rarely just about resources. Whatever are the ostensible terms of conflict, resource conflicts have a tendency to spill over into many other issues. They are struggles over labour, livelihoods, cultural identity, inter-ethnic relations, historical struggles and many other possibilities. As a result, the actors implicated by resource conflicts are often diverse and wide-ranging, as are their forms and levels of engagement. Hence, our analyses of resource conflict should necessarily be multi-dimensional, multi-level and multi-sited.

At the same time, conflict should not be seen strictly as an indicator of problems. Conflict may also reflect new levels of recognition of those who were previously invisible, new forms of dialogue, or other emergent possibilities for politics. As Anthony Bebbington has argued, the push and pull of conflict, even when among highly asymmetric power relations, can co-produce alternative development outcomes.[25] Hence, one of the challenges to understanding conflict is being able to excavate their messy social, cultural and historical lineages to understand how they might challenge or contest existing political and economic orders.

Either way, China's global investments in natural resources will continue. Since 2010, China has been the world's largest energy consumer, having been the world's second largest oil consumer since 2003.[26] China is also the largest global consumer of iron ore, coal, zinc, lead, tin, nickel, copper and aluminum.[27] It consumes more than 80 per cent of the world's zinc and tin, more than half of its iron ore and coal, and around 40 per cent of its copper, aluminum, lead and nickel, and nearly one third of its tungsten.[28] Both the global acquisition, as well as domestic production, of these resources have been vital to China's own "resource-led economic development strategies".[29] Hence, examining China's global investments and development cooperation from a resource lens is appropriate.

A resource lens urges us to employ more critical and relational understandings on China's rise in Southeast Asia. It requires us to trace contingent multi-level networks of uneven power relations, examine strategies of territorialization as key mechanisms of struggle within those networks, and conflict as an open terrain with multiple meanings and possibilities for a diverse set of actors. Not only does a resource lens provide critical insight into an important part of China's international development cooperation, it is also an approach to examining China's rise in a region that it has long characterized as its own backyard.

China's Return to Southeast Asia

A new wave of Chinese investment has spread across Southeast Asia. The levels of investment have been as impressive as their implications have been wide-ranging. They have brought to Southeast Asia new economic opportunities, regional cooperation and development, but they have also included allegations of land grabbing, environmental degradation, cultural conflict, political bullying, and distrust. Furthermore, underlying these tensions are long intertwined histories characterized by both trust and suspicion. A recent volume on China in the Mekong Region captures these sentiments in its descriptions of China as an awakening "imperial dragon ... seeking to reassemble the 'Middle Kingdom'".[30] To complicate matters further, the new wave of investments is happening against a backdrop of heightened geopolitical tensions surrounding China's territorial claims in the South China Sea.

That China has become such an important investor in Southeast Asia is remarkable, recalling that a little more than three decades ago most countries in the region maintained distant, if not adversarial, relations with China. During the Cold War, the capitalist countries of Southeast Asia saw communist China as an existential threat. For communist Vietnam and, to a lesser degree, Laos, China fell on the wrong side of the Sino–Soviet split and its support of the Pol Pot regime led to the China–Vietnam border war in 1979.

Not until the 1990s, more than a decade after China had begun its transition to a market economy, did major powers in Southeast Asia begin to normalize relations with China, among them Singapore, Malaysia, Indonesia, Vietnam and the Philippines. However, China boosted its candidacy as a new regional leader for Southeast Asia during the Asian Financial Crisis of 1997 and 1998. As western capital took flight, China stepped in to provide much needed financial assistance and committed to maintaining its currency value for the sake of regional stability. These gestures of Chinese largesse helped pave the way for new political and economic agreements with the ten ASEAN member states, including agreements to remove tariffs in 2004 and create "the world's largest free trade area by 2010".[31]

Since then, China has established itself as the most important economic trading partner of nearly every Southeast Asian country. China is now the first or second most important trading partner for every major economy in Southeast Asia (see Table 1.1).[32] Chinese direct investment into the region has also grown rapidly from US$120 million in 2003 to US$6.3 billion in 2012, representing a twofold increase in China's total FDI to the region from 4 to 8 per cent.[33] And even as 8 per cent might reflect a minor share of China's total outgoing investments, it translates into a very big number for most Southeast Asian countries. For poorer countries, such as Cambodia and Laos, China is their number one foreign investor.[34]

TABLE 1.1
Chinese Two-Way Trade with Southeast Asia

| | Imports from China | | | Exports to China | | | Total Trade with China |
	Rank	Per Cent	US$ Billion	Rank	Per Cent	US$ Billion	US$ Billion
Brunei	3rd	15%	0.6	—			
Cambodia	2nd	22%	3.3	—			
Timor-Leste	5th	10%	*0.0	—			
Indonesia	1st	18%	32.5	2nd	11%	20.8	53.3
Lao PDR	2nd	26%	3.9	1st	45%	1.7	5.6
Malaysia	1st	18%	36.4	2nd	12%	32.6	69.0
Myanmar	1st	36%	4.3	1st	42%	8.8	13.1
Philippines	1st	24%	19	1st	17%	13.8	32.8
Singapore	1st	13%	44.3	**2nd	12%	33.2	77.5
Thailand	1st	18%	37.3	1st	12%	29.5	66.8
Vietnam	1st	30%	44.7	2nd	11%	17.5	62.2

Notes: * Rounded from US$15 million
       ** First is ranked as Hong Kong (13%, 34.1B)
Source: Compiled by the author using economic data from <http://atlas.media.mit.edu>.

Where Chinese direct investment has been less significant, Chinese economic activity has found other channels for pouring into Southeast Asia. For example, Chinese contractors are believed to hold up to 90 per cent of Engineering, Procurement and Construction (EPC) contracts in Vietnam's thermal power sector.[35] By 2010, China held US$10–15 billion worth in project contracts, making Vietnam China's largest project contracting market in Southeast Asia.

Along with trade and investment, Chinese aid to Southeast Asia has also grown. Through a mix of preferential credit, concessional loans and direct grants, China is now one of the most important sources of economic development assistance to the region. Between 2002 and 2007, China directed US$7.4 billion worth of economic assistance to Southeast Asia, of which US$7.1 billion were concessional loans.[36] Cambodia, Laos, Myanmar and Vietnam each received around US$500 million in Chinese economic assistance between 2004 and 2008.[37] Cambodia alone received US$9.17 billion between 1994 and 2012, mainly concentrated in hydropower and transport infrastructure.[38]

China has also been a major source of investment and infrastructural development, especially roads, in under-funded regions such as Laos.[39] Growing regional cooperation will only strengthen these trends through such initiatives as the China–ASEAN Investment Cooperation Fund announced in 2008 (to which China committed US$15 billion in loans), the ADB's Greater Mekong Subregion Economic Cooperation Program, and, more recently, China's One Belt, One Road programme to support regional transport and communication infrastructure, including a "maritime silk road" in Southeast Asia, and the US$100 billion Chinese-led Asian Infrastructure Investment Bank.

However, China's growing economic influence has also been accompanied by new waves of Chinese migration, cross-cultural conflicts, and geopolitical tensions. Increasing waves of Chinese traders, investors and workers moving into the region has had mixed effects on national and local businesses, communities and inter-ethnic relations, especially in the more volatile hinterland regions.[40] An expanding demographic presence has also given the Chinese government new reasons to sponsor Chinese cultural developments, such as building Confucian Institutes across the region. Yet such initiatives have not always been appreciated by everyone and some have regarded them as "civilizing missions".

Growing geopolitical tensions in the South China Sea has only heightened the potential for conflict on all of the above-mentioned issues. China claims up to 90 per cent of the South China Sea through its famously ambiguous ten-dashed line. The claimed area extends from the southern coast of Hainan Island along the entire eastern coast of Vietnam, the northern edge of Malaysian Borneo and Brunei, and the western islands of the Philippines. Over the past ten years, China has become increasingly bold in asserting its claims to the South China Sea and has engaged in ambitious efforts to reclaim land and build infrastructure, including military installations, which has sparked tensions both within and among Southeast Asian claimant countries.

The mix of economic development together with cross-cultural, socio-demographic and geopolitical challenges makes assessing the implications of China's rise in Southeast Asia necessarily complex. Any final analysis that describes these developments as singularly, or even fundamentally, productive or destructive, cooperative or conflictive, beneficial or detrimental is necessarily limited. Such assessments will necessarily vary according to where developments are taking place, from whose perspective they are viewed, and on what specific issues they are being assessed. Identifying, recognizing and learning to work within these multiple and contradictory tendencies is the challenge that this volume puts out to its readers.

China's Resource Investments in Southeast Asia

An important part of China's growing economic cooperation in Southeast Asia has been in the natural resource and energy sectors. Over the past two decades, Chinese FDI has expanded rapidly in the region's resource rich countries. In the ten years between 2003 and 2012, Chinese FDI increased from US$0.8 million to US$809 million in Laos, from US$12.8 million to US$349 million in Vietnam, from US$2.0 million to US$199 million in Malaysia, from US$26 million to US$592 million in Indonesia, and from US$4.1 million (2004) to US$749 million in Myanmar.[41] Chinese investment in Myanmar has been primarily in the natural resource and energy sectors, where all three of China's national oil companies are active.[42] No less than 99 per cent of China's total FDI in Indonesia went to the mining sector in 2014, which, together with India, has replaced Japan and Korea as Indonesia's main energy market (Zhao Hong and Sambodo, this volume). In all of Southeast Asia, the proportion of Chinese FDI in the mining sector increased from 9.7 per cent in 2008 to 28 per cent in 2012, which is similar to the proportion of China's mineral investments in Africa (i.e., 31 per cent). Even despite vagaries in global market prices for resource commodities, the sheer size of the Chinese economy ensures that its long-term demand for resources in global markets will continue.

This volume addresses China's growing interest in Southeast Asia's resources by bringing together diverse researchers and scholars from different disciplines, institutions and backgrounds. Among them are resource economists and policy analysts, international relations specialists, human geographers, political ecologists, a historian, a sociologist, and an anthropologist. The authors also hail from different countries and backgrounds in Southeast Asia, China and beyond. However, what each of us share in common is a deep and long-standing interest in Southeast Asia and its evolving relations with China. The volume also addresses the topic on multiple levels by professively zooming in from a regional view and state perspectives through to multi-level country studies and ethnographies of conflict.

The first four chapters begin with the views of resource economists and policy analysts who demonstrate that, despite the challenges of coordinating state behaviour and the many actors involved, resource investments can be strategic at the highest levels. In their chapter on "Mixed Motivations, Mixed Blessings: Strategies and Motivations for Chinese Energy and Mineral Investments in Southeast Asia", Philip Andrews-Speed, Mingda Qiu and Christopher Len examine the different rationales and logics at work among the various Chinese and Southeast Asian actors engaged in "resource deals" in Southeast Asia. They note that Chinese resource companies currently have the largest footprint among all Asian energy and resource investors in Southeast Asia. Over the past twenty years, they have committed tens of billions of dollars to resource and energy sector projects across the region. Where "traditional" or western multinational companies have been leaving the region for larger deposits or more attractive fiscal and regulatory regimes elsewhere, Chinese investors have been taking their place.

Andrews-Speed et al. further note that a dominant proportion of the new Chinese resource investments are through companies wholly or partly owned by the Chinese state. They may be owned at central or provincial levels, but almost invariably they arrive with "substantial workforces and access to generous financing" (p. 49). However, Andrews-Speed et al. present a complex picture of the mixed state and corporate motivations that drive these investments. They apply Dunning's model of overseas investment to describe complex structures of motivations and investment strategy, as well as how they vary according to resource type. These authors reject the notion of any single narrative driving Chinese resource investments in the region. They demonstrate that resource investments serve multiple agendas among different actors, which affects how resources are extracted and produced and the types of contestation they may generate.

Yu Hongyuan's chapter on "Mineral Resources in China's 'Periphery' Diplomacy" also takes a regional view of Chinese resource investment strategies. However, it does so from a Chinese policy perspective on China's political and economic relations with its "peripheral regions", which is a policy designation for the countries and regions surrounding China. The Chinese government has identified forty-five "strategic mineral resources" as vital to the country's development, of which twenty-seven are considered to be in short supply or without guarantee of adequate supply. More than half of them, however, are present in Southeast Asia and Australia. They include almost all of China's nickel imports (Indonesia and the Philippines), more than 70 per cent of its aluminum ore (Indonesia), more than half of its tin (Myanmar until 2011), and 83 per cent of its coal (Indonesia, Australia, Vietnam, Mongolia and other neighbouring countries), and they have yet to reveal the potential importance of underexplored countries, such as Laos. These numbers underline the importance of Southeast Asia to China's domestic development.

From a Chinese policy perspective, a key conundrum is how to secure long-term strategic resources amid diverse social, economic and political contexts, and a near complete range of diplomatic relations with China. In Southeast Asia, these concerns are compounded by existing geopolitical competition in the South China Sea, as well as China's concerns about the Malacca Straits "chokepoint" through which more than 80 per cent of its oil and gas imports.[43] Yu Hongyuan examines these dilemmas through the Chinese lens of "unbalanced development", spectres of resource nationalism, and the potential influence of major powers outside the region, most notably the United States. He shows that even as Chinese resource investments may be driven by profits, they cannot be neatly disentangled from Beijing's economic, political or geopolitical policy concerns.

The next two chapters offer complementary case studies of Chinese and Indonesian bilateral resource cooperation. Indonesia is China's largest resource trading partner and investment destination in Southeast Asia, largely due to the massive coal trade between the two countries. As Cecilia Springer describes, their bilateral coal trade was at one time sizable enough to influence global market prices, production levels, and production agendas for coal.

In her chapter on "Energy Entanglement: New Directions for the China–Indonesia Coal Relationship", Springer provides a close-up examination of China–Indonesia bilateral relationship through the lens of "old King coal". Taking us through three distinct phases of the relationship, she demonstrates how domestic patterns of energy consumption and transport logistics forced China to depend increasingly on Indonesian coal, despite China's own substantial coal production in its interior regions. After the 2008 Global Financial Crisis, the two countries effectively established a global duopoly on coal trade. However, trade began to diminish as China, on the one hand, increasingly dissociated itself from coal (reflecting wider global trends in the sector) and Indonesia, on the other, enacted policies to restrict the export of unprocessed mineral ores.

Yet even as coal trade has declined, new possibilities have emerged to reinforce bilateral relations, notably through Chinese investment in Indonesia's coal production infrastructure. While the future of the Chinese–Indonesian coal relationship may still be an open question, especially as India rapidly emerges as an important coal supplying alternative for China, Springer's analysis demonstrates how the strength of the bilateral coal trade has at times proven to be a bulwark against turbulent relations.

In the next chapter on "Indonesia–China Energy and Mineral Ties: The Rise and Fall of Resource Nationalism?", Zhao Hong and Maxensius Tri Sambodo combine perspectives of a Chinese and an Indonesian resource policy analyst, respectively, to examine Indonesian resource policy in a "context of fraught political and economic relations" (p. 105). More specifically, these authors question a suite of Indonesian policy measures that aim to protect national resource industries — and, more poignantly, specific resource companies and well-worn networks of resource patronage. These authors point out that not only are such "resource nationalist" measures often ineffective as economic policy, they can increase conflict at local levels in contexts of weak governance and a lack of transparency.

Zhao Hong and Sambodo demonstrate that both China and host countries use the tools of statecraft to protect resource rents and supply, whether this is within the national territory, as in Indonesia's case, or by acquiring resources from overseas, as in China's case. These are merely different forms of resource nationalism, which are visible in varying degrees across the region. As such, these authors show how bilateral relations both shape and have been shaped by resource cooperation. They emphasize that states continue to occupy important nodes in transnational resource production, although the particular role played by any one government may vary according to local contexts.

The following six chapters provide more grounded views and multi-level country studies of Chinese resource investments in Southeast Asia. The first three examine what happens when resource policies and investment strategies conceived in one place (such as government offices, company headquarters, or policy research institutes) hit the ground in other, often very different, places. Alvin Camba's chapter on "The Directions, Patterns, and Practices of Chinese Investments in Philippine Mining" challenges popular notions of Chinese resources companies dominating by their sheer size and state-backed resources. In contrasting them with "traditional" multinational resource companies, notably Canadian and Australian mining companies, Camba provides a different picture of how Chinese companies pursue mining sector projects in the Philippines.

As Camba demonstrates, rather than pursuing large-scale mining investments through central level agencies, Chinese mining companies tend towards small-scale mining. Small-scale mining enables Chinese companies to avoid national regulators and negative limelight, which has been especially important in a context of open geopolitical conflicts with China's encroachments in the South China Sea. Chinese companies also employ a wider range of methods for capital accumulation than "traditional" mining investors, which include labour-intensive extraction, household and community labour, and more flexible infrastructure. Being able to connect with diaspora networks further enables Chinese companies to work through local elites rather than national-level policymakers.

Camba emphasizes that these distinctions are not hard and fast, neither can they be strictly attributed to *Chinese* way of doing business. Rather they have historical roots in Japanese forms of non-hierarchical production overseas, where the foreign investing firm own the highest node of the production and then outsources the rest to local and regional actors. In this regard, Camba gives evidence of the important place-based differences that shape Chinese resource investment strategies and practices across the region.

In "Develop Cooperation with Chinese Characteristics: Opium Replacement and Chinese Rubber Investments in Northern Laos", Juliet Lu's study of the China–Laos Opium Replacement Program (ORP) examines a Chinese homegrown "model of development" as it travels to Laos. For Chinese investors, rubber is the "miracle crop" that developed and modernized the impoverished uplands of Yunnan, which borders with northern Laos. The topographic and ecological similarities of northern Laos to Yunnan, the low levels of socio-economic development among the Lao upland people (including several of the same ethnic groups as in the Yunnanese uplands), and global economic prospects of rubber appeared to set the stage for another development miracle. Yet, unfortunately, that miracle has yet to appear for most Lao rubber farmers.

In trying to understand why Chinese development worked in one place but not another, Lu highlights key differences in the levels of state support to the Chinese rubber industry, which protected it from global market volatility, allocated tenure rights to smallholders, provided market access services to producers, and oversaw relations with agribusiness companies. In contrast, Chinese rubber investors in Laos were hit hard when prices plunged in 2011. For Lao rubber farmers, weak tenure arrangements and a lack of regulation over rubber investors resulted in their livelihoods becoming more vulnerable and uncertain. Lu illustrates how models for resource development are forged in geographically specific histories, which are not easily transported from one place to the next. In Laos, the application and rhetoric of a Chinese development model was, at best, simplistic and, at worst, a disingenuous way of enabling Chinese rent-seeking in the Lao uplands.

Chinese investment and development cooperation is arguably more important to Cambodia than any other Southeast Asian country. Recently, China's annual FDI in Cambodia has been more than double the next largest investing country. An important part of these investments has been in agricultural and resource sector projects. As Siem Pichnorak argues in his chapter on "The High Cost of Effective Sovereignty: Chinese Resource Access in Cambodia", Chinese companies have been the largest foreign holders of economic land concessions, mining licenses and hydropower construction projects. The development benefits of these investments are important, but they have also led to allegations of land grabbing, deforestation and human rights abuses. In turn, incidents and allegations related to these investments generated conflict and protests at local, national and international levels.

To make sense of these competing processes, Siem argues that Chinese resource cooperation helps to enhance Cambodia's "effective sovereignty", which refers to the state's ability to effectively claim control over and govern its national territory. In other words, by conceding certain territories to foreign investors, the Cambodian state is able to reinforce its claims to sovereignty through land administration and a rhetoric of development in outerlying areas.

However, in the process, these partnerships can exacerbate already weak systems of land and resource governance, as well as lead to devastating social and environmental consequences. Siem makes clear that these dilemmas of resource cooperation offer no simple solutions. Rather, Siem argues for complex understandings of Cambodian land and resource governance, and the state's capacity for enacting and enforcing its own regulations.

The next three chapters focus more specifically on conflict and contestation, emphasizing how struggles over resources are often embedded within geographically specific histories of conflict. Myanmar is a fitting place to begin, as massive protests led the Myanmar government to suspend the Chinese-backed Myitsone hydroelectric dam in 2011, sending shock waves through China–Myanmar relations, as recounted in Diane Tang-Lee's chapter on "Complex Contestation of Chinese Energy and Resource Investments in Myanmar". For decades, China had been Myanmar's most reliable political ally and its top foreign investor. To China, Myanmar was an important region for securing natural resource supplies, as well as offering a strategic overland route for oil and gas from the Middle East and Africa to shortcut the Malacca "chokepoint". Some of China's largest investments in Myanmar have been in the resource sector, including the Myitsone hydroelectric dam, the Letpadaung copper mine, and both Sino–Myanmar oil and gas pipelines. Yet domestic protests surrounding each of these projects shows that local and civil society concerns do matter, despite the best laid plans of states and corporations.

Through ethnographic research with Chinese companies and Myanmar civil society actors, Tang-Lee identifies a complex mix of factors that hinder more meaningful engagement between these two sets of actors. They include the restrictive legal and political context for civil society in Myanmar, organizational limitations, and dismissive attitudes by company representatives, often based on their past experiences with civil society organizations in China. Not only have these problems hindered the development of more formal and regular communications, they have also undermined the efforts of Chinese companies to build trust and reputation with the wider Myanmar public. Tang-Lee's analysis questions China's pretenses of "non-interference" in a context where the central government has little support or trust among local or regional populations.

Vietnam has perhaps the longest history of political and economic cooperation with China among all Southeast Asian countries, as it has passed through nearly all variations of cooperative and conflictive relations.[44] Since the 1990s, however, Chinese trade and investment in Vietnam have been expanding rapidly. However, popular attitudes towards Chinese investment changed radically after the global financial crisis of 2008. In their chapter on "Anti-Chinese Protest in Vietnam: Complex Conjunctures of Resource Governance, Geo-politics and State-Society Deadlock", Jason Morris-Jung and Pham Van Min trace this shift in domestic attitudes by providing an overview of Chinese economic activity and the popular concerns that have emerged around it. They then examine two case studies of anti-Chinese protest in relation to resource sector projects as complex conjunctures of domestic conflict and tension.

Morris-Jung and Pham Van Min demonstrate how protests over Chinese-backed resource projects have been used to express popular grievances towards the government in a context where open criticism of political leaders is personally risky and routinely suppressed. The controversy over Chinese involvement in bauxite mining exposed recurring problems in Vietnam's land and resource governance, economic management, and government decision-making. However, these developments also reached a disturbing crescendo in 2014 when riots and violent attacks on Chinese workers at a steel factory in Central Vietnam broke out in response to Chinese encroachments in what Vietnam calls the Eastern Sea. Morris-Jung and Pham Van Min's analyses of these events show how these conjunctures of China, resources and governance have also generated new openings and risks for political contestation in Vietnam.

The Philippines is another country where resource conflict has been deeply entangled with territorial conflicts on what the Philippines calls the West Philippine Sea. Continuing with Camba's earlier discussion on Chinese involvement in small-scale mining, Menandro Abanes' chapter on the "Complexities of Chinese Involvement in Mining in the Philippines" focuses on the local level conflicts that have emerged around it. He begins with a paradox on why an injunction ordered by the Philippine Supreme Court on ninety-four small-scale mines in 2013 was popularly understood as an anti-China measure, rather than a pro-environment or even an anti-mining one.

After providing an overview of the Philippine mineral sector and its legal context, Abanes examines the complex local realities of mining conflict. In particular, he demonstrates the strong correlation between regional poverty incidence and mining sector activities, whose percentage rate, between 1988 and 2009, went from half to surpassing that of agriculture. Mining is the only sector where regional poverty rates have actually increased. While Abanes warns against deducing a clear causal relationship from statistics alone, he notes that they nonetheless challenge claims that mining reduces poverty. Amid the power asymmetries between transnational companies and local communities, Abanes highlights how enduring problems with resource governance, divided loyalties among local officials, historical conflicts with indigenous peoples, and challenges in regulating small-scale mining companies have combined with a broader malaise towards Chinese actors as a result of territorial disputes. In conclusion, Abanes suggests that Chinese companies' greater cognizance of national laws, responsiveness to local needs, and operations in more locally inclusive and transparent ways could improve the current situation.

The chapters in this volume disabuse us of notions that resource production is inherently cooperative or conflictive. Rather, they draw our attention to how one often presupposes the other, and how any assessment of resource production must be attuned to local history, geography and socio-economic conditions. As importantly, they show that shortcomings in Chinese companies' own resource, environmental or social management are also co-produced with a host of local and national level actors, including government regulators, partner companies, local leaders and local communities. That these conflicts are playing out against a backdrop of territorial conflict in the South China Sea adds further fuel to the fire. At the same time, such conflicts can also generate new channels for citizens and communities to express their grievances to their political leaders. Examining China's return to Southeast Asia from a resource lens highlights the wide cross-section of actors involved, the many interrelated issues often at stake, and both the diversity and complexity of their outcomes.

NOTES

1. Gavin Bridge, "Contested Terrain: Mining and the Environment", *Annu. Rev. Environ. Resour.* 29 (2004): 205–59.
2. Suisheng Zhao, "The China Model: An Authoritarian State-Led Modernization", in *Handbook on China and Developing Countries*, edited by Carla P. Freeman (Cheltenham, UK: Edward Elgar Publishing, 2015); Yos Santasombat, *Impact of China's Rise on the Mekong Region* (New York: Palgrave Macmillan, 2015).
3. Zhao, "The China Model: An Authoritarian State-Led Modernization"; Wenran Jiang, "Fuelling the Dragon: China's Rise and Its Energy and Resources Extraction in Africa", *The China Quarterly* 199 (2009): 585–609.
4. Elizabeth Economy and Michael Levi, *By All Means Necessary: How China's Resource Quest is Changing the World* (Oxford: Oxford University Press, 2014); David Shambaugh, *China Goes Global: The Partial Power* (New York: Oxford University Press, 2013); Dambisa Moyo, *Winner Take All: China's Race for Resources and What It Means for Us* (UK: Penguin, 2012); Chris Alden, Daniel Large, and Ricardo Soares de Oliveira, *China Returns to Africa: A Rising Power and a Continent Embrace* (New York: Columbia University Press, 2008); Chris Alden, *China in Africa* (London: Zed Books, 2007).
5. Economy and Levi, *By All Means Necessary*; Moyo, *Winner Take All*.
6. Julia Ebner, "The Sino–European Race for Africa's Minerals: When Two Quarrel a Third Rejoices", *Resources Policy* 43 (March 2015): 113.
7. Ruben Gonzalez-Vicente, "Mapping Chinese Mining Investment in Latin America: Politics or Market?", *The China Quarterly* 209 (March 2012): 37.
8. Mel Gurtov, "China's Third World Odyssey: Changing Priorities, Continuities, and Many Contradictions", in *Handbook on China and Developing Countries*, edited by Carla P. Freeman (Cheltenham, UK: Edward Elgar Publishing, 2015), p. 17.
9. Deborah Bräutigam, *The Dragon's Gift: The Real Story of China in Africa* (Oxford: Oxford University Press, 2009), p. 17.

10. Bräutigam, *The Dragon's Gift*; Deborah Bräutigam and Kevin P. Gallagher, "Bartering Globalization: China's Commodity-Backed Finance in Africa and Latin America", *Global Policy* 5, no. 3 (1 September 2014): 346–52.
11. Alden, *China in Africa*, p. 90.
12. Ian Scoones, Kojo Amanor, Arilson Favareto, and Gubo Qi, "A New Politics of Development Cooperation? Chinese and Brazilian Engagements in African Agriculture", *World Development, China and Brazil in African Agriculture* 81 (May 2016): 1–12; Ebner, "The Sino–European Race for Africa's Minerals"; Alden, Large, and Oliveira, *China Returns to Africa.*
13. Alden, Large, and Oliveira, *China Returns to Africa.*
14. Julia C. Strauss and Martha Saavedra, eds., *China and Africa*, vol. 9, *The China Quarterly Special Issues* (Cambridge: Cambridge University Press, 2010); Ching Kwan Lee, "se Managers, African Workers and the Politics of Casualization in Africa's Chinese Enclaves", *The China Quarterly* 199 (September 2009): 647–66; Bräutigam, *The Dragon's Gift.*
15. Ariel C. Armony and Julia C. Strauss, "From Going Out (*Zou Chuqu*) to Arriving In (*Desembarco*): Constructing a New Field of Inquiry in China–Latin America Interactions", *The China Quarterly* 209, no. 1 (March 2012): 4.
16. Deborah Bräutigam and Sigrid-Marianella Stensrud Ekman, "Briefing Rumours and Realities of Chinese Agricultural Engagement in Mozambique", *African Affairs* 111, no. 444 (1 July 2012): 10.
17. Jin Zhang, "China's Oil Industry, International Investment and Developing Countries", in *Handbook on China and Developing Countries*, edited by Carla P. Freeman (Cheltenham, UK: Edward Elgar Publishing, 2015), p. 298.
18. Rubén Gonzalez-Vicente, "Development Dynamics of Chinese Resource-Based Investment in Peru and Ecuador", *Latin American Politics and Society* 55, no. 1 (2013): 47.
19. Halina Ward, "Resource Nationalism and Sustainable Development: A Primer and Key Issues" (London: IIED, May 2009), available at <http:// pubs.iied.org/G02507/>; Ian Bremmer and Robert Johnston, "The Rise and Fall of Resource Nationalism", *Survival* 51, no. 2 (1 May 2009): 149–58.
20. Jody Emel, Matthew T. Huber, and Madoshi H. Makene, "Extracting Sovereignty: Capital, Territory, and Gold Mining in Tanzania", *Political Geography* 30, no. 2 (2011): 70–79.

21. Emel, Huber, and Makene, "Extracting Sovereignty"; John Agnew, "The Territorial Trap: The Geographical Assumptions of International Relations Theory", *Review of International Political Economy* 1, no. 1 (1 April 1994): 53–80.

22. Robert David Sack, *Human Territoriality: Its Theory and History* (Cambridge: Cambridge University Press, 1986), p. 19.

23. Ken MacLean, "Unbuilt Anxieties: Infrastructure Projects, Transnational Conflict in the South China/East Sea, and Vietnamese Statehood", *TRaNS: Trans-Regional and -National Studies of Southeast Asia* 4, no. 2 (July 2016): 365–85.

24. Nancy Lee Peluso and Peter Vandergeest, "Genealogies of the Political Forest and Customary Rights in Indonesia, Malaysia, and Thailand", *The Journal of Asian Studies* 60, no. 3 (2001): 761–812; Peter Vandergeest and Nancy Lee Peluso, "Territorialization and State Power in Thailand", *Theory and Society* 24, no. 3 (1995): 385–426.

25. Anthony Bebbington, Leonith Hinojosa, Denise Humphreys Bebbington, Maria Luisa Burneo, and Ximena Warnaars, "Contention and Ambiguity: Mining and the Possibilities of Development", *Development and Change* 39, no. 6 (2008): 887–914; Anthony Bebbington, "Underground Political Ecologies: The Second Annual Lecture of the Cultural and Political Ecology Specialty Group of the Association of American Geographers", *Geoforum*, Themed Issue: Spatialities of Ageing 43, no. 6 (November 2012): 1152–62.

26. Ibid.

27. Gonzalez-Vicente, "Development Dynamics of Chinese Resource-Based Investment in Peru and Ecuador".

28. Gonzalez-Vicente, "Mapping Chinese Mining Investment in Latin America", p. 37; Yu Hongyuan, this volume.

29. According to Jeffrey Wilson, China currently maintains or produces 45 per cent of the world's iron ore, 18 per cent of its bauxite, 53 per cent of its coal, 5 per cent of its crude oil, and 3 per cent of its natural gas. However, China's resource production has mainly provided outputs for downstream Chinese industries rather than export markets. As Wilson notes, resource industries contributed 6.9 per cent to Chinese GDP, but only 3.1 per cent to exports. Jeffrey D. Wilson, "Resource Powers? Minerals, Energy and the Rise of the BRICS", *Third World Quarterly* 36, no. 2 (1 February 2015): 223–39.

30. Santasombat, *Impact of China's Rise on the Mekong Region*, page unavailable.

31. Hsing-Chou Sung, "China's Geoeconomic Strategy: Toward the Riparian States of the Mekong Region", page unavailable.

32. The only exceptions are the region's two smallest states, Brunei and Timor-Leste. China is the number one source of imports for all of the region's major economies, except Laos (where it is second) and Cambodia. It is also the first or second most important destinations for their exports.

33. Zhao Hong, *China's Quest for Energy in Southeast Asia: Impact and Implications*, Trends in Southeast Asia 2015 #01 (Singapore: Institute of Southeast Asian Studies, 2015).

34. Bien Chiang and Jean Chih-yin Cheng, "Changing Landscape and Changing Ethnoscape in Lao PDR: On PRC's Participation in the Greater Mekong Subregion Development Project", in *Impact of China's Rise on the Mekong Region* (New York: Palgrave Macmillan, 2015), available at <http://www.palgrave.com/br/book/9781137476210>; Sung, "China's Geoeconomic Strategy: Toward the Riparian States of the Mekong Region".

35. Nguyen Van Chinh, "China's 'Comrade Money' and Its Social-Political Dimensions in Vietnam", in *Impact of China's Rise on the Mekong Region* (New York: Palgrave Macmillan, 2015), available at <http://www.palgrave.com/br/book/9781137476210>; Sung, "China's Geoeconomic Strategy: Toward the Riparian States of the Mekong Region"; Le Hong Hiep, "The Dominance of Chinese Engineering Contractors in Vietnam", *ISEAS Perspective* 2013 #04 (Singapore: Institute of Southeast Asian Studies, 17 January 2013).

36. Thomas Lum, Hannah Fischer, Julissa Gomez-Granger and Anne Leland, "China's Foreign Aid Activities in Africa, Latin America, and Southeast Asia", *Congressional Research Service: Report for Congress* (25 February 2009).

37. Zhu 2009, as cited in Terence Chong, "Chinese Capital and Immigration into CLMV: Trends and Impact", *ISEAS Perspective* 2013 #50 (Singapore: Institute of Southeast Asian Studies, 29 August 2013).

38. Pal Nyiri, *New Chinese Migration and Capital in Cambodia*, Trends in Southeast Asia 2014 #03 (Singapore: Institute of Southeast Asian Studies, 2014).

39. Danielle Tan, "'Small is Beautiful': Lessons from Laos for the Study of Chinese Overseas", *Journal of Current Chinese Affairs* 41, no. 2 (10 July 2012): 61–94.

40. Kevin Woods, "Ceasefire Capitalism: Military–Private Partnerships, Resource Concessions and Military–State Building in the Burma–China Borderlands", *Journal of Peasant Studies* 38, no. 4 (2011): 747–70; Tan, "'Small is Beautiful'".

41. Zhao Hong, *China's Quest for Energy in Southeast Asia*.

42. Fan Hongwei, "Enmity in Myanmar against China", *ISEAS Perspective* 2014 #08 (Singapore: Institute of Southeast Asian Studies, 17 February 2014).
43. Sara Hsu, "China's Energy Insecurity Glaring in South China Sea Dispute", *Forbes*, 2 September 2016, available at <http://www.forbes.com/sites/sarahsu/2016/09/02/china-energy-insecurity-south-china-sea-dispute/>.
44. Brantly Womack, *China and Vietnam: The Politics of Asymmetry* (Cambridge: Cambridge University Press, 2006).

2

MIXED MOTIVATIONS, MIXED BLESSINGS
Strategies and Motivations for Chinese Energy and Mineral Investments in Southeast Asia

Philip Andrews-Speed, Mingda Qiu and Christopher Len

Chinese energy and mineral companies have been investing overseas for more than twenty years, and the quantity and size of their projects have been growing steadily. Although Southeast Asia is not a preferred region for these investments, they remain significant on account of their relative value to the host country and because of the geostrategic importance of the region for China. Most of the companies making the investments or undertaking the projects are wholly or partly owned by the Chinese government at either central or local levels. As a result, the motivations for their investment activities reflect a mix of corporate and state objectives. Corporate objectives include securing energy or resource supply chains, increasing or diversifying their asset base, and enhancing their profits or market share. Government motivations range from direct support to companies for purposes of industrial strategy and resource security to indirect support through development assistance, diplomacy and regional strategic positioning.

This chapter presents an overview of the scores of investments and major infrastructure projects undertaken by Chinese companies in Southeast Asia in oil and gas, coal, hydroelectricity and metalliferous mining, showing how the mix of motivations for these activities varies between industries and, to a lesser extent, between host countries. While the economic benefits of these investments to host countries are in most cases evident, there are some risks. The lack of transparency and low operating standards that characterize some projects can weaken the social license to operate, creating risks for both the Chinese companies and the host governments

Introduction

China's economic engagement in Southeast Asia has grown markedly in recent years, and this includes the energy and mining industries. In addition to importing energy and mineral commodities from the region, Chinese enterprises have been investing in primary resources and constructing energy infrastructure. While the objectives of these companies are often primarily commercial, they are also multi-faceted and involve the ambitions of both the enterprises and the Chinese government. In addition, this mix of objectives varies between the different resource industries. The aims of this chapter are to identify the specific motivations of Chinese enterprises and government in their engagements with Southeast Asia's energy and mineral resources, and examine some of the implications from their mix of corporate and state drivers.

The analysis builds on the framework developed by John H. Dunning to explain the logic of overseas investment, while also extending application of Dunning's framework by incorporating motivations of Chinese and Southeast Asian governments. The following section provides background on the reasons and conditions for China's growing engagement in Southeast Asia's energy and mineral resources. Next, we explain Dunning's framework and then apply them to examine the drivers for the internationalization of Chinese companies in general and Chinese energy and mineral companies, more specifically. The subsequent section examines the nature of these engagements in Southeast Asia and specific motivations. In conclusion, we highlight implications for transparency, operating standards and social license in Southeast Asia.

China's Growing Engagement with Southeast Asian Resources

China is one of the largest producers of energy and mineral raw materials in the world.[1] It produces nearly 50 per cent of the world's coal and is the largest producer of non-energy minerals. Although its oil production amounts to just 5 per cent of the global total, it is the fourth largest producer after Russia, Saudi Arabia and the United States. China's gas production continues to rise and it is now the sixth largest producer.

Despite its status as a major producer of raw materials, China's sustained and rapid economic growth — combined with the resource intensive nature of this growth — has led to a dramatic rise in the country's import requirement for raw materials of all types, including oil, natural gas and non-energy minerals, as well as a small but growing quantity of electricity. Net imports of oil have risen steadily since the country became a net importer in 1993. They now account for nearly 70 per cent of domestic consumption. Gas is playing an increasing role in the national energy mix. Imports provide about 30 per cent of gas supply, a proportion that rises each year. In respect of coal, China fluctuates between being a net importer and a net exporter, depending on conditions in the domestic coal market. Imports of iron ore, copper, bauxite and nickel all grew rapidly from 2002 when the economy accelerated,[2] though the level of imports has declined since 2013 as economic growth has slowed. China also imports small amounts of electricity from Russia and Myanmar.

The growing requirements for importing raw materials triggered the internationalization of many of China's energy and mineral companies, for a variety of reasons which will be described in the following sections. In the oil and gas industry, the great majority of these investments, in terms of both the number of projects and their aggregate value, has been carried out by the four national oil companies (NOCs), namely the China National Petroleum Corporation (CNPC)/PetroChina, Sinopec, China National Offshore Oil Corporation (CNOOC) and Sinochem.[3] All four NOCs are owned at the central government level. Overseas investment in minerals has involved a much wider range of companies including those owned by subnational governments and by private investors.[4] However, state-owned enterprises (SOEs) owned at national or provincial levels hold the largest number of overseas projects that are directly controlled by the Chinese enterprise.[5] The total value of these overseas energy and mineral investments probably lies between US$100 billion and US$200 billion but, even so, Chinese companies account for only a small share of energy and mineral production outside China.[6]

A combination of economic growth and population increase will drive rising demand for all forms of primary energy and mineral ores in Southeast Asia and the wider Asian region. In Southeast Asia alone, energy demand may grow by 80 per cent between 2011 and 2035 (see Figure 2.1). The absolute quantity of energy used and the energy mix will depend greatly on policy decisions taken by governments to improve energy efficiency and reduce the share of coal and other fossil fuels in the energy mix. A further challenge will be to provide electricity and clean cooking energy to an estimated 130 million people in Southeast Asia who lack access to electricity and 280 million who cook using traditional biomass.

Southeast Asia possesses only a small share of the world's proven recoverable fossil fuel reserves, about 1 per cent for oil and between 3 per cent and 4 per cent for gas and coal.[7] It is currently a net importer of oil and these imports are likely to continue growing for the foreseeable future (see Figure 2.2). In contrast, the more generous reserves of natural gas and coal allow the region to be a net exporter, and production of these fossil fuels will continue to rise over the next two decades. Great potential exists for hydroelectricity, especially in the Greater Mekong Subregion.[8] A wide variety of mineral resources also exist in Southeast Asia,[9] but only Indonesia's bauxite is produced on a major scale, accounting for nearly 20 per cent of the global total.[10] Nonetheless, significant potential exists to raise the output of different mineral resources in Southeast Asia and to refine these ores within the region.[11]

In order to address Southeast Asia's energy requirements, it is estimated that investment in energy infrastructure between 2013 and 2035 needs to reach US$2.0 trillion, or about US$100 billion per year in 2012 US dollar terms.[12] Much of this investment may have to come from outside Southeast Asia, notably from international financial institutions, multinational enterprises, and foreign SOEs and their home governments. A similar argument applies to the mining sector.[13]

In the past, multinational energy and mining companies played a significant role in exploring and exploiting Southeast Asia's energy

FIGURE 2.1
ASEAN Primary Energy Demand by Source, IEA New Policies Scenario

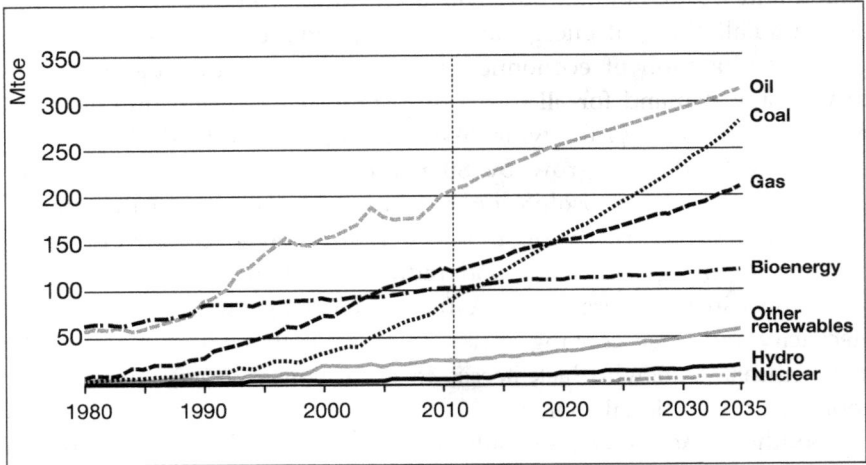

Source: International Energy Agency (2013).

FIGURE 2.2
ASEAN Fossil Fuel Production and Trade

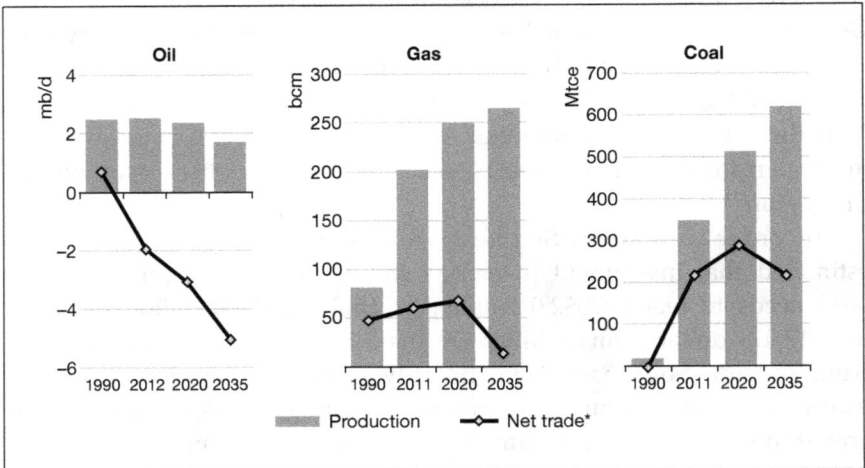

* Positive values are exports; negative values are imports.
Source: International Energy Agency (2013).

and mineral resources. However, their involvement in the region has steadily declined as larger deposits have become available for exploitation elsewhere and under more attractive fiscal and regulatory regimes. Their place is currently being filled by state-owned and state-backed enterprises from different Asian countries, among which China is the leading actor.[14] Chinese energy and mining companies have built a significant presence across Southeast Asia over the last twenty years, spending and committing tens of billions of US dollars on projects to extract, transform and transport energy and minerals.

The Logic of Overseas Investment: Dunning's Model of Foreign Direct Investment Strategies

Analyses of the drivers for international investment in the business studies literature start by identifying the motivations of the corporate actors, and include government motivations as part of the investment environment. In simple terms, when a company decides whether or not to invest outside its home territory, it has to address three key questions:

- What are the overall objectives of the investment strategy?
- To what sort of assets should investment be directed?
- What countries should be targeted for investment?

To address these questions in regards to Chinese energy and mineral companies, we take as our framework the work of John H. Dunning.[15]

Dunning identified four primary and three secondary objectives for overseas investment. The primary objectives concern the search for resources, markets, efficiency and strategic assets. Resource-seeking investments are made to secure resources, such as raw materials, that are of higher quality or lower cost than in the investor's home country. Firms may also invest abroad to gain access to new or larger markets for their products, or benefit from local conditions to improve efficiencies. Companies may also try to gain strategic assets such as technology, skills or managerial expertise. Dunning further outlines three other secondary motivations for companies to invest abroad: to escape from restrictive policies and legislation of the home, to provide support for their affiliates or subsidiaries, and capital gains (see Table 2.1).

TABLE 2.1

Simplified Classification of Range of Possible Motivations for Investor and Host Governments and Companies in the Energy and Resource Sectors

Goals of Investor Companies	Goals of Investor Governments	Goals of Host Governments
RESOURCE • Supply of commodity for own supply chain • Reserves replacement (future production)	ENERGY/RESOURCE STRATEGY: • Acquire sources of supply • Diversify sources of supply • Build relationship with suppliers • Enhance security of supply	ENERGY/RESOURCE STRATEGY: • Need for investment/ technology/labour/ skills to produce the energy/resources • Diversify sources of investment • New markets for resources/ commodities
MARKETS • New investment opportunities and profits; limited opportunities at home • New markets for technologies, equipment, labour, etc. EFFICIENCY • Not relevant? STRATEGIC ASSETS • Access to technology and skills	ECONOMIC POLICY Industrial policy: • Support for internationalization of key industries • Markets for technology/skills • Spin-off opportunities for other companies Employment policy: Labour export Financial: • Foreign exchange generation • Additional tax revenue (if profits are made) • Use of Sovereign Wealth Funds	ECONOMIC POLICY Industrial policy: • Need for technology, skills • Enhance or develop new/key industries (technology/skills) Employment policy: Labour import Infrastructure: • Associated infrastructure investments Financial: • Need for foreign investment • Tax revenue • Foreign exchange generation
OTHERS • Escaping regulatory and pricing constraints at home • Support subsidiaries/ affiliates	DIPLOMATIC/ STRATEGIC POLICY • Build/enhance bilateral relationship • Build/enhance regional influence • Aid policy	DIPLOMATIC/ STRATEGIC POLICY • Build/enhance bilateral relationship • Need for aid policy

Dunning also developed a framework for understanding the extent, pattern and foreign composition of overseas production, which he termed the "eclectic paradigm", and he regularly updated this framework over the course of his life to take into account the changing international environment. The approach focuses on three factors which take the form of specific advantages possessed by the firm or its potential investment opportunity: ownership, location and internalization. This "Ownership-Location-Internalization (OLI)" framework reflects the contexts of the firm, the home and host countries, and the specific industry and market in which the firm operates.

The ownership advantages of a firm include the possession of or exclusive access to specific assets (such as resources or technologies) and the ability to coordinate efficiently the management of these assets. A further type of asset takes the form of competitive advantages arising from the institutional context either in the company's home state or in the host country. In this respect, the attitudes and actions of the home and host governments can play decisive roles. Specific host countries may possess locational advantages (such as immobile natural resources and man-made endowments) or policy and economic characteristics that are attractive to foreign investors. The final dimension of the OLI framework is internalization, which addresses the question of whether or not the internalization or ownership of the asset yields a net benefit to the firm. In the terminology of transaction cost economics, if the transaction costs of using arms-length markets are greater than using internal hierarchies, then the company should invest in internalizing the asset. In principle, the more perfect the market, the lower the transaction costs and the less the need for such investment. But this ignores the desire of firms to gain access to resources, capabilities, markets, or assets.

Using Dunning's model of companies' foreign direct investment (FDI) strategies and the OLI framework to analyse the nature of those investments, we can now begin to untangle the complex mix of motivations that drive Chinese resource and mineral companies to invest in Southeast Asia. In the following sections, we discuss the reasons for the boom in the internationalization of Chinese companies in recent decades and then examine more specifically the overseas investment strategies and motivations of Chinese energy and mineral companies in Southeast Asia.

The Internationalization of Chinese Companies

Chinese companies are relatively new actors in the global economy compared to the more established transnational corporations from North America and Europe. Overseas investment by Chinese enterprises has grown ten times since 2005 and, in 2014, the value of outward direct investment (ODI) exceeded 90 per cent of the value of inward FDI.[16] By 2015, the value of China's ODI had overtaken that of inward FDI, which is the normal state of affairs in the United States, for example.

Dunning's model for FDI strategies can been applied to analyse the motives and nature of China's ODI. Although China is rich in many primary energy and mineral resources, growing demand has turned it into a net importer. Its search for primary natural resources has become an increasingly important driver of ODI over the last twenty years. As a consequence, resource companies account for a disproportionately large share of China's cumulative ODI.[17] Manufacturing companies have moved overseas to seek new markets, enhance or exploit their competitive advantages,[18] and acquire strategic assets, such as knowledge, technology and skills.[19] A final important objective for Chinese firms is to escape restrictive policies and market conditions at home, such as price controls and high barriers to entry.[20] Senior managers may also be trying to escape controls on their ability to obtain private benefits from corporate activities.[21]

Dunning's OLI framework can also help elucidate the nature of these investments. An important key to understanding the nature of China's ODI, especially those investments undertaken by SOEs, lies under the heading of ownership. Direct ownership by government provides SOEs with many competitive advantages both at home and abroad. At home, government policies help SOEs to have a strong market position, close ties to the government and suppliers, receive favourable treatment from banks, and have a high level of savings due to weak requirements for dividend payments.[22] Among the many reasons for the government to support overseas investment by SOEs, two are particularly important. The first one, as stated in the "Going Out" policy formulated in the year 2000, is to produce globally competitive enterprises.[23] This advantage is complemented by the government's willingness to use these companies as tools of diplomacy.[24] In order to particularize these two sets of objectives, the government periodically publishes guidelines and catalogues that explicitly identify those industries and countries for which outward FDI is encouraged.[25] These policies, combined with other non-commercial objectives, have provided a strong push for overseas investment by China's SOEs.

In the actual execution of investments, SOEs also receive government support in several forms, such as access to finance and foreign exchange, favourable tax treatment, and protection under inter-governmental agreements.[26] These benefits are coordinated between ministries and banks.[27] State support helps to compensate for the SOEs' relative lack of international experience and it is particularly important for companies investing in developing countries with weak institutions and high political risk.[28] In these contexts, China's SOEs also have the advantage of being accustomed to complex and opaque regulatory systems.[29]

When choosing a location to invest, aside from the issue of resource abundance, Chinese companies aggregated across all sectors appear to have been attracted to countries with various combinations of good governance, close trade ties to China, large markets, and rapid GDP growth.[30] While Chinese resource companies have gained a reputation for investing in countries with poor governance, this is not representative of Chinese companies as a whole. Investments also tend to be associated with countries with close cultural, geographic and institutional proximity.[31] The guidelines and catalogues published by the Chinese government, along with less visible government steering, also play significant, but not necessarily decisive, roles in determining the location of investments.

The relatively strong ability and willingness of Chinese companies to internalize their overseas activities over the last twenty years arise from the combination of their motives and their ownership characteristics, along with their being relative newcomers to outward FDI and feeling the need to catch up. The most important of these ownership characteristics is the strong political and economic support provided by China's government, which lessens the risks and costs of internalization through investment compared to those faced by conventional multinational companies. However, this support can be a double-edged sword, for it can also lead to inadequate assessment of investment risks and poor project management, which, in turn, cause tension and conflict with host governments and local communities. These problems have become particularly apparent in the resources industries, as discussed next.

Mixed Motivations for China's International Energy and Mineral Investments

In this section, we use the Dunning model to examine the complex motivations of their complex mix of motivations driving Chinese resource investments in Southeast Asia. We focus here on Chinese resource and mineral corporations, Chinese government, and Southeast Asian host governments to provide an analytical basis for discussing specific industries and resource projects in the next section.

Corporate Motivations

Within the overall objective of building the business and making profits, the main corporate motivation for Chinese overseas investments in energy and mineral resources is to gain access to resources. There are two components to the quest for resources (see Table 2.1). The first is the desire to diversify and secure supplies of the commodity for their own use back in China. In other words, it is a form of upstream vertical integration to acquire feedstock for industrial processes by companies facing a shortage of the required resources at home in China, notably in the oil, petrochemical and metallurgical sectors. Significant differences exist, however, between the mineral and petroleum industries. Mineral companies tend to ship their overseas product back to China to feed into their supply chains,[32] while also hedging against market control by dominant multinational corporations.[33] In contrast, most oil and gas produced overseas is sold on international markets. In this case, the main corporate purpose for securing overseas resources is a mix of replacing declining reserves, building or enlarging their reserve base, and vertical integration, depending on the company involved.[34]

Besides searching for resources, China's companies also want to build themselves into truly international corporations with a strong presence in international markets and able to secure their long-term profitability and survival. As part of this strategic move to internationalize, the companies have also made strategic acquisitions to gain access to technology and managerial skills.[35] Examples in the oil and gas industry include the acquisition of companies or assets in very deep offshore waters or in unconventional oil and gas fields.

In addition to the core activity of extracting resources, these resource projects also support the subsidiaries and affiliates of the Chinese companies through the provision of contracts for services, labour and technology.[36] The final motivation is to escape regulatory constraints at home in China. These include regulated prices for oil products and natural gas, as well as controls on private rent seeking.[37] Although concrete evidence for private rent seeking is rarely exposed, one recent extreme example involved the PetroChina Daqing Oilfield Company paying US$85 million for an oil field in Indonesia that was already depleted, suggesting that the management of the Chinese company had secured a private financial gain.[38]

Motivations of the Chinese Government

While Dunning's model was devised to explain company strategies, government policies and motivations form an important element of the company's investment environment. China's government has a strong interest in overseas activities, especially in the case of oil and gas which are seen as commodities of strategic importance. Its "Going Out" policy aims to build a number of international corporations among selected SOEs to be able to compete with the best in the world. Formally initiated in the year 2000, this policy built on the earlier drive in the 1990s to create "pillar industries"[39] by providing positive support for companies to go overseas in search of resources and markets.[40] Since the first catalogue was issued in 2004, oil, gas and minerals have featured prominently in official documents relating to outward investment.[41] Securing resources is also an important motivation for the energy and mineral sector. The government applies a mix of economic and diplomatic actions to manage the risk of supply disruptions.[42] This approach has been described as "strategic",[43] "neo-mercantilist"[44] or "hedging".[45]

In addition, overseas investment by energy and mineral companies addresses other economic goals, such as providing employment and generating foreign exchange and, possibly, taxable profits. The large scale and long duration of commitments related to some of these projects also provide China with diplomatic advantages, especially if the investments are backed up by loans and other economic and political engagement. Nevertheless, events can conspire to undermine such strategies, for example a change of leadership in the host government or poor operating practices of the Chinese company.[46] The following sections discuss how this close involvement of the government and the wide scope of engagement can create suspicion and resentment among the host country population, especially if the Chinese labour force is large or the companies' practices are poor.

Host Government Motivations

Host government policies and motivations are evidently important dimensions of the company's investment environment. The motivations for host government to seek or support investment by Chinese energy and mineral companies are highly diverse (see Table 2.1). Many energy- and mineral-rich nations encourage foreign investment in the extraction of their resources to bring in capital, technology or skills. At one extreme are countries which have very equitable and transparent policies for inward investment, under which Chinese companies are treated in the same way as companies from most other countries. At the other extreme are those nations which are experiencing sanctions from the international community (mainly western countries) and are happy to receive investment from any state. In between are governments that want to diversify sources of investment away from the long-standing multinational corporations. They may see China as a new market for their resource commodities, or have been unsuccessful at attracting investment from other parties for various economic or political reasons.

The host country will also try to gain a number of other economic advantages such as tax revenues, development of indigenous skills, technology and businesses, and possible infrastructure investments associated with the projects. In addition, the government may also be seeking to enhance its diplomatic relations with China for a mix of strategic and economic reasons, including the supply of development aid or military hardware. Corrupt practices can also play a role in motivating government officials to conclude contracts with Chinese companies. Such practices appear to be particularly prominent in resource-rich countries with poor governance characteristics.[47]

Nature of the Engagement

Applying the terminology of Dunning's OLI framework, the preceding sections have shown that China's energy and mineral companies possess a number of ownership advantages, most notably in the range of support they receive from their government and state-owned banks. In respect of choice of location, three factors appear to be important. First, and most important, is that the resource must be present, preferably at a large scale. As a result, Chinese investments in oil and gas have tended to target Central Asia, the Middle East, Africa and North and

South America,[48] while most metallic mining investment has targeted Australia, Canada, Latin America and Africa.[49]

Second, the Chinese company must have the opportunity to acquire rights to the resource. As a result, some Chinese investments, especially in the oil and gas industry, have been directed to countries which either had a poor resource base or were out-of-bounds for western companies for political reasons. Iran, Sudan and Myanmar are examples of the latter category.[50] In this way Chinese resource companies have gained a reputation for investing in countries with poor governance.[51]

Finally, all other things being equal, there appears to be a preference among some companies for geographic proximity to China on account of reduced transport costs. In some cases this includes neighbouring countries, such as in Southeast Asia, as this provides the opportunity to transport the commodities to China overland and thus avoid the sea lanes where security is controlled by the United States.[52]

Although China imports large quantities of oil, gas and minerals through international markets, decisions to internalize transactions through investment arises from a combination of the objectives of the Chinese companies and the government described above, and the ownership advantages the companies have through the support they receive. In these cases, the company may either acquire a share of the rights to a particular resource deposit, or may acquire a company that already owns those rights. The main exception to the practice of engagement in major international projects without internalizing the asset is found in the hydroelectricity sector where Chinese companies tend to provide the construction and engineering services, rather than take ownership. The reason for the different approach in the case of hydroelectric dams is that the output of the project, electricity, is usually sold into a domestic market at a price controlled by the host government, as opposed to an international market as is the case of oil and minerals. This reduces the likely profitability of the project, as electricity prices tend to be relatively low in most developing countries.

Specific Motivations in the Case of Southeast Asia's Energy and Mineral Resources

Southeast Asia offers distinct locational advantages to China's energy and mineral companies and to the government. These countries are close to or even immediately adjacent to China and, in some cases, they have political, economic and ethnic ties which date back centuries. The region has significant reserves of different forms of energy and mineral resources which Chinese companies can exploit and, in some cases, ship back to China or earn revenues through project construction and other services. In recognition of these advantages, the Catalogue guiding outward investment issued in 2004 lists a number of energy and mineral resources in Southeast Asian countries (see Table 2.2). The updates of the Catalogue issued since 2004 have not included any additional countries or resources in this region. For China, Southeast Asia is a region of strategic importance, not least on account of its sea lanes through which much of the nation's trade passes. As a result, the government has enhanced its political and economic engagement with the region, both bilaterally with individual countries as well as with the Association of Southeast Asian Nations (ASEAN) as a group.[53]

TABLE 2.2
Summary of Relevant Information from the
2004 Foreign Investment Industrial Guidance Catalogue

	Energy Resources	**Non-Energy Minerals**
Thailand		Potash, tungsten, antimony
Laos	Electricity	Potash
Myanmar	Oil, gas	Tungsten, nickel, copper, gems
Vietnam	Coal, electricity	Bauxite, iron, chromium
Singapore	Oil refining	
Philippines	Electricity	Copper, nickel
Malaysia		Gold
Indonesia	Oil, gas, electricity	
Brunei	Oil, gas	
East Timor	Oil, gas	
Papua New Guinea	Oil, gas	Copper

Source: Ministry of Commerce and Ministry of Foreign Affairs, *Foreign Investment Industrial Guidance Catalogue, Country Directory*, 2004.

Oil and Gas

Chinese investment in the oil and gas sector in Southeast Asia can be
dated back to 1993, the year that China first became a net importer
of oil. Between 1993 and 1995, CNOOC gained a share of the Malacca
oilfield in Indonesia, while CNPC, through its subsidiary PetroChina,
won blocks in Thailand and Papua New Guinea. The next phase of new
acquisitions in exploration and production began in 2001 and continued
for about ten years as CNOOC and CNPC built modest portfolios
across the region, although mainly in Indonesia and Myanmar (see
Table 2.3). The China Petrochemical Corporation (Sinopec), Sinochem
and the CITIC Group have also built up small positions. Information
on the companies' oil and gas reserves and production in Southeast
Asia is not readily available. In the case of Indonesia, by 2010, CNPC
had accumulated a total output of 5.8 million tonnes of oil and gas

TABLE 2.3
Estimated Number of Chinese NOC Investments in Southeast Asia

	CNPC/ PetroChina	Sinopec	CNOOC/ CNOOC Ltd.	Others	Subtotal
Brunei				1*	1
Cambodia			1	1	2
Indonesia	8	1	9	3	21
Myanmar	9**	1	4		12
Papua New Guinea	(2)		3		3
Philippines			1		1
Singapore	1*				1
Thailand	3				3
Total	19	2	18	5	44

Notes:
(i)   Some projects include multiple licenses or blocks.
(ii) Numbers in brackets () refer to projects that are known to have been relinquished or sold.
      These are not included in totals. Other projects may also have been relinquished or sold.
(iii) *Refers to a downstream (refining) project
(iv) **Includes two pipelines
Sources: Include Kong (2010), the websites of Chinese national oil companies, international
news agency articles, policy papers by international think-tanks, the Chinese press, and
various Chinese language websites.

equivalent,[54] while CNOOC had more than 15 million tonnes of oil reserves.[55] These are small numbers in proportion to the total output and reserves of the companies.

Since 2009, the largest share of overseas investment by China's NOCs has been directed to North and South America, the Middle East, and Australia, where the remaining reserves of oil and gas are much larger than in Southeast Asia.[56] In contrast, the NOCs in Southeast Asia have been directing their attention at mid-stream and downstream projects. For example, in 2009 PetroChina purchased a 96 per cent share of the Singapore Petroleum Corporation (SPC). In addition to a small number of upstream assets in Indonesia, Cambodia and Vietnam, SPC has interests in oil refining, pipelines, oil storage and bunkering, and service stations. In this way, the acquisition of SPC gave PetroChina a number of advantages. The company can enhance its relatively modest portfolio of refineries, gain access to Southeast Asian markets, learn advanced refining technology, and escape the highly regulated downstream market at home.

However, of much greater significance than upstream and downstream projects is CNPC's involvement in two pipelines running the length of Myanmar. In 2009, China and Myanmar agreed that CNPC would construct one gas and one oil pipeline from Myanmar's deep-water port of Kyauk Phyu to Kunming in China's Yunnan Province. The gas pipeline was commissioned in 2013 and carries the output from the offshore Shwe gas field, operated by Korea's Daewoo Corporation. Under the current arrangement, Myanmar can offtake 20 per cent of the throughput which is intended to rise gradually to a maximum of 20 billion cubic metres per year. The balance will flow to Yunnan. The oil pipeline was commissioned in January 2015 with an annual capacity of 20 million tonnes. It is fed with oil unloaded from tankers at Kyauk Phyu. Myanmar may tap up to 2 million tonnes per year, with the rest going to China.[57] The total cost for the two pipelines is reported to be as high as US$5 billion.[58]

China's motivations for these two pipelines are largely strategic in nature. The gas pipeline adds to the existing routes for importing natural gas which is seen as a key source of relatively clean energy to substitute for coal. The oil pipeline provides a marginal reduction in China's dependence on sea lanes for its oil imports and CNPC will receive feedstock for its refineries which are under construction in Southwest China. Despite these gains, the pipeline projects have run into a number of controversies on social and environmental grounds. In addition, a strong feeling exists in Myanmar that the country should receive a larger share of the pipelines' throughput.[59]

Hydropower

Chinese companies have two main objectives in Southeast Asia's hydropower sector: to win construction contracts and thus sustain their businesses, given the slowing down of dam construction in China; and, in some cases, to gain access to the resource to transmit the electricity back to China. The Chinese government supports these initiatives with financing through the state policy banks, for a mix of economic and diplomatic reasons.

The highest level of activity is in continental Southeast Asia, where hydropower resources are abundant in Myanmar and in the Mekong River Basin (especially Laos and Cambodia). These rivers provide excellent opportunities for Chinese companies to win construction contracts, for the Chinese government to build influence in its immediate neighbourhood, and, in a few cases, for the electricity to be transmitted to China.

Myanmar has the largest number of dams with Chinese involvement. The six plants built between 1996 and 2006 played an important role in boosting national power generating capacity.[60] We have documented 49 dam projects with Chinese involvement, including those completed (probably 10–15 in total), under construction, planned and suspended. More than twenty of these have a scale of 500 MW or above. The largest completed dam with Chinese involvement is the Yeywa dam at 790 MW, which was commissioned in 2011. In many cases, Chinese involvement is limited to construction services provided by companies such as SinoHydro. In other cases, the financing also comes from China, normally through the China Exim Bank or the China Development Bank. If the power is flowing back to China, enterprises in Southwest China, such as power generating companies, provide a significant share of the funding. An example is the 600 MW Shewli-1 dam completed in 2008 that sells most of its power to China.[61]

As well as projects under construction with a capacity of 1,000–1,600 MW, there are a small number of very large projects with a scale of several 1,000 MW. Most of these projects have been suspended due

TABLE 2.4
Estimated Number of Hydropower Projects with Chinese Involvement

	Subtotal	Large (> 500MW)	Key Projects
Brunei	1		
Cambodia	8	1	*Sambor Hydropower Project*: with designed capacity of 2,600 MW, proposed on the Mekong River. The feasibility study has received approval in 2012.
Indonesia	3		
Laos	30	4	*Pak Lay Dam*: with designed capacity of 1,320 MW, to be built on the Mekong River. China Exim Bank is providing US$270 million loan to Laos government for the dam.
Malaysia	14	4	*Baleh Hydropower Project*: with designed capacity of 1,295 MW, proposed on the Sejang River in Sarawak. The project is a crucial part for Malaysia's Sarawak Energy Corridor.
Myanmar	49	22	*Myitsone Dam*: with designed capacity of 6,000 MW, to be built on the Irrawady, N'Mai and Mali Rivers in Kachin State. *Tasang Dam*: with designed capacity of 7,100 MW, to be built on the Salween River in Karen State. *Hagyi Dam*: with designed capacity of 1,360 MW, to be built on the Salween River in Karen State. The total investment is projected as US$2.4 billion.
Papua New Guinea	1		
Philippines	4		
Thailand	3		
Vietnam	9		
Total	122	31	

Sources: Include the websites of Chinese investing companies, international news agency articles, reports by NGOs, policy papers by international think-tanks, the Chinese press, and other Chinese language websites.

to protests from local communities and wider civil society. The most notorious of these is the 6,000 MW Myitsone dam where construction work was suspended by Myanmar's government in 2011. In addition to concerns over social and environmental impact, there was rightful indignation that the previous government had agreed that 90 per cent of the electricity generated by the dam flow to China.[62] This dam is just one example of a number of projects that have undermined the economic relationship between China and Myanmar in recent years on account of the non-transparent relationship between the various parties and the practices of the Chinese enterprises.[63]

The dams built and planned in Laos and Cambodia tend to be on a smaller scale, with a few exceptions, such as the Sambor dam in Cambodia and the Pay Lak dam in Laos. Both of these proposed large dams have encountered opposition on social and environmental grounds and, as a result, construction had not started as of 2015. Overall, the engagement of Chinese enterprises in constructing large hydroelectric dams in Southeast Asia continues to encounter opposition. This resistance arises from a combination of poor management on the part of the Chinese companies and weak governance on the part of the host government. That being said, large-scale dams are controversial in most parts of the world, whoever is constructing them.

Mining

Unlike the oil and gas sector, the overseas investments in mining (both coal and metallic minerals) in Southeast Asia are carried out by both SOEs and private enterprises. The SOEs are further categorized into central SOEs and local SOEs. Central SOEs are directly invested and, to a certain extent, managed by the central government; whereas local SOEs are organized and invested by local, mainly provincial, governments. The most prominent central SOEs are very large enterprises, such as the Aluminium Corporation of China (Chinalco), the China Minmetals Corporation, the China Nonferrous Metal Mining Corporation (CNMC), the Shenhua Corporation, and the Guohua Corporation. They are also among the main Chinese companies supplying imported minerals to the domestic market. In addition to the formally established mining enterprises with officially granted mining licenses, there exists in Myanmar a poorly documented informal mining sector driven by Chinese investment which targets mainly precious and semi-precious stones.[64]

As explained in section above, China has a large import requirement for certain categories of minerals, and this determines the pattern of

ODI by the nation's mining companies, including in Southeast Asia. These countries have significant deposits of nickel, bauxite, copper and coal, which are the main targets for Chinese investors (see Table 2.5). The increase in investment in the region's mineral resources is in line with the increase of China's overseas mining investment globally. Compared to the oil and gas sector, which is generally profitable, some of the Chinese mining companies are incurring huge losses. Chinalco, for example, declared a loss of 8.2 billion yuan in its 2012 annual report. As a result, these enterprises are heavily dependent on state funding through the policy banks, namely the China Development Bank and the China Exim Bank. This evidence supports the contention that the mining companies are, to a certain extent, executing central government policy through their investments in Southeast Asia, both to supply resources for the domestic market and to enhance diplomatic relations. However, given the close relationship between government leaders and SOE executives, the exact nature of the principal–agent relationship is hard to unravel, a feature common to China's energy and minerals industries.[65]

TABLE 2.5
Mining Projects Invested by Chinese Companies

	Documented Projects	Minerals	Investing Companies
Indonesia	11	Nickel, iron, coal, bauxite	SOEs & private
Laos	5	Gold, bauxite, copper	SOEs & private
Malaysia	1	Iron ore	SOE
Myanmar	11	Nickel, copper, coal, zinc	Mainly SOEs
Papua New Guinea	1	Nickel, cobalt	SOE
Philippines	7	Nickel, coal	SOEs
Vietnam	3	Bauxite, copper	SOEs

Sources: Include the official website of the China Mining Association (<www.chinamining.org>), Chinese mining companies, articles by international news agencies, reports by NGOs, the Chinese press, and other Chinese language websites.

As has been the cases with oil and gas pipelines and hydroelectric dams, Chinese mining companies have been encountering challenges in their Southeast Asian operations. In Myanmar, the Letpadaung copper mine involves an investment of about US$1 billion. The Chinese company Wanbao (a subsidiary of Norinco) took over the project in 2011 after Ivanhoe of Canada withdrew. Local protests arose in 2012 from perceptions that compensation had been insufficient and that measures to protect the environment were inadequate. Production was suspended later that year. While these problems probably originated with Ivanhoe, the Chinese company has had to bear the legacy. A Myanmar government commission investigated the complaints and published a report in 2013. The report argued that production should restart subject to certain conditions being met. Despite Wanbao taking steps to address these conditions, the violence continues, including kidnappings and shootings.[66]

Vietnam has rich resources of bauxite (aluminium ore). In 2006, Chinalco was reported to have signed an agreement to invest US$1.6 billion in two mining and processing projects in the Central Highlands region, which would have given it a dominant position in this mining region. Most of the processed alumina would be exported to China for aluminium smelting. This arrangement was, however, subsequently downgraded to engineering, procurement and construction (EPC) contracts with a Chinalco subsidiary company. Two years later, major domestic opposition to these projects started to appear. The reasons for opposition were multiple and complex. They included the large influx of Chinese workers, which was perceived as an unwelcome intrusion on local populations and as a national security threat by such prominent individuals as the retired General Vo Nguyen Giap.[67] There were also strong environmental objections on account of the large quantity of toxic waste produced from processing alumina. Despite these and other objections, the Vietnamese government has been determined to carry on with the projects.[68] Recently, however, it has emerged that the projects are encountering technical problems and incurring substantial financial losses, which affect both Chinalco and its Vietnamese partner, VINACOMIN.[69]

In Papua New Guinea, the Metallurgical Corporation of China (MCC) has encountered a number of obstacles in bringing its US$2 billion Ramu nickel and cobalt mine into production, having taken over operation of the project in 2005. The preference of both the Papuan and Chinese governments for conducting discussions at a national level precluded the involvement of local governments, land-owners and communities. Disputes have arisen over such issues as the disposal of waste and tailings, the large size of the Chinese labour force during construction, low wages paid to local workers, and land access.[70] These disputes delayed the commissioning of the mine to 2012. However, MCC appears to have learned many lessons and the mine was operating at 72 per cent of its nameplate capacity at the end of 2014,[71] despite a violent attack on the mine by local people in August of that year.[72]

Perhaps the misfortune of Chinese mining companies, from a business perspective, is that they have internationalized during a period when the behaviour of the international mining industry is being increasingly scrutinized on environmental, social and wider sustainability grounds. The failure of Chinese enterprises to recognize these challenges and to adapt their practices from those they deploy at home has resulted in tension and conflict with host governments and local populations in Southeast Asia and in other parts of the world.

Conclusions

Among Asian energy and resource companies, Chinese companies have the largest footprint in Southeast Asia. These are generally, but not always, large state-owned companies with substantial workforces and access to generous financing. In this chapter we have applied John Dunning's framework to examine corporate motivations for these investments. These drivers vary between industries, but the overwhelming objective among them is to gain access to resources, either to send back to China or boost the individual companies' production. Overseas investment also gives them access to international markets, technology and skills, and allows them to escape domestic regulatory constraints. Although not the richest region in the world in terms of energy and mineral raw materials, Southeast Asia does have a range of resources of interest to Chinese companies and has the advantage of being close or immediately adjacent to China with strong historic ties.

Home and host governments of these companies play a very important role in these undertakings, a factor on which Dunning's framework does not place sufficient weight. The Chinese government provides these energy and resource enterprises with policy support through financial and diplomatic channels for a variety of reasons. Principally, it wants to promote the internationalization of large SOEs, to enhance the security of supply for critical raw materials, and to use this economic engagement for diplomatic purposes. This latter objective has particular resonance in Southeast Asia, a region of particular strategic importance to China. For host governments, Chinese companies with home government support bear the promise of resource development, infrastructure construction, fiscal revenues and, in some cases, raw material or energy supply. The close relationships between governments and companies, however, may also create opportunities for rent seeking and corruption.

This rapid surge to extract the resources of its neighbours has not met with unalloyed success on account of the limited international experience of the Chinese enterprise and poor standards of governance on the part of the host nations. A small number of notorious cases have led to disputes, project delays, financial losses and occasional violence, all of which have led to growing levels of distrust and loss of reputation for China. Such cases can also undermine trust between communities and their own governments.

Two-way trade between China and ASEAN is expected to reach US$500 billion by 2015 and a target has been set to boost trade to US$1 trillion by 2020.[73] While Chinese investments patterns have been evolving over time to increasingly feature the real estate and transport sectors in Southeast Asia,[74] the energy, gas and mining sectors will continue to feature highly among Chinese businesses looking for overseas growth opportunities.

From a broader perspective, the increasing involvement of Chinese energy and resource companies may prove a mixed blessing to Southeast Asia on account of state-centric motivations, lack of international experience, and occasionally poor operating practices. Although these enterprises bring much needed investment to the region, there is a risk that strong strategic motivations will undermine Southeast Asia's development by failing to deliver sufficient benefits for host nations and their peoples.

NOTES

1. Magnus Ericsson, "Mineral Supply from Africa: China's Investment Inroads into the African Mineral Resource Sector", *The Journal of the Southern African Institute of Mining and Metallurgy* 111 (July 2011): 497–500; BP, *Statistical Review of World Energy 2015* (London: BP, 2015).

2. David Humphreys, "New Mercantilism: A Perspective on How Politics is Shaping World Metal Supply", *Resources Policy* 39 (2013): 341–49.

3. Julie Jiang and Jonathan Sinton, *Overseas Investments by Chinese National Oil Companies: Assessing the Drivers and Impact* (Paris: OECD/IEA, 2011); Julie Jiang and Chen Ding, *Update on Overseas Investments by China's National Oil Companies: Achievements and Challenges since 2011* (Paris: OECD/IEA, 2014).

4. Ericsson, "Mineral Supply from Africa".

5. Ruben Gonzales-Vicente, "Mapping Chinese Mining Investment, with a Focus on Latin America", paper presented at the China–Latin America meeting at UCLA Asia Institute, 15–16 April 2011.

6. Philip Andrews-Speed and Roland Dannreuther, *China, Oil and Global Politics* (London: Routledge, 2011); Ericsson, "Mineral Supply from Africa".

7. BP, *Statistical Review of World Energy 2015*.

8. International Energy Agency, *Southeast Asia Energy Outlook 2013* (Paris: OECD/IEA, 2013).

9. Antimony, bauxite (aluminium), chromium, cobalt, copper, gold, iron, nickel, tin, tungsten and zinc.

10. British Geological Survey, *World Mineral Production 2009–2013* (Nottingham: British Geological Survey, 2014).

11. Institute for Essential Services Reform, *The Framework for Extractive Industries Governance in ASEAN* (Jakarta: Institute for Essential Services Reform, 2014).

12. International Energy Agency, *Southeast Asia Energy Outlook 2013*.

13. Institute for Essential Services Reform, *The Framework for Extractive Industries Governance in ASEAN*.

14. Jeffrey D. Wilson, "Northeast Asian Resource Security Strategies and International Resource Politics in Asia", *Asian Studies Review* 38, no. 1 (2014): 15–35; Zhao Hong, *China's Quest for Energy in Southeast Asia: Impact and Implications*, Trends in Southeast Asia 2015 #01 (Singapore: Institute of Southeast Asian Studies, 2015); Philip Andrews-Speed, "China and Russia's Competition for East and Southeast Asia Energy Resources", in *The EU–China Relationship: European Perspectives. A Manual for Policy Makers*, edited by Kerry Brown (London: Imperial College Press, 2015).

15. John H. Dunning, "The Eclectic Paradigm as an Envelope for Economic and Business Theories of MNE Activity", *International Business Review* 9 (2000): 163–90; John H. Dunning, "Towards a New Paradigm of Development:

Implications for the Determinants of International Business Activity", *Transnational Corporations* 15, no. 1 (2006): 173–227; John H. Dunning and Sarianna M. Lundan, "Institutions and the OLI Paradigm of the Multinational Enterprise", *Asia Pacific Journal of Management* 25 (2008): 573–93.

16. United Nations Conference on Trade and Development, *World Investment Report 2015, Annex Tables*, available at <http://unctad.org/en/Pages/ DIAE/World%20Investment%20Report/Annex-Tables.aspx> (accessed 4 April 2015).

17. Peter J. Buckley, L. Jeremy Clegg, Adam R. Cross, Xin Liu, Hinrich Voss and Ping Zheng, "The Determinants of Chinese Outward Investment", *Journal of International Business Studies* 38 (2007): 499–518; Xiaoxi Zhang and Kevin Daly, "The Determinants of China's Outward Investment", *Emerging Markets Review* 12 (2011): 389–98; Filip De Beule and Daniel van den Bulcke, "Locational Determinants of Outward Foreign Direct Investment: An Analysis of Chinese and Indian Greenfield Investments", *Transnational Corporations* 21, no. 1 (2012): 1–34.

18. Daphne W. Yiu, Chungming Lau and Gary D. Bruton, "International Venturing by Emerging Economy Firms: The Effects of Firm Capabilities, Home Country Networks, and Corporate Entrepreneurship", *Journal of International Business Studies* 38 (2007): 519–40.

19. Mark Yaolin Wang, "The Motivations Behind China's Government-Initiated Industrial Investments Overseas", *Pacific Affairs* 75, no. 2 (2002): 187–206; Randall Morck, Bernard Yeung and Minyuan Zhao, "Perspectives on China's Outward Foreign Direct Investment", *Journal of International Business Studies* 39 (2008): 337–50; Ping Deng, "Why Do Chinese Firms Tend to Acquire Strategic Assets in International Expansion?", *Journal of World Business* 44 (2009): 74–84; De Beule and van den Bulcke, "Locational Determinants of Outward Foreign Direct Investment".

20. Deng, "Why Do Chinese Firms Tend to Acquire Strategic Assets in International Expansion?"; Yadong Luo, Qiuzhi Xue and Binjie Han, "How Emerging Market Governments Promote Outward FDI: Experience from China", *Journal of World Business* 45 (2010): 68–79.

21. Morck et al., "Perspectives on China's Outward Foreign Direct Investment".

22. Yiu et al., "International Venturing by Emerging Economy Firms"; Morck et al., "Perspectives on China's Outward Foreign Direct Investment".

23. Deng, "Why Do Chinese Firms Tend to Acquire Strategic Assets in International Expansion?".

24. Wang, "The Motivations Behind China's Government-Initiated Industrial Investments Overseas".

25. Ministry of Commerce and Ministry of Foreign Affairs, *Foreign Investment Industrial Guidance Catalogue, Country Directory* (Beijing: August 2004); National Development Reform Commission et al., *2006 Catalogue of*

Industries for Guiding Outward Investment (Beijing: 2006); Ministry of Commerce et al., *Foreign Investment Industrial Guidance Catalogue, Country Directory* (Beijing: 2007); Ministry of Commerce et al., *Foreign Investment Industrial Guidance Catalogue, Country Directory* (Beijing: 2011).

26. Luo et al., "How Emerging Market Governments Promote Outward FDI: Experience from China"; Jiangyong Lu, Xiaohui Liu, Mike Wright and Igor Filatotchev, "International Experience and FDI Location Choices of Chinese Firms: The Moderating Effects of Home Country Government Support and Host Country Institutions", *Journal of International Business Studies* 45 (2014): 428–49.

27. Luo et al., "How Emerging Market Governments Promote Outward FDI: Experience from China"; Henry Sanderson and Michael Forsythe, *China's Superbank. Debt, Oil and Influence — How China Development Bank is Rewriting the Rules of Finance* (Singapore: John Wiley, 2013); Elizabeth C. Economy and Michael Levi, *By All Means Necessary: How China's Resource Quest is Changing the World* (Oxford: Oxford University Press, 2014).

28. Buckley et al., "The Determinants of Chinese Outward Investment".

29. Morck et al., "Perspectives on China's Outward Foreign Direct Investment".

30. Zhang and Daly, "The Determinants of China's Outward Investment"; De Beule and van den Bulcke, "Locational Determinants of Outward Foreign Direct Investment"; Julan Du, Kai Wang and Yongqin Wang, "Political Determinants of the Location Choice and Entry Mode of Chinese Outward FDI", paper presented at the 6th Annual Joint Workshop on Socio-Economics co-sponsored by Fudan University, University of Paris 1 and FERDI, Paris, June 2014.

31. Buckley et al., "The Determinants of Chinese Outward Investment"; De Beule and van den Bulcke, "Locational Determinants of Outward Foreign Direct Investment".

32. Ericsson, "Mineral Supply from Africa"; Humphreys, "New Mercantilism".

33. Michael Komesaroff, "Screwing Up in Foreign Climes", *China Economic Quarterly* (June 2012): 9–11.

34. Bo Kong, *China's International Petroleum Policy* (Santa Barbara: Praeger Security International, 2010); Andrews-Speed and Dannreuther, *China, Oil and Global Politics*; Jiang and Sinton, *Overseas Investments by Chinese National Oil Companies*; Hongyi Lai, Sarah O'Hara and Karolina Wysoczanska, "Rationale of Internationalization of China's National Oil Companies: Seeking Strategic Natural Resources, Strategic Assets or Sectoral Specialization?", *Asia Pacific Business Review* 21, no. 1 (2014): 77–95.

35. Economy and Levi, *By All Means Necessary*; Lai et al., "Rationale of Internationalization of China's National Oil Companies".

36. Kong, *China's International Petroleum Policy*.

37. Andrews-Speed and Dannreuther, *China, Oil and Global Politics*; Economy and Levi, *By All Means Necessary*.

38. Charlie Zhu, David Lague and Fergus Jensen, "Depleted Oil Field is Window into China's Corruption Crackdown", Reuters, 19 December 2014, available at <http://www.reuters.com/article/2014/12/19/us-china-corruption-indonesia-specialrep-idUSKBN0JX00720141219> (accessed 14 January 2015).

39. Peter Nolan, *China and the Global Business Revolution* (Basingstoke: Palgrave, 2001).

40. Duncan Freeman, "China's Outward Investment: Institutions, Constraints, and Challenges", Brussels Institute of Contemporary China Studies, *Asia Paper* 7, no. 4 (May 2013).

41. Guidelines and catalogues 2004, 2006, 2007, 2010. See endnote 25.

42. Kong, *China's International Petroleum Policy*; Monique Taylor, *The Chinese State, Oil and Energy Security* (Basingstoke: Palgrave Macmillan, 2014).

43. Philip Andrews-Speed, Xuanli Liao and Roland Dannreuther, "The Strategic Implications of China's Energy Needs", *Adelphi Paper* 346 (2002).

44. Kenneth Lieberthal and Mikkal Herberg, "China's Search for Energy Security", *NBR Analysis* 17, no. 1 (2006).

45. Oystein Tunsjo, *Security and Profit in China's Energy Policy: Hedging Against Risk* (New York: Columbia University Press, 2013).

46. Andrews-Speed and Dannreuther, *China, Oil and Global Politics*.

47. Ivar Kolstad and Arne Wiig, "What Determines Chinese Outward FDI?", Chr. Michelsen Institute, Working Paper No. 2009/3 (2009); Economy and Levi, *By All Means Necessary*.

48. Jiang and Sinton, *Overseas Investments by Chinese National Oil Companies*; Jiang and Ding, *Update on Overseas Investments by China's National Oil Companies*.

49. Ericsson, "Mineral Supply from Africa".

50. Andrews-Speed and Dannreuther, *China, Oil and Global Politics*.

51. Kolstad and Wiig, "What Determines Chinese Outward FDI?".

52. Andrews-Speed and Dannreuther, *China, Oil and Global Politics*; Economy and Levi, *By All Means Necessary*.

53. See, for example, Prashanth Parameswaran, "Beijing Unveils New Strategy for ASEAN–China Relations", The Jamestown Foundation, *China Brief* XIII, no. 21 (2013): 9–12; Zhao Hong, "China's FDI into Southeast Asia", *ISEAS Perspective* 2013 #08 (Singapore: Institute of Southeast Asian Studies, 31 January 2013).

54. China National Petroleum Corporation, "CNPC in Indonesia", available at <http://www.cnpc.com.cn/en/cnpcworldwide/indonesia/PageAssets/Images/CNPC%20in%20Indonesia.pdf> (accessed 21 May 2015).

55. China National Offshore Oil Corporation, "Key Operating Areas — Indonesia", available at <http://www.cnoocltd.com/encnoocltd/AboutUs/zygzq/Overseas/1639.shtml> (accessed 21 May 2015).

56. Jiang and Ding, *Update on Overseas Investments by China's National Oil Companies*.

57. Zhao Hong, "The China–Myanmar Energy Pipelines: Risks and Benefits", *ISEAS Perspective* 2013 #30 (Singapore: Institute of Southeast Asian Studies, 15 May 2013).

58. Yingjie Sun, "Sino–Myanmar Oil and Gas Pipelines: An Expensive Lesson for China?", *New York Times*, 11 July 2013, available at <http://cn.nytimes.com/china/20130711/cc11myanmar/> (accessed 2 March 2015).

59. Leslie Hook, "China Starts Importing Natural Gas from Myanmar", *Financial Times*, 29 July 2013, available at <http://www.ft.com/intl/cms/s/0/870f632c-f83e-11e2-92f0-00144feabdc0.html#axzz43VwWZ15P> (accessed 17 November 2013).

60. Toshihiro Kudo, "Myanmar's Economic Relations with China: Who Benefits and Who Pays?", in *Dictatorship, Disorder and Decline in Myanmar*, edited by Monique Skidmore and Trevor Wilson (Canberra: ANU E Press, 2008).

61. Toshihiro Kudo, *China's Policy Toward Myanmar: Challenges and Prospects* (Chiba: IDE-JETRO, 2012).

62. International Rivers, "The Myitsone Dam on the Irrawaddy River: A Briefing", 28 September 2011, available at <http://www.internationalrivers.org/resources/the-myitsone-dam-on-the-irrawaddy-river-a-briefing-3931> (accessed 9 February 2015).

63. Su-Ann Oh and Philip Andrews-Speed, *Chinese Investment and Myanmar's Shifting Political Landscape*, Trends in Southeast Asia 2015 #16 (Singapore: Institute of Southeast Asian Studies, 2015).

64. Oh and Andrews-Speed, "Chinese Investment and Myanmar's Shifting Political Landscape".

65. Economy and Levi, *By All Means Necessary*; Janet Xuanli Liao, "The Chinese Government and the National Oil Companies (NOCs): Who is the Principal?", *Asia Pacific Business Review* 21, no. 1 (2015): 44–50.

66. *The Economist*, "Chinese Miner Tries to be Nice", 24 May 2014, available at <http://www.economist.com/news/business/21602719-chinese-miner-tries-be-nice-kidnapped> (accessed 23 April 2015); *The Irrawaddy*, "Protests Continue Against Letpadaung Copper Mine", 19 January 2015, available at <http://www.irrawaddy.org/burma/protests-continue-letpadaung-copper-mine.html> (accessed 12 March 2015).

67. *The Economist*, "Bauxite Bashers: The Government Chooses Economic Growth over Xenophobia and Greenery", 23 April 2009, available at <http://www.economist.com/node/13527969> (accessed 28 March 2015);

Le Hong Hiep, "The Dominance of Chinese Engineering Contractors in Vietnam", *ISEAS Perspective* 2013 #04 (Singapore: Institute of Southeast Asian Studies, 17 January 2013).

68. *Vietnam Business Forum*, "Bauxite Project Still Effective", 21 April 2015, available at <http://vccinews.com/news_detail.asp?news_id=32009> (accessed 17 May 2015).

69. Hunter Marston, "Bauxite Mining in Vietnam's Central Highlands: An Arena for Expanding Civil Society?", *Contemporary Southeast Asia* 34, no. 2 (2012): 173–96; Claire Mai Colberg, "Catching Fish with Two Hands: Vietnam's Hedging Strategy Towards China", a thesis submitted to the Interschool Honors Program in International Security Studies, Stanford University, June 2014.

70. Yingjie Guo, Shumei Hou, Graeme Smith and Selene Martinez-Pacheco, "Chinese Outward Directed Investment: Case Studies of SOEs Going Global", in *Law and Policy for China's Market Socialism*, edited by John Garrick (London: Routledge, 2012).

71. Highlands Pacific, "Ramu Nickel", available at <http://www.highlandspacific.com/current-projects/ramu-nickel> (accessed 27 May 2015).

72. Sonali Paul, "China's Ramu Nickel Mine in PNG Restarts after Attacks", Reuters, 7 August 2014, available at <http://in.reuters.com/article/2014/08/07/papua-nickel-ramu-idINL4N0QD0GY20140807> (accessed 27 May 2015).

73. Xinhua News Agency, "ASEAN–China Trade Expected to Reach 500 bln USD by 2015", 14 November 2014, available at <http://news.xinhuanet.com/english/china/2014-11/14/c_133788265.htm> (accessed 17 March 2015).

74. Sarah Oliver and Kevin Stahler, "Can Japan Tell Us Where Chinese FDI is Going in ASEAN?", Peterson Institute for International Economics, 3 July 2014, available at <http://blogs.piie.com/china/?p=3944> (accessed 29 April 2015).

3

MINERAL RESOURCES IN CHINA'S "PERIPHERY" DIPLOMACY

Yu Hongyuan

This chapter analyses key challenges that China currently faces in securing supplies of strategic mineral resources from its neighbouring or "periphery" regions. It provides some perspectives on Chinese state interests in ensuring long-term mineral supply from its neighbouring mineral-rich countries, as well as offer policy recommendations on how China can improve on resource diplomacy in its "peripheries". The chapter does not aim to be comprehensive in its review of these minerals, but instead surveys some important minerals to reflect on how Chinese strategies for sourcing minerals from peripheral regions are relevant to China's mineral investments in Southeast Asia.

Introduction

"Strategic minerals" can be defined as natural resources that are important to a state's economic development, national security or people's livelihoods. The US Geological Survey (USGS) includes additional criteria for defining "strategic minerals" as commodities that are not easily replaceable yet they are extracted mainly from foreign sources.[1] Among industrialized nations, competition over the exploitation, production and accumulation of strategic minerals have come under international media attention.

As China's economy has been developing rapidly since economic reforms that started in 1979, its consumption and import of strategic mineral resources has also increased significantly. According to the Chinese Academy of Engineering's definition of forty-five strategic minerals, China is in short supply or cannot guarantee adequate supply of twenty-seven different kinds of strategic minerals. More than half of them are currently imported from Southeast Asia and Australia, mineral-rich regions proximate to China. Southeast Asia is also China's most important supplier for nickel, aluminum ore (i.e., bauxite) and coal. As China's demand for strategic minerals is likely to keep increasing, based on projected slowing but continuing growth, China will need to consider several important factors to ensure accurate and realistic assessments of availability and accessibility for long-term resource supplies, especially from its neighbouring regions such as Southeast Asia.

This chapter analyses the key challenges China currently faces in securing supplies of strategic mineral resources from its neighbouring or "periphery" regions (translated as "zhoubianguojia" in Chinese *hanyu pinyin* romanization), which refer to those countries and regions surrounding the Chinese overland borders and coastal areas. The "periphery" is a necessarily loose categorization, but it has particular significance because of the particular geopolitical and close economic relations that China shares with these countries and regions historically. Southeast Asia is a necessarily important part of this periphery, where national and local-level protests against Chinese investment and escalating geopolitical tensions in the South China Sea make for complex political and economic relations. This chapter provides perspective on Chinese state interests in ensuring long-term mineral supply from its neighbouring mineral-rich countries, as well as offers policy recommendations on how China can improve on resource diplomacy in its "peripheries". The chapter does not aim to be comprehensive in its assessment of all strategic minerals, but rather provides a broad survey of key minerals to reflect on how Chinese strategies for sourcing minerals from its peripheral regions are relevant to China's mineral investments in Southeast Asia.

China's Growing Demand for Strategic Mineral Resources from Its "Peripheral" Regions

Over the past couple of decades, the rise and development of the Chinese economy has provided an important stimulus to the global economy. Much of it has come from the rise of China, often characterized as the "World Factory" for a large percentage of the world's consumer goods. In particular, China's consumption of natural resources has reached new and unprecedented levels during this period (the reform era from 1979 to the twenty-first century), which has boosted commodity markets and natural resource production around the world. It has been a windfall for many mineral-rich economies. According to Lin Yifu, ex-chief economist and senior vice president at the World Bank and pioneering director of the China Center for Economic Research, in the next ten years or so, the Chinese economy will continue to develop at an average annual growth rate of 8 per cent, although this figure is now realistically pegged by the Chinese government as between 5–7 per cent. (The Chinese government has started using a range instead of a definitive single digit to characterize and project its economic growth from recent times.) In this context, China's demand for resources will only keep on growing at a substantial scale given the growing base of its consumption, even despite slowing economic growth. Therefore, China will continue to play, along with other large emerging economies like India, a vital role in the consumption of global strategic mineral resources. In other words, the global outlook for these mineral-dependent industries and economies remains strong despite short-term price fluctuations.

In 2011, Chinese consumption of zinc and tin was almost 90 per cent of total global consumption, while its consumption of copper and aluminum was near or over 40 per cent (see Figure 3.1).

China's levels of strategic mineral resources security can also be measured by its external dependence and import concentration. External dependence is the ratio of net imports of a kind of mineral resource to the consumption of it. In other words, the closer to 100 per cent the higher the dependence. China's dependence on external sources are high for the following metals (figures from 2010): iron 51 per cent; copper 72 per cent; aluminum 48 per cent; and lead 26 per cent, although there are no universally-accepted current

FIGURE 3.1
Consumption of Strategic Minerals as Percentage of Total Global Consumption, 2011

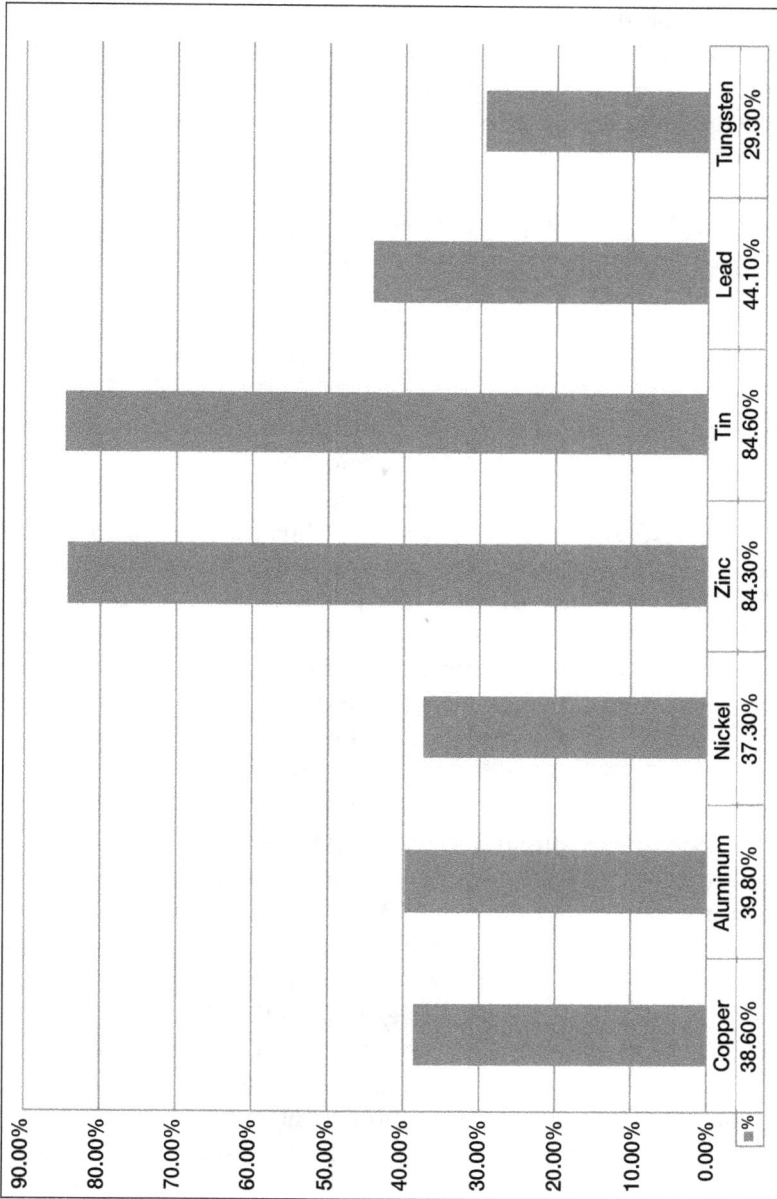

Source: The data is summarized by the author according to the report from US Department of the Interior, US Geological Survey, *Mineral Commodity Summaries 2013* (English translations and *Hanyu Pinyin* romanization of the minerals are, from left to right, copper (*tong*), aluminium (*lu*), lead (*qian*), zinc (*xin*), nickel (*nie*), tungsten (*wu*), tin (*xi*) respectively).

benchmarks for the definition of mineral overdependence.[2] In 2011, China's external dependence on chrome was 97.4 per cent, iron 73.4 per cent, nickel 72.6 per cent, and copper 65.7 per cent.

A better measure of mineral dependence is perhaps utilizing the concept of "import concentration". "Import concentration" refers to the number of source countries upon which a country is dependent for its mineral supply. The more countries upon which China depends, the higher is its percentage of import concentration. At present, China's import concentration for more than one third of its strategic mineral resources is over 80 per cent, while aluminum ore and nickel import concentrations are about 95 per cent, iron ore is 90 per cent, and potash fertilizer is 85 per cent.

According to import statistics collected by the Chinese Customs Network in 2012, China's peripheral region countries have become the most important suppliers of such Chinese mineral imports, notably in nickel, aluminum ore, and coal from Southeast Asian countries. In 2011, 97 per cent of nickel imports were from Indonesia and the Philippines, amounting to 52 per cent and 45 per cent of China's total nickel imports, respectively. Aluminum ore imports were 78 per cent from Indonesia and 11 per cent from Australia. Coal imports were 83 per cent from Indonesia, Australia, Vietnam, Mongolia and other neighbouring countries. Iron ore imports were 43.3 per cent from Australia, 10.7 per cent from India, and 2.3 per cent from Russia. Potassium ores were 44 per cent from Russia and Belarus. Copper ore imports were 20 per cent from Mongolia and Australia. In 2010, 45.3 per cent of uranium ore imports were from Thailand and Malaysia (see Figure 3.2).

Furthermore, other neighbouring countries that currently provide only a small proportion of strategic mineral resources to China have a huge potential. For example, the Laotian Ministry of Energy and Mining worked with the Government of China to construct a railway that transports raw materials from Vientiane to Yunnan with China coming up with a US$7 billion loan and the projection that, by 2020, China can have access to 5 million annual metric tons per year (Mt/yr) of mineral commodities through these lines (minerals such as bauxite, copper, gold, iron ore, lead, potash, zinc, agricultural and timber products).[3]

FIGURE 3.2
China Mineral Import, 2011

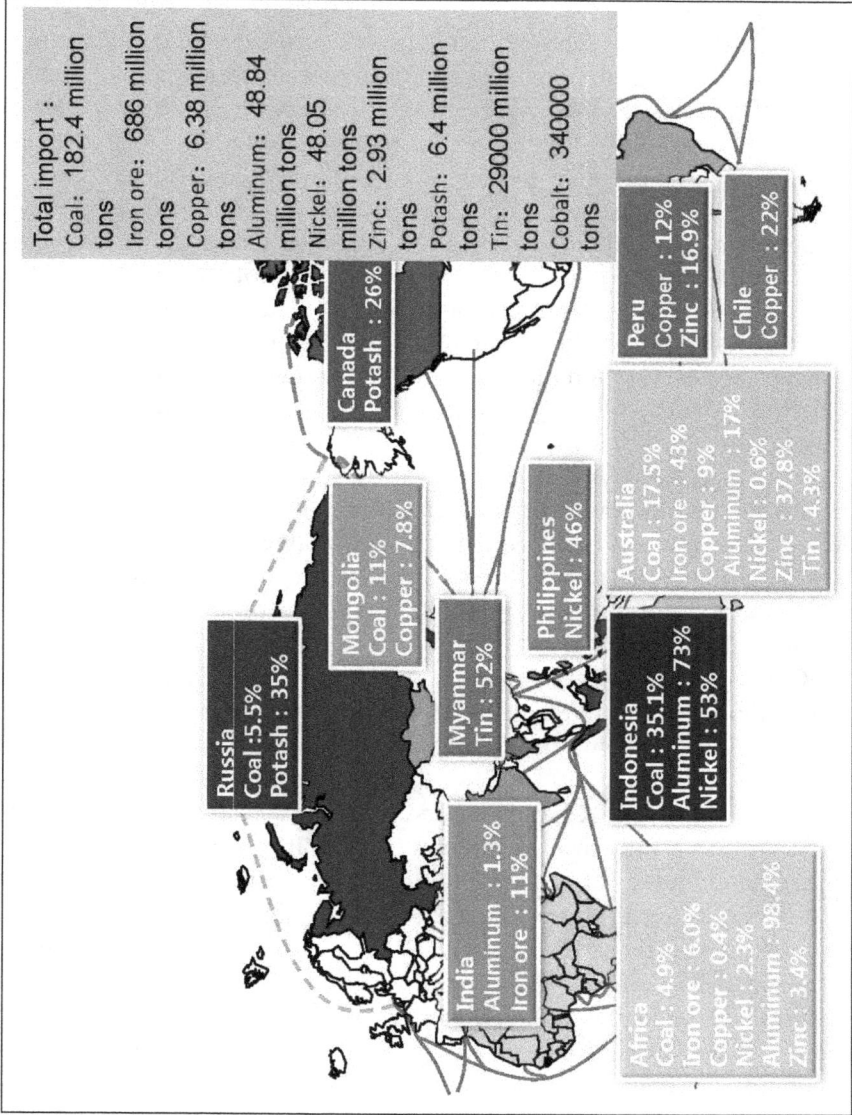

Total import :
Coal : 182.4 million tons
Iron ore : 686 million tons
Copper : 6.38 million tons
Aluminum: 48.84 million tons
Nickel: 48.05 million tons
Zinc : 2.93 million tons
Potash : 6.4 million tons
Tin : 29000 million tons
Cobalt: 340000 tons

Canada
Potash : 26%

Peru
Copper : 12%
Zinc : 16.9%

Chile
Copper : 22%

Russia
Coal :5.5%
Potash : 35%

Mongolia
Coal : 11%
Copper : 7.8%

Myanmar
Tin : 52%

India
Aluminum : 1.3%
Iron ore : 11%

Philippines
Nickel : 46%

Australia
Coal : 17.5%
Iron ore : 43%
Copper : 9%
Aluminum : 17%
Nickel : 0.6%
Zinc : 37.8%
Tin : 4.3%

Indonesia
Coal : 35.1%
Aluminum : 73%
Nickel : 53%

Africa
Coal : 4.9%
Iron ore : 6.0%
Copper : 0.4%
Nickel : 2.3%
Aluminum : 98.4%
Zinc : 3.4%

Source: Author, based on Chinese Customs data, available at <http://www.haiguan.info/>.

Challenges for Resource Imports from Peripheral Regions

Yet it is not uncommon for strategic minerals to be at the centre of political and military tensions. Strategic mineral-exporting nations may use these resources to advance political or economic goals based on national interests. For example, major oil suppliers — from Russia to Iran and Venezuela — have been increasingly able and willing to use their energy resources to pursue strategic and political objectives.[4] At the same time, it is also important to take a long-term perspective, deepen strategic mineral cooperation, increase energy efficiency, and facilitate the development and use of new technologies.[5]

From a scientific–historical viewpoint, well-known energy historian Daniel Yergin[6] and political scientist Paul Kennedy[7] have argued that strategic mineral domination is a prerequisite condition for significant structural changes in the international system. In their arguments, the colonization of strategic resources in the nineteenth and twentieth centuries and neo-colonial dominance by multinational resource firms have amounted to power, prestige and wealth in the current international system. Yet many of the world's major strategic mineral producing regions are also locations of geopolitical tension, and possibilities exist of unexpected supply disruptions. Instability in producing countries is the biggest challenge China faces.

China is currently the world's largest mineral resource consumer, which drives its ambitions to secure supplies of strategic mineral resources. A steady supply of strategic mineral resources will not only ensure a solid foundation for China's continued economic development, but it will also help China to pursue an independent foreign policy and have more influence in international politics. China's dependence on overseas sources of energy like oil and the mineral resources stated earlier in this writing are viewed by policymakers as vulnerabilities, especially when those resources (such as the case of oil from the major suppliers in the Middle East) have to pass through chokepoints like the Strait of Malacca. China's military and government are concerned about the blockade of the Straits in the event of a conflict with other major powers active in that region. In the past, Western media and analysts have cited several Chinese initiatives as ways to circumvent this resource delivery vulnerability, including developing the overland One Belt One Road (OBOR) economic initiative, organizing the so-called "string of pearls" (a term used in the Western media and strategic circles but not in China) strategy of developing access points in Myanmar, Iran and Pakistan for delivery of energy resources without the need to enter the Strait of Malacca. Other plans include developing strategic reserves of commodities and resources to offset price fluctuations and political instabilities in resource-supplying economies and states.

These issues are especially important for countries in China's peripheral regions because of their geopolitical implications, perhaps nowhere more so than in Southeast Asia. In general, as China has little bargaining power in the trade of strategic minerals, it is vulnerable to export price fluctuation and political or economic instability in the source country.[8] For example, from 2000 to 2010, prices of iron ores, copper and nickel rose 3.8 times, 3.2 times and 1.5 times, respectively, leading to significant negative economic repercussions in China.

Therefore, when China deals with neighbouring countries for the import of strategic mineral resources, it must consider the following three aspects: unbalanced development in regional political and economic relations; emerging resource nationalism in host countries; and the influence of powers outside the region and China, notably the United States. Furthermore, these three aspects are not mutually exclusive, but overlapping and interrelated. The following subsections will explore these themes as they generally concern China's peripheral regions.

"Unbalanced Development" in Peripheral Regions

"Unbalanced development" is a literal translation of the Chinese term *bupingheng de waijiaoguanxi*. The term is derived from the indigenously-developed ideas of the relations China maintains with different states, including those in the surrounding regions. It is a unique Chinese classification that pegs countries like Pakistan as *tiegermen* ("iron clad brothers"), an unofficial term that refers to strong alliances, to *youhao* or friendly countries like Russia and Kazakhstan. China generally does not believe in alliances because it would suggest equal partnership of a senior partner (i.e., China) with a junior counterpart. Therefore, these unofficial descriptive phrases provide the rationale to distinguish friendly countries from ones that are less close.

The "imbalances" in the intensity of friendliness and hostilities caused by regional political and economic relations can be divided into two main categories. The first concerns China's political friends and allies (i.e., the *youhao* countries), with whom China can deepen strategic mineral ties through mutually beneficial political and economic relationships. For instance, China's relations with Russia and Kazakhstan belong to this category. In fact, Kazakhstan was seen as the beachhead for the overland component of the OBOR initiative running through Central Asia. The second one concerns those nations that are "not friendly" with China, which can threaten the stability of strategic mineral imports. Indonesia, India and Mongolia fall into this category. These countries are not rivals or enemy states, but they may not enjoy the same level of closeness and preferential treatment as *youhao* countries in the Chinese worldview. Nevertheless, these countries can still be useful and productive business partners, particularly in the sector of supplying mineral resources.

Russia is a prime example of China's mineral sourcing from a *youhao* country. Russia is abundant in iron ore resources, and it is also one of the larger steel-producing countries in the world. According to the US Mining Bureau, Russia's iron ore reserves reached 64 billion tons in 1995 (equal to 14.5 per cent of the global reserve), while the reserve base was 78 billion tons. The term "reserve base" is a component of a resource that conforms to the minimal physical and chemical conditions relevant to contemporary mining and production benchmark practices, including the criteria of grade, quality, thickness and depth.[9] The definition of "reserves" is the component of the reserve base that can be extracted economically or is suitable for production purposes.[10] Furthermore, the reserve of ferrous metals was 24 billion tons and the ferrous metal reserve base was up to 29 billion tons.[11]

China and Russia enjoy some points of complementarities in the security and economic realms. For example, subject of an economic embargo from the West, Russia is finding China and other East Asian states as alternatives to garner investments. Together with exceptional geopolitical advantages underpinned by common interests between China and Russia, the comprehensive strategic cooperative partnership signed in 1996 and growing bilateral political and economic relations have been the foundation for a flourishing trade in strategic minerals between the two countries. One sector where there is potential to expand further is in iron ore minerals. Chinese import of iron ores was 9.98 million tons,

1.4 per cent of the total imports (686 million tons).[12] Australia and Brazil had been China's traditional suppliers of iron ore but, starting from at least 2011, Russia had been considered as a possibility for forging a strategic relationship in the iron ore mineral sector.[13]

Similarly, Kazakhstan has an abundant supply of solid minerals. It has more than 90 different kinds of minerals, 1,200 raw mineral materials, and more than 500 mineral territories of black, coloured, rare and precious metals. Also, its proportion of reserves for wolfram, uranium, chromium, lead, zinc, copper, and iron is among the highest in the world. However, the trade volume of the strategic minerals between China and Kazakhstan is little. Therefore, there exists great potential for cooperation between China and Russia, and between China and Kazakhstan, in strategic mineral resources.

Kazakhstan has closer relations with China based on common and complementary systems of authoritarian governments. Both countries also understand the necessity of accommodating Russia's backyard interests in Central Asia and they are members of the Shanghai Cooperation Organization (SCO), an organization that bands together the common strategic interests of China, Russia and Central Asia.

In contrast, India and Mongolia are not in the same category as strategic Russia or Kazakhstan. However, neither are they considered as rivals or enemies. Beijing perceives India as a strongly non-aligned country with a fiercely independent stance on world affairs. Landlocked Mongolia depends on Russian and Chinese ports to trade with the outside world and it is also close to Japan in economic exchanges. Because Mongolia and India are rich in minerals, the Chinese are interested to upgrade economic exchanges with them, especially as labour costs in both countries are considered low.

India has about 8 per cent of the world's mineral ore. India is an important country of origin of iron ores for China that once captured 25 per cent share of Chinese iron ore imports in the 1990s. However, due to the domestic politics of nationalism and resource protection as well as the rise of the Indian economy (at the point of writing, India's economic growth is outstripping China's maturing economy), iron ore exports to China has declined considerably since 2012.

As one of the most important neighbours of China, Mongolia shares more than 4,000 kilometres of common border with China. Located in the middle eastern part of the polymetallic metallogenic belt rich with the bronze from East to West in Central Asia, Mongolia is also abundant in mineral resources. With plentiful mineral resources, Mongolia also hopes to use its resources to its advantage when dealing with the major powers in the region, including China. In particular, it has a strategic goal of securing greater independence from China and preventing overdependence on Chinese acquisition of minerals. For example, Japan is interested in developing the rare earth resources in Mongolia to diversify rare earth supply sources away from their industry's dominance by China.

These factors are also relevant to Southeast Asia. For example, Indonesia also aspires to be a non-aligned country. In fact the Bandung conference that gave rise to the non-aligned movement (NAM) was hosted by Indonesia. Indonesia aspires to be a major naval power in Southeast Asia and has, in the recent past, blown up Chinese fishing boats (along with those from Malaysia and Vietnam) to demonstrate their resolve in punishing transgressions into their maritime boundaries. Indonesia is also putting surface to air missiles on the Natuna Islands, showing its determination to protect their island territories.

In other areas, however, the two countries have strong economic partnerships. Indonesia is working closely with China on the medium-speed railway construction. Indonesia is also amongst the first applicants to obtain funds from China's multilateral lending agency, the Asian Infrastructure Investment Bank (AIIB). A caveat to Chinese economic cooperation is the challenges of Chinese minerals and natural resources development in Southeast Asia in general. Maritime disagreements and unbalanced focus on infrastructure development related to the commodities trade and Chinese resource needs complicate overall economic cooperation.

Emerging Resource Nationalism and Historical Boundary Disputes

China is one of the countries that shares its borders with the largest number of neighbouring countries. There are fourteen countries that share a land border with China and the total length of these boundaries lines is 22,000 kilometres. Furthermore, there are six countries located directly across the sea from China's 18,000 kilometre coastline. Some of these countries share strong cultural and historical bonds with China, while others are characterized by a diversity of political and economic systems, as well as important social and cultural differences. Furthermore, a history of border disputes and growing geopolitical tensions in the region have made relations between China and its peripheral regions increasingly complicated. In turn, these tensions have also propelled protests against Chinese companies and sentiments of resource nationalism.

One key example is China's relations with India. Boundary disputes left over from historical relations between China and India have left constant mistrust in their contemporary political and economic cooperation. Historically, the two countries fought a brief but bloody boundary war in 1962. The Indian scholar S.B. Asthana has suggested that with China's growing strength, military capacity, and its plan to develop offshore power projection, nuclear strike capacity and military modernization, China will become a potential security threat to India in the near future.[14] In the recent past, accusations have also been made against China about economic development projects having political or geopolitical motivations *vis-à-vis* India. For example, these narratives suggest the possibility of Chinese forces breaching the dam on the Yarlung Zangbo River so that "all the northeastern areas in India will probably be flooded"; that India's "drought, the flood and even domestic water are under the control of China"; and the flood waters from Yarlung Zangbo in India "was created on purpose by the Chinese military", which were used as "Chinese ecological weapons against India". This is one aspect of Sino–Indian relations. On the other hand, India is one of the first applicants to apply for AIIB funding for coal-fired power plants and other infrastructure projects. Both countries are also trying to promote intellectual and people-to-people exchanges by reviving an ancient Nalanda University, which used to be a Buddhist religious linkage between the two countries on the ancient Silk Road.

The remnant worries (particularly from the 1962 Sino–Indian conflict) about China as a geopolitical threat influences commercial trade in mineral resources between China and India. India once was an important country for producing iron ores, among the top three iron ore producers in the world together with Australia and Brazil. With the influx of iron ore imports from India, China started a bulk market for it, also known as spot ore.

Recently, however, India has focused on developing the steel industry on its own to satisfy domestic demand, as well as raised tariffs and transport fares to limit ore exports. From January to September 2012, only 32.35 million tons of iron ore were imported from India to China, a decrease of 47 per cent compared with the same period last year, and falling to 5.8 per cent of the overall share of total iron ore import by the Chinese. These statistics are similar to Chinese imports from South Africa, which was 31.13 million tons of iron in the corresponding period and indicates an increase of 17.8 per cent from the same period last year. Moreover, from June to September 2012, the iron ore exports from India to China were decreasing constantly.

However, India is not an exception when it comes to border tensions with China. Even comprehensive strategic partners like Russia have outstanding issues with China. Similar tensions also exist between China and Russia. For instance, the Russian mass media reported that China tried to control water resources in Siberia, causing more than one million people and dozens of agricultural enterprises to face the danger of hydropenia (absence of water from living tissues, a severe form of dehydration) and stopping farm production. Some regions in Russia and their policymakers also worry that Russia will become a "resource dependent" country *vis-à-vis* China due to the hollowing out of their manufacturing industries, as large quantities of Chinese manufactured consumer products enter into Russia and compete directly with their own industries. These concerns have also limited trade development in strategic mineral resources between China and Russia. Both Russian and Indian case studies are illustrative of Chinese geopolitical barriers in reaching out to its neighbouring states. The case of India demonstrates how historical tensions and contemporary geopolitical anxieties have hindered China's access to strategic mineral resources in its peripheral region and causes China to seek them further abroad.

In Southeast Asia, uncertainty over China's rise as a regional hegemon and growing geopolitical tensions in the South China Sea have also complicated regional mineral trade. Southeast Asia is an important region for exporting mineral resources to China. Amongst the Southeast Asian states, Indonesia is a globally important mineral producer, whose production of tin, coal, copper and nickel are among the top in the world. In terms of volume, its tin production is ranked second in the world, while copper and nickel are fifth. The copper reserve in the Philippines is fourth in the world, while nickel is fifth and chrome is sixth. China's main sources of nickel from Southeast Asia are from the mines in Indonesia and the Philippines. In 2011, 52 per cent of China's nickel imports came from Indonesia and another 45 per cent from the Philippines, amounting to 97 per cent of its total nickel import (45 per cent) (see Figure 3.3). China's main source of aluminum ore is Indonesia, comprising 78 per cent of its total aluminum ore imports (see Figure 3.4).

FIGURE 3.3
Top Three Sources of China's Nickel Imports, 2011

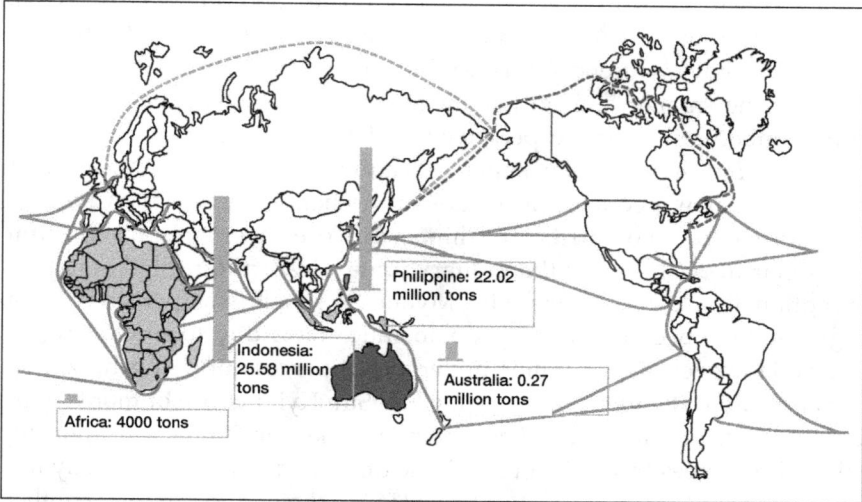

Source: The data is summarized by the author according to the statistics on the main bulk commodity from China Customs in 2012.

FIGURE 3.4
Top Three Sources of China's Aluminum Ore Imports, 2011

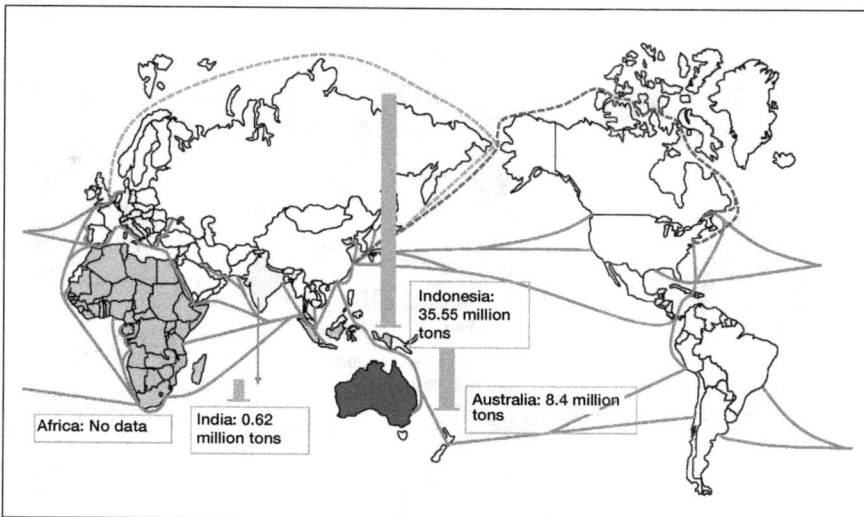

Source: The data is summarized by the author according to the statistics on the main bulk commodity from China Customs in 2012.

However, recent tensions over maritime sovereignty in the South China Sea have negatively affected China's trade in strategic mineral resources with Vietnam, Malaysia, the Philippines, and Brunei Darussalam. Most of the Southeast Asian countries named above, including the original ASEAN 5, have a friendly businesslike relationship with China, while some of the smaller and former Indo–Chinese socialist states like Laos and Cambodia tend to be perceived as having closer ties with China. Amongst Southeast Asian states, Myanmar is probably the clearest example of how geopolitical fortunes can change rapidly for China.

More than two thirds of China's tin ores were imported from Myanmar in 2011. Under the military junta's leadership, China enjoyed exceptionally close relations with Myanmar, which was perceived as a rogue state and embargoed by democratic states due to human rights issues. However, as Myanmar undergoes a process of democratization under released dissident leader Aung San Suu Kyi, the rise of more "anti-China" forces domestically has potential risks for China in importing mineral resources from Myanmar. Some of these domestic forces may not be necessarily "anti-China", but the interests they advocate may conflict with Chinese economic ones. For example, Burmese environmentalists may take issue with Chinese-funded hydropower dam projects and cause delays (see Figure 3.5). China will have to navigate carefully

FIGURE 3.5
Top Three Sources of China's Tin Imports, 2011

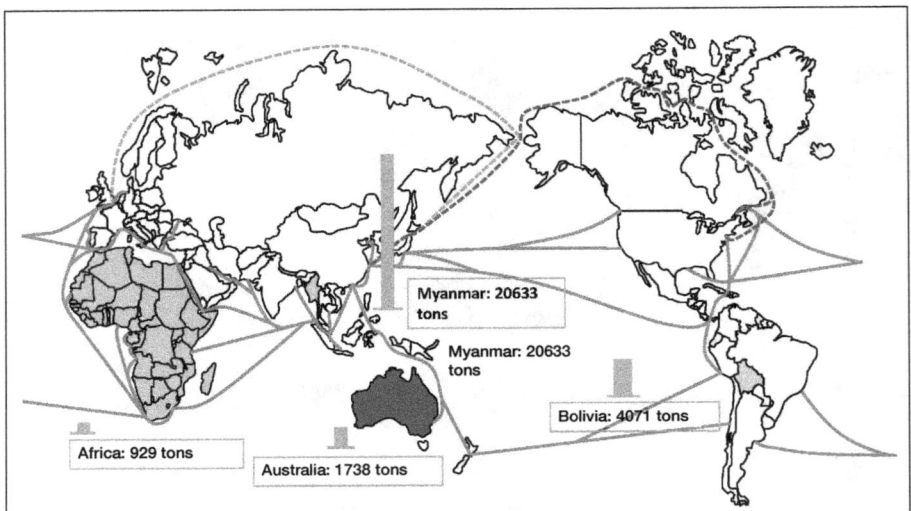

Myanmar: 20633 tons

Myanmar: 20633 tons

Bolivia: 4071 tons

Africa: 929 tons

Australia: 1738 tons

Source: The data is summarized by the author according to the statistics on the main bulk commodity from China Customs in 2012.

through the complicated domestic politics of diverse Southeast Asian political systems and confront the reality of major groups in those countries that are ideologically or policy-wise incompatible with Chinese national interests.

Influence of Powers outside the Region and China

In Chinese strategic conceptualization, the "first island chain" strategy is a US-led alliance consisting of Japan, the Philippines, and network partners like Taiwan (specializing more on intelligence gathering) to encircle and prevent further Chinese naval expansion into the Pacific. This is then followed by a "second chain" of islands, such as Guam and Hawaii as a second perimeter of defence against Chinese blue-water naval expansion. In an address to Georgetown University, US diplomat Susan Rice declared that resources needed by Asian economies in future decades will greatly affect world energy supply and climate security.[15]

There is also an "Asia-Pacific Crisis Arch" that stretches from resource-rich Sakahlin island under dispute between Japan and Russia through to energy-poor South Korea, North Korea and Japan and the southeast coast of China with its fast-growing and energy intensive manufacturing economies. The "crisis arch" covers a number of major maritime disputes in East Asia and major economies dependent on overseas energy sources, especially oil. The "arc" is an informal term that is also utilized by some Western analysis to geographically describe the 3,000 mile area of territorial disputes stretching from Senkaku/Diaoyu Islands in East China Sea to the Spratlys disputes in South China Sea.[16]

Further west from East Asia, the Indian Ocean lane is another centre of US resource strategy. China's dependence on the Indian Ocean lane makes it a security issue because of the potential threats faced by key transportation routes for Chinese resources and it affects military strategy. In the name of its "pivot" back to the Asia Pacific, the US considers East Asia as one of the flashpoints of energy geopolitics and a strategic chokehold to control strategic passage of energy resources. At present, the islands in the South China Sea have a great potential for conflicts and big power competition. Chinese hawks, realists and conservatives believe that the island conflicts reflect the US and its allies' aspirations to dominate the South China sea lanes (along with India in the Indian Ocean) and, by doing so, "contain" the development of China. Controlling the South China Sea would mean controlling three shipping lanes to China:

(1) the Middle East route through the Persian Gulf — the Strait of Hormuz–Strait of Malacca–Taiwan Strait–China;

(2) the African route through: North Africa–Mediterranean–Gibraltar Strait–Cape Horn–the Strait of Malacca–Taiwan Strait– China;

(3) Southeast Asia routes: the Strait of Malacca–the Taiwan Strait–China.

That is to say, except for the minerals and energy resources that China imports from South American countries such as Venezuela, China's resource imports coming from all other directions must go through the Strait of Malacca. Hawkish Chinese perspectives believe that, while China actively built relationships with global resource producers to ensure its stable energy supply, this proactive resource diplomacy has caught the attention of the United States and Japan, whose increasing concerns about China's proactive diplomacy and growing demand for energy could lead to future conflicts. In contrast, the moderates in Chinese foreign policy believe that China can work out its differences with other claimants in these islands due to the deep economic interdependence between their economies with Chinese consumption, production and investments.

Concluding Section: Integrating the Sourcing of Strategic Mineral Resources with Peripheral Diplomacy

In its economic development strategies to expand trade routes across Central Asia through the "New Silk Road Economic Zone" and the waters of Southeast Asia with "the 21st Century Maritime Silk Road", China wants to emphasize trade in strategic mineral resources in Central Asia, Southeast Asia and Australia. These world regions are the main importing regions for energy resources (e.g., coal, oil and gas) and minerals for Chinese state enterprises. Chinese state companies believe that collaborative development of strategic mineral resources with these countries can be used to promote mutual economic development and regional stability.

At the current Chinese outreach to Russia and Central Asia, China aims to improve strategic mineral imports from Russia and Kazakhstan. In particular, China and Russia should grasp the opportunity to promote regional cooperation through initiatives like the Conference on Interaction and Confidence-Building Measures in Asia (CICA) Summit, a twenty-six-member state multilateral organization that brings together nations to discuss security issues openly in the interest of peace and stability. Chinese state companies and government could also offer to rebuild the network of gas and oil pipelines constructed by Central Asian countries and Mongolia during the Soviet period into an integrated trans-regional network of natural gas pipelines. Given the comprehensive strategic partnership with Russia, China may have to communicate such ideas with its SCO counterparts first.

In Southeast Asia, the State Grid Corporation of China, a state-owned enterprises (SOE) that is also the world's largest utilities company, can promote its expertise in the integration of power supply in Southeast Asia, based on cooperation with existing energy infrastructure, such as the proposed Sino–Cambodia pipeline. These two projects are highly controversial and, while they may enhance the national interests of some states in Southeast Asia as well as some stakeholders in China, they may meet with some resistance from other stakeholders within Southeast Asia. For example, China Pipeline Bureau, whose parent company is the state-owned China National Petroleum Corporation, is keen to have an oil pipeline from Sihanoukville to Phnom Penh but they run into a competing bid from Indonesia's Pertamina oil and natural gas state firm.[17] Therefore, unless they find ways to work together, such commercial rivalries between state companies may complicate other deals.

China wants to promote the integration of infrastructure development in this area. In some ways, this is already happening with the dispensing of AIIB and other bilateral loans to poverty eradication in Indonesia, Laotian railway projects, and the Kuantan port in Malaysia. China should develop public diplomacy in energy security in the direction of greater transparency, to eliminate the West's suspicions of the resource investment, trade, mineral acquisition activities of China. As for its existing partnerships with Russia as well as the producers in Middle East, China should enhance its information release channels in various kinds of public communication in energy, build up informal dialogue mechanism with high standards for transparency, and reduce the distrust as much as possible. To ally suspicions, it should take great efforts to create the harmonious atmosphere of win-win collaboration, and its state firms should comply with the local laws (such as those pertaining to the environment) and institutions, and strengthen existing cooperative partnerships.

Second, China should help to improve the development of energy and mineral resource pricing regulatory mechanisms in APEC. "The Declaration on Asia-Pacific Partnership on Clean Development and Climate" and "the Sydney Declaration" are the outcomes of such multilateral intentions to improve the APEC energy mechanism. The usual criticism of APEC is that it lacks regulatory teeth as a loose coordination mechanism that is not legally binding, therefore inherent problems remain in energy resource cooperation. For instance, a proposal for increased cooperation in the field of energy and mineral resources at the APEC conference in 2014 was mentioned only in an annex.[18] Meanwhile, in the framework of the Asia-Pacific Partnership (APP) on Clean Development and Climate, the countries in the Asia Pacific set an example in institution of energy cooperation. According to the Asian Development Bank (ADB) website, APP has eight public–private sector task forces overlooking the following sectors of aluminum, buildings and appliances, cement, cleaner fossil energy, coal mining, power generation and transmission, renewable energy and distributed generation, and steel and each task force looks into "clean development and environmental issues and build action plans for both immediate and medium-term activities".[19] China should continue to emphasize on openness and tolerance in energy cooperation mechanism and promote the mechanism construction based on it.

China can choose to use the existing cooperation mechanism including APEC, Sino–Russian cooperation in border areas, the SCO, cooperation in the Greater Mekong Subregion, ASEAN–China Free Trade Area, China–Japan–South Korea Free Trade Area, Sino–Russia–North Korea–South Korea–Mongolia the Tumen River Cooperation and others to develop a stronger strategy for the import of strategic mineral resources from peripheral regions. China should consider developing different design strategies and policies in the South, North and East China for promoting trade in strategic mineral resources in their adjacent regions. For adjacent regions in the North, China should emphasize economic development to improve cooperation potential. For adjacent regions in the South, China should recognize the needs of neighbouring Southeast Asian countries. China can also be in constructivist consultations with the US to promote win-win economic cooperation policies amidst the fraught geopolitical situation in the South China Sea. For the neighbouring regions in Northeast Asia, China should consider carefully the implications of resource development for territorial conflicts and other politically sensitive issues.

Third, much room still remains to improve cooperation between China and countries like Russia, Kazakhstan, South Africa, Brazil, and so on. China should enrich the content of cooperation in strategic mineral resources at the multilateral and bilateral levels with member states in BRICS and SCO, as well as promote the cooperation and win-win in strategic mineral resource.

However, China may have to avoid transforming economic cooperation into the singular focus of resource imports. Starting from AIIB, the Silk Road Fund and the others, China should strengthen mechanisms for cooperation in the trade of strategic mineral resources. In October 2014, Chinese President Xi Jinping pointed out that China should operate the AIIB and the Silk Road Fund in accordance with "innovative thinking". "Innovation" in this sense is related to the idea that China initiates and cooperates with some countries to establish AIIB to provide funds for infrastructure development to promote economic cooperation among countries along the OBOR. Part of this OBOR initiative has been to set up the Silk Road Fund to support the construction in OBOR directly with Chinese funds.

NOTES

1. Jan Ishee and Alex Demas, "Going Critical: Being Strategic with Our Mineral Resources", 13 December 2016, available at <https://www2.usgs.gov/blogs/features/usgs_top_story/going-critical-being-strategic-with-our-mineral-resources/> (accessed 1 June 2016).
2. Geeta C. Goled, "Urban Mining: A Solution to China's Resource Crisis?", *Yale Environment Review*, 29 October 2015, available at <https://environment.yale.edu/yer/article/urban-mining-a-solution-to-chinas-resource-crisis#gsc.tab=0> (accessed 1 June 2016).
3. Yolanda Fong-Sam, "The Mineral Industry of Laos", in *2012 Minerals Yearbook* (US: US Department of the Interior, US Geological Survey, November 2014), p. 151.

4. In the area of improving energy efficiency, the Report on China Mineral Resources 2015 published by the Ministry of Land and Resources, People's Republic of China indicated that new stratigraphic charts of China were revised in 2014 incorporating the latest research in this area. New remotely manipulated vehicles that can explore deep oceans were also developed. New technologies including geophysical exploration information processing equipment from the air were also completed, and China has improved rare earth magnetic separation technologies. For more, see Foreign Relations Council, "National Security Consequences of U.S. Oil Dependency", available at <http://www.cfr.org>.

5. The Editorial Committee of the Ministry of Land and Resources, People's Republic of China, *2015 China Mineral Resources* (Beijing: Geological Publishing House, 2015), pp. 42–44.

6. Daniel Yergin, "Ensuring Energy Security", *Foreign Affairs* (1 March 2006): 69–77.

7. Paul Kennedy, *The Rise and Fall of the Great Powers* (New York: Vintage, 1968).

8. Willis, "Mining Market Review", Spring 2012, available at <https://www.willis.com/naturalresources/pdf/MiningMarketReview2012.pdf>.

9. US Geological Survey (USGS), "Appendix C", 13 December 2016, p. 191, available at <http://minerals.usgs.gov/minerals/pubs/mcs/2009/mcsapp2009.pdf> (accessed 1 June 2016).

10. Ibid., p. 192.

11. 张孟伯:《2012年,俄罗斯铁矿石资源研究》,载《矿床地质》2012年第31卷增刊,第54页.

12. Zhang Qi, "China Collecting New Sources Overseas to Provide Iron Ore", *China Daily*, 7 September 2011, available at <http://shandong.chinadaily.com.cn/e/2011-09/07/content_13636834.htm> (accessed 1 June 2016).

13. Ibid.

14. S.B. Asthana, "The People's Liberation Army of China: A Critical Analysis", *Combat Journal* 30, no. 2 (1992): 1–8.

15. National Security Advisor Susan Rice's Address at Georgetown on "America's Future in Asia", The White House Office of the Press Secretary, 20 November 2013, available at <http://www.voltairenet.org/article181088.html>.

16. Alan Dupont and Christopher G. Baker, "East Asia's Maritime Disputes: Fishing in Troubled Waters", *The Washington Quarterly* 37, no. 1 (1994): 79.

17. Kali Kotoski and Sor Chandara, "Royal Group in Talks with China, Indonesia on Proposed Oil Pipeline", *Phnom Penh Post*, 11 April 2016, available at <http://www.phnompenhpost.com/business/royal-group-talks-china-indonesia-proposed-oil-pipeline> (accessed 1 June 2016).

18. Asia Pacific Economic Cooperation (APEC), "Annex C – APEC Accord on Innovative Development, Economic Reform and Growth", available at <http://www.apec.org/Meeting-Papers/Leaders-Declarations/2014/2014_aelm/2014_aelm_annexc.aspx> (accessed 1 June 2016).

19. Asia Regional Integration Center, Asian Development Bank (ADB), "Regional Public Goods Asia-Pacific Partnership on Clean Development and Climate", 2015, available at <https://aric.adb.org/initiative/asia-pacific-partnership-on-clean-development-and-climate> (accessed 1 June 2016).

4

ENERGY ENTANGLEMENT
New Directions for the China–Indonesia Coal Relationship

Cecilia Han Springer

Coal is the dominant source of energy in Asia, and China and Indonesia are its leading producers and consumers. Understanding the interconnection between their coal industries also illuminates their shifting geopolitical relationship, which has become stronger through resource-based economic ties despite the occasional contest over issues on sovereignty. China and Indonesia have maintained a huge and mutually beneficial coal trade for many years, and this coal-based relationship has been nationally and globally influential. However, the China–Indonesia coal trade dropped precipitously in 2015 when shifting political and economic contexts drove both countries to decrease their levels of exchange. China's domestic air pollution and climate mitigation policies led to a large decrease in demand for coal, while Indonesia introduced new policies, including new restrictions on coal exports, to promote economic development and nationalization of its energy industry. However, the changing incentive structure resulted in new directions for the China–Indonesia coal relationship, further strengthening the economic ties that undergird their geopolitical relationship. The China–Indonesia coal relationship is no longer a consumer–producer relationship defined by trade and export, but is instead shifting toward Chinese investment in coal production within Indonesia. China is becoming more involved in coal-related infrastructure in Indonesia, especially electric power generation. These new directions will be critical for understanding the environmental and economic impacts of coal.

Introduction

Southeast Asia is one of the only regions in the world in which coal will constitute a greater share of the energy mix in the future.[1] China is the world's top producer and consumer of coal, while Indonesia holds these same titles within Southeast Asia, producing 89 per cent of the coal in Southeast Asia.[2] Indonesia is also the world's largest coal exporter. The coal industry has major political sway in both countries. China and Indonesia have become politically and economically entangled through coal trade and investment. For the past several years, their intimate import–export relationship has dominated global steam coal trade in both volume and market power.[3] Although other regional partners, namely Australia, are major suppliers of coal to China, this chapter focuses on Indonesia due to the growing importance of the China–Southeast Asia relationship for global resource governance. Southeast Asia has a huge natural resource endowment, including biodiversity and carbon stocks in its forests, major fossil fuel reserves, minerals, and ores. The development of these resources requires large amounts of capital, technology, and planning, and many Southeast Asian countries turn to external sources of finance and development expertise. China, in particular, is a major consumer of Southeast Asian products and provider of development finance in the region. China's growing involvement in Southeast Asia's resource development projects can set global precedents for resource management and development, particularly for coal, arguably the region's most important source of energy.

Extraction and combustion of coal cause health and environmental damages at local and global levels. Coal extraction can cause occupational hazards for miners, local air pollution from coal dust and transport, noise pollution, water and soil contamination, and even coal fires that threaten surface infrastructure. Coal combustion is also associated with air pollution and public health issues. Worldwide, air pollution from coal combustion causes 210,000 deaths, almost 2 million serious illnesses, and over 151 million minor illnesses per year, though some estimates of the burden of disease in coal-heavy economies indicate higher rates of illness among the local population.[4] Globally, coal-fired power plants are the largest source of carbon emissions, which drive global climate change. Climate change also drives social and environmental upheaval through increased frequency and severity of natural disasters, habitat loss, and sea level rise.

These externalities have politicized the use of coal in many countries, with environmental and health advocates around the world pushing for policies to reduce coal combustion. In the United States, air quality and carbon emissions regulations have made new coal-fired power plants uneconomical.[5] In Western countries, a number of institutions, including governments, private banks, development finance institutions, and individual companies are cutting financial ties to the coal industry under pressure from environmental advocates, as well as because of financial evidence of the industry's decline.[6] As a result, the construction of new coal-fired power plants has essentially stopped in developed countries.[7]

Yet this deliberate dissociation from the coal industry has yet to take hold in the developing world. Coal continues to be the fuel of choice for many countries in Asia and especially in Southeast Asia, where supply and demand are ample, prices are low, and coal operations are partially or fully state-owned. The growth of coal use in Asia stands in stark contrast to coal's decline in other parts of the world. Coal dominates the energy scene in Asia due to its abundance, reliability, and affordability when compared to oil and natural gas in the region. Asia accounts for nearly all the growth in demand for coal in developing countries, with demand coming primarily from coal combustion for electricity generation. In Asia, around 70 per cent of electricity comes from coal, compared with a 40 per cent global average.[8]

Why should readers care about what the coal industries in China and Indonesia do? From an economic perspective, market dominance can leave other coal-using countries dependent on the China–Indonesia relationship to set the price — and the governance policies — for coal in the global market. As the 1st and 4th most populous countries in the world (comprising more than 20 per cent of the global population), local impacts of coal production in China and Indonesia are important for a range of global social and environmental problems. The health impacts discussed above can be exacerbated on global and local levels by resource decisions in Indonesia and China. Coal burned in China and Indonesia contributes to the global impacts of climate change, and China is the world's largest carbon polluter. Both countries also face local problems associated with coal production and use. Difficult geological conditions result in roughly 90 per cent of China's mines being constructed underground, where there are higher risks of accidents and exposure.[9] China's labour-intensive coal mining has a long history of dangerous working conditions and high worker mortality in underground mines.[10] Coal combustion in China is also a major source of air pollution that has been notorious in recent years as a visible symbol of the consequences of China's rapid development.

In Indonesia, while coal deposits are close to the surface of the earth, mining often requires stripping away carbon-rich and biologically diverse ecosystems, such as peatlands or tropical rainforests. The transformation of land through coal mining and infrastructure displaces indigenous communities and transforms local ecologies and economies. For example, flooding in East Kalimantan in the late 2000s, which were exacerbated by deforestation from coal mining, had a major toll on communities there. The cost of rebuilding and preventing floods far exceeded the local income from coal.[11] In addition, Indonesia has a large number of small and illegal mines, many of which are located in forest reserves and largely unregulated. In some cases, their presence hinders economic development for local communities.[12]

Given the local and global impacts associated with the coal industry in Asia, it is important to examine the dynamics between the largest coal players in the region — China and Indonesia. This chapter first describes how the powerful coal industries in China and Indonesia became tightly interconnected through trade, to the point that they were together able to influence prices and production in the global market in a duopolistic manner. The chapter then elucidates the reasons behind a recent downturn in their coal trade, which health and environment advocates have welcomed. However, the chapter goes on to suggest that the China–Indonesia coal relationship will shift towards investment in Indonesia's coal sector as political factors in both countries make them see a more investment-oriented relationship as mutually beneficial.

China and Indonesia: The Era of Duopolistic Entanglement

Coal is a pillar of both the Chinese and Indonesian economies, which, following the global financial crisis in 2008, became entangled into a formidable duopoly on the global coal market. A duopoly refers to the ability of two actors to dominate the market through setting prices, production levels, or agendas.

China has plentiful coal reserves, primarily located in its northern and western provinces. China's coal industry operates on a massive scale. It produces nearly half of the world's coal, the vast majority of which is consumed domestically.[13] As coal is one of China's few truly abundant natural resources, centralized state planning, since the founding of the People's Republic of China, has prioritized the development of heavy industry fuelled by cheap energy inputs. Even after the Chinese reform period in the 1980s, when general commodity pricing became more liberalized and despite disgruntled coal producers, China maintained low coal prices.[14] This policy enabled the Chinese economy to take off at an unprecedented rate through manufacturing and heavy industry.

Coal is also plentiful in Indonesia, primarily in deposits in Kalimantan and Sumatra. Indonesia's coal reserves may be two to three times larger than what is currently proven, as coal is currently so abundant as to not require additional geological exploration to find new reserves.[15] Historically, Indonesia has oscillated between nationalizing its energy industry and opening it up to foreign investment. Since the Sukarno era, when the government had centralized the entire mining industry, Indonesia has increased protections for foreign investors and allowed investment in the mining sector from companies like Shell and Rio Tinto.[16] Indonesia further expanded its coal industry after petroleum became scarce domestically in the 1990s.[17] Growth in Indonesia's mining industry is driven by coal, the production value of which has an annual growth rate of roughly 10 per cent.[18] Indonesia is also the world's largest exporter of coal.

These powerful but relatively separate coal economies became increasingly entangled in the late 2000s. 2009 was a momentous year for the international coal market. China transitioned from being a net coal exporter to a net coal importer, meaning that coal imports exceed coal exports. This moment marked a major milestone in the entanglement of China and Indonesia through coal, as coal imports from Indonesia to China jumped dramatically (see Figure 4.1).

A number of trends in both countries prior to 2009 set the stage for their increased entanglement in the coal trade. In China, coal supply and demand are geographically distant. China's ample coal reserves and major mines are concentrated in its northern and western provinces.[19] Yet the Pearl River Delta in southeastern China is a major centre of energy demand due to large population (over 100 million people) and a heavily industrialized economy. Major cities in the Pearl River Delta

FIGURE 4.1
Coal Volumes Traded from Indonesia to China, 2000–13

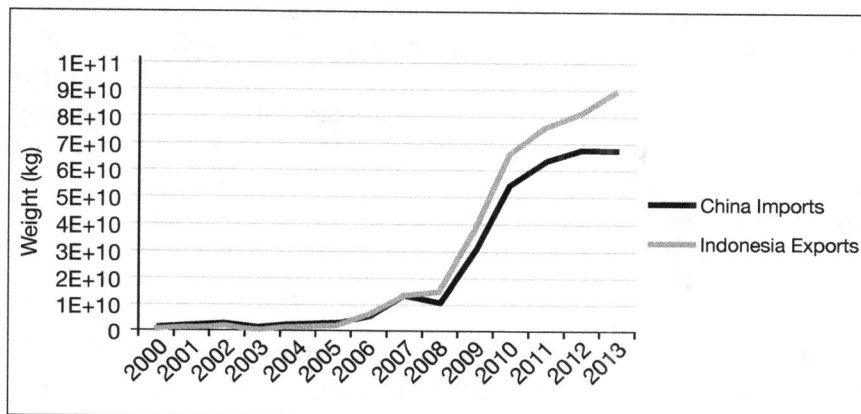

Source: UN Comtrade Database. Since there is a discrepancy between reported data from China and from Indonesia, both series are displayed.

include such manufacturing hubs as Guangzhou, Shenzhen, Dongguan, and Foshan. Because transporting coal to the Pearl River Delta from Northwest China by rail or truck is prohibitively expensive, coal is instead moved by rail to ports on the northeastern coast and then shipped by sea to southeastern China. However, these rail-to-sea links are still expensive and can account for 50–60 per cent of the price of coal in southeastern China.[20] Furthermore, China's rail and freight truck transport infrastructure has struggled to modernize, in part due to the monopoly power of its central government regulator, the Ministry of Railways.

In contrast, China's ports were opened up to competitive investment in the 1990s and they have since enjoyed relatively greater construction investments proportional to coal transporting capacity.[21] Importing coal through coastal routes avoids further pressure on China's railways, reducing the need for costly government infrastructural investment.[22] Thus, coastal import of coal has become an increasingly attractive option in China. Unless China builds massive transmission infrastructure from its supply regions to bypass rail transport bottlenecks, imported coal for electricity will continue to be cheaper to transport than domestic coal.[23]

Furthermore, electricity generation is the top source of demand for coal in China, responsible for just over half of coal consumption.[24] Disputes between Chinese coal companies and electricity companies over contract prices for steam coal (i.e., the type of coal used to produce electricity, as well as the primary type of coal found in Indonesia) led the utility companies to move towards imports.[25] In China's electricity industry, contract prices for steam coal were maintained at artificially low levels until 2006, when China's National Development and Reform Commission began to deregulate. However, due to electricity price controls, utilities had trouble passing on the rising cost of coal to customers, and began to turn to imports as a way to continue to purchase low-cost coal.[26]

In addition to these physical and regulatory limitations, China's strategic resource policies also paved the way for major coal imports. The Chinese government has encouraged the import of resources over domestic sourcing when imports are cheaper through the "two markets, two resources" policy. This policy was first promulgated in the 1990s and has since been applied to energy, agricultural, and mineral sourcing decisions,[27] and it holds true for coal as well.[28] In addition to economic benefits, China promotes this policy in order to avoid other externalities associated with resource extraction. For example, the coal production process is highly water-intensive. Northern and western China, where most coal reserves are located, are arid regions facing extreme pressure on water resources. Indonesia has more fresh water per capita than China, and a more optimal ratio of fresh water to coal production than China.[29] China can alleviate its water scarcity by importing water-intensive resources like coal. Thus, China has both economic and environmental motivations for reaching overseas for coal.

The global recession following the 2008 financial crisis, however, was the tipping point for China's flip from being a net coal exporter to a net coal importer. The recession caused flagging demand for coal in many markets, but not in China. Freight costs dropped in other countries, but not as much in China, which weathered the recession relatively well. This meant that coal purchasers in China could take advantage of recession-hit countries, where transport costs had dropped, such as Indonesia, rather than face the relatively expensive Chinese coal transport system.[30]

Taken together, these trends lowered the cost of importing coal over sourcing it domestically in China, demonstrating why China became a net importer of coal in 2009 despite the abundance of domestic coal reserves (see Figure 4.2). In the span of a year, Chinese demand came to account for 15 per cent of internationally traded coal.[31] Even though imports only make up about 10 per cent of China's coal supply, China is now the world's largest coal importer.

FIGURE 4.2
Coal Production, Import, and Export in China, 2000–12

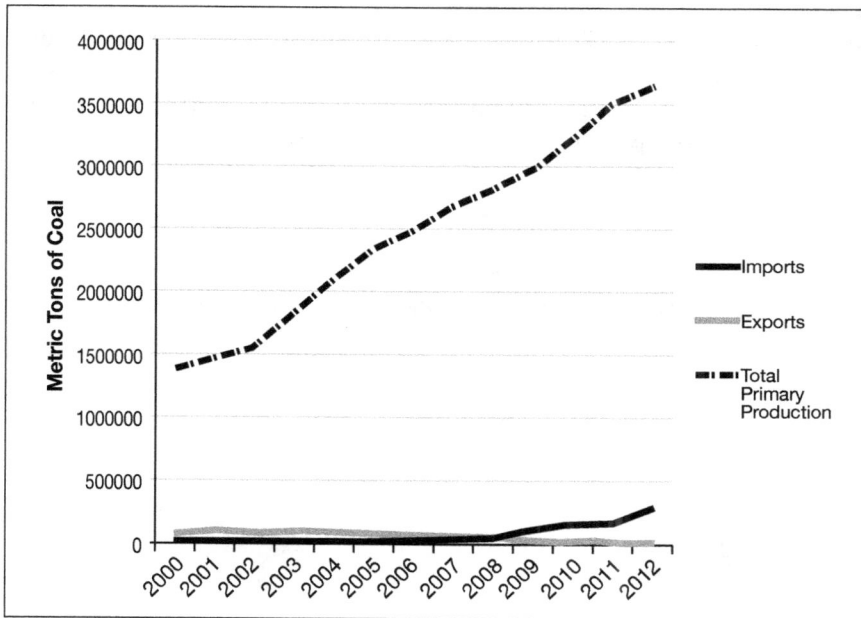

Source: Energy Information Administration.

In 2009, Indonesia's export-oriented coal industry was well equipped to satisfy China's new appetite for overseas coal. Following China's flip from net exports to net imports, Indonesia emerged as China's biggest overseas supplier of coal. Indonesia has a competitive advantage over other countries that export coal to China, primarily Australia. Indonesia's geographic proximity to China and historically low freight costs make imports of Indonesian coal cheaper than coal from other countries. In 2013, freight costs for coal shipped from Indonesia to southeastern China were $6–10/ton, while freight costs for Australia were $10–18/ton.

Ninety-five per cent of Indonesian coal that is exported to China is steam coal, which is burned to produce electricity. Indonesian steam coal is particularly cheap due to its high ash content and low energy content, which further stoked Chinese purchasing. The other primary type of coal is metallurgical coal, which is mostly used for industrial processes and steel production. The steam coal and metallurgical coal markets are separate, and metallurgical coal has a much higher market price (justifying the higher shipping costs from Australia). Indonesia has exceeded its historical competitor, Australia, in exports of steam coal to China (see Figure 4.3). However, certain types of coal in Indonesia, such as lignite, have recently fallen out of favour in China due to environmental concerns, as discussed in the following section.

FIGURE 4.3
China Coal Imports by Country and Type of Coal, 2013

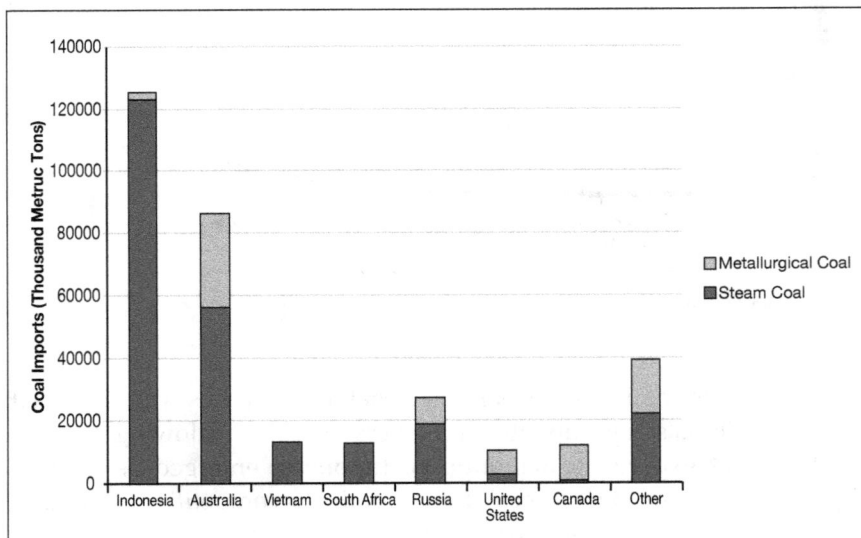

Source: International Energy Agency, Coal Information 2014.

The China–Indonesia Coal Duopoly Under Strain

More recently, however, the duopolistic relationship between China and Indonesia in the global coal market has begun to weaken. The primary reasons for this has been falling demand for coal in China, as it too begins to show signs of slowly dissociating itself from the coal industry, as well as growing demand and a resurgent resource nationalism in Indonesia.

The air pollution crisis in China had a major impact on China's domestic demand for coal. Dire environmental health has forced the government to act. In northern China, where coal is burned in the winter for heating, air pollution has reduced life expectancy by over five years.[32] In November 2015, air pollution levels in Shenyang reached levels 50 times higher than World Health Organization recommendations, while activists cited these levels as the highest ever recorded in China.[33]

As a result, China has begun to devote major political will to reducing domestic air pollution, and a major part of that strategy has been to lower coal burning. Several cities and provinces have capped coal use, while China recently declared that coal use nationwide would peak by 2020. This will require significant policy action. In addition, China has been considering for some time now a ban on the import of low-quality coal with sulfur and ash content above a certain quantity, such as the lignite type commonly found in Indonesia.[34] China has begun testing the quality of imported coal and refusing high ash-content coal at some ports. The Chinese government also requested that utilities stop importing coal in late 2014.[35]

These environmental policies, together with an ongoing structural transition away from heavy industry in the Chinese economy, led coal consumption in China to drop by 4.7 per cent in the first quarter of 2015. Imports fell by 11 per cent in 2014 from the previous year.[36] To match decreased demand for coal, China closed down small domestic coal mines responsible for about 200 million tons of local production. China then decided to cut imports rather than continue to shut down domestic mines. By February of 2015, Chinese coal imports from Indonesia were down 65 per cent compared to February 2014.[37]

While Indonesia's coal economy is still highly export-dependent, exposing it to the energy politics of its trade partners, Indonesia also has its own motivations for decreasing coal exports to China. In

Indonesia, coal users (primarily utilities) are highly nationalized, but coal producers are mostly a mix of small and large private companies that the Ministry of Energy and Mineral Resources regulates through mining laws and policies.[38] The Ministry's 2009 Mining Law aimed to streamline these companies' operations and increase the added value of Indonesian mining industries by using more domestic commodities, services, technologies, and labour. One way of doing this was by keeping more coal for Indonesia itself. The Mining Law stipulated that Indonesian coal producers must sell at least 21.47 per cent of their coal to the domestic market, beginning in 2011, thus restricting the percentage of coal available for export.[39] This domestic market obligation is set each year based on projected domestic demand, and is a policy instrument unique to Indonesia, indicating a form of expressing nationalism through resource control.[40] The policy ensures that even foreign contractors producing coal in Indonesia must give some of their product back to Indonesia. The Ministry is also considering banning exports of unprocessed coal, in addition to other commodities.[41]

The Indonesian government is also trying to reign in illegal coal mining and exports, which have reached a volume equalling nearly 10 per cent of legal exports. The government wants to control supply in order to boost prices and reduce royalties lost through illegal mining and exports, as illegal coal exports are estimated to have a value of $1–4 billion per year.[42] The Indonesian government has scaled up monitoring at ports, and it is also developing fourteen state-controlled coal terminals to aid with monitoring and export restriction.[43] Finally, in late 2014, the Indonesian government began requiring export licenses in order to better manage export levels and taxes.[44]

These export restrictions, however, are driven in large part by energy security concerns. Energy demand in Indonesia is growing faster than infrastructure is expanding, particularly in the electricity sector. The electricity sector is the largest source of coal consumption in Indonesia, and it is poised only to grow further. Currently, 20 per cent of Indonesians lack access to electricity, though the government aims to achieve universal electricity access by 2020.[45] As the Indonesian grid expands, more people will have access to electricity, and as standards of living rise, so will electricity consumption.

Currently, the electricity grid faces frequent issues with insufficient generation and power shortages. Sources of energy for electricity besides coal are limited. Domestic oil production has been declining since its peak in 1991, and Indonesia already must import expensive foreign oil to meet domestic demand.[46] Coal is less expensive than diesel and fuel oil for electricity.[47] Although Indonesia has large gas reserves, they are difficult to access, largely located in frontier regions that lack infrastructure and regulatory certainty.[48] Indonesia is unable to afford the massive up-front capital cost of nuclear power infrastructure, and public opposition to nuclear power is strong in the wake of the Fukushima accident.[49]

In contrast, coal in Indonesia is plentiful and cheap, and coal-fired power plants use established technology and can be built rapidly. Thus, as the government promotes electrification, Indonesia seeks to retain more of its own coal to bolster the domestic energy supply. Indonesia's National Energy Policy aims to make coal a core energy source over oil and gas by increasing the share of coal for primary energy from roughly 25 per cent to 30 per cent by 2030.[50] The Indonesian government fast-tracked a plan to add 10 gigawatts of coal-based generation by the end of 2015, an increase of 20 per cent over current installed capacity for all types of fuel. Indonesia has largely achieved this goal, and aims to install 20 more gigawatts of coal power by 2019.[51]

In sum, the recent slump in the coal import–export relationship between China and Indonesia can largely be attributed to policy and political changes within each country driven by separate domestic needs. To rebalance decreased overseas demand and increased domestic demand, Indonesia is attempting to consolidate coal production, boost domestic coal prices, and reduce exports, potentially motivated by nationalism as well as energy security concerns. As the China–Indonesia coal trade slumped in 2014, India overtook China to become the largest importer of Indonesian coal.[52] Although the China–Indonesia coal duopoly ended as a result of these trends, the two countries' coal relationship has continued through other channels.

A New Direction Towards Coal Investment

While a high-volume coal trade from Indonesia to China is no longer a mutual priority, the downturn does not mean that the China–Indonesia coal relationship is over. In fact, China's investment in the Indonesian coal industry is increasing, signalling that the two countries' coal relationship is taking a new direction. A mix of political and economic benefits among a range of Chinese and Indonesian actors is driving these investments.

Although the coal industry in Asia is not facing the same structural decline as in the West, markets are increasingly volatile. Monthly and spot purchasing — in which a commodity is purchased, delivered, and paid for at roughly the same time — are slowly displacing longer-term contracts. A wider range of coal qualities is becoming available on the market.[53] As discussed above, prices in the internationally traded coal market have dropped rapidly in recent years.[54] Amidst this atmosphere of change, exchange of financial products in the form of foreign aid and investment, rather than direct trade of commodities, represents a more long-term and stable mode of engagement between the two coal giants of Asia.

The Chinese government recognizes the political and strategic benefits of supporting economic and especially resource development in other developing countries. China's economic diplomacy abroad often emphasizes resource development. In particular, Chinese state investment allows the Chinese government to accumulate raw materials, transcending the benefits of trade by securing long-term access to resources.[55] Development finance also helps China build political good will. The New Development Bank (headquartered in Shanghai) and the Asian Infrastructure Investment Bank will channel South–South public finance to developing countries, much of it in energy and mining. Currently, Indonesia is the 6th greatest recipient of foreign aid and government-sponsored investment activities from China, and the top recipient in Asia (countries ahead of Indonesia, like Iran, Venezuela, and Nigeria, are top recipients primarily because of multi-billion dollar oil-related agreements).[56] In addition to public finance, China's "Going Global" or "Going Out" policy since 1999 has opened up a wave of investment from private and state-owned firms in China, usually coupled with loans from the state-owned China Exim Bank (Exim) and the China Development Bank. In 2013, China's Ministry of Commerce reported that 23 per cent of outbound Chinese investment went to the mining sector.[57]

Chinese finance will be increasingly important for Indonesia's coal industry, notably as other international sources of finance are drying up. As discussed above, many Western countries are using less coal and financial institutions within these countries are actively divesting from coal. Over the past few decades, international financial institutions were a major source of funding for coal-fired power plants in Indonesia, focusing specifically on more efficient coal generation technologies.[58] Yet since 2014, the World Bank, the European Investment Bank, the US Export-Import Bank, several national governments, and a host of other banks and financial institutions have ended financial support for coal-fired power plants in other countries. While these policies will have little effect on countries like China, where private capital for domestic coal plants is ample, Indonesia lacks replacement sources of development finance, a gap that Chinese financing could fill. According to some sources, China is now the largest provider of overseas public funding for coal power plants.[59]

Furthermore, China's foreign investment goals align with Indonesia's current needs for infrastructure development capital. The country's infrastructure index remains very low. Government infrastructure spending in 2011 fell as a share of GDP by nearly 50 per cent, when compared to the first half of the 1990s, and it is well below that of other faster growing Asian economies. Interestingly, the Indonesian coal industry first took off after the post-Suharto regime opened the industry to foreign investment, yielding, for the first time, coal production growth rates of 30 per cent each year throughout the 1990s. Most of the first-generation licensees to produce coal in Indonesia were foreign or joint ventures.[60]

Before that, in the first half of the twentieth century, due to the legacy of colonialism, Indonesia's electricity utilities were 100 per cent foreign-owned.[61] These companies were nationalized in the 1950s. Today, the state-owned electricity utility Perusahaan Listrik Negara (PLN) and domestic independent power producers work together to build and deliver power supply. The particular openness to foreign investment in Indonesia's coal and coal power industries can be seen as one indication of a return to a less nationalized framework in the Indonesian energy sector.

Meanwhile, Indonesia is actively encouraging investment from its regional partners while also restricting its export of raw materials. President Jokowi visited China and met with President Xi Jinping in March 2015 as part of a diplomatic trip to encourage further investment from China in Indonesia. Investments from China will play a pivotal role in achieving the Indonesian government's target of accelerating and expanding infrastructure development in the country.[62] However,

Indonesia's government is also attempting to coerce investment by restricting the export of raw materials.

Historically, Indonesia has been a natural resource export-based economy, which has left it vulnerable to the "resource curse", which is when resource-rich countries underperform in long-term economic growth, in part due to an over reliance on raw material exports over more value-added and productivity enhancing industries, such as manufacturing. The Indonesian government has considered banning the export of unprocessed coal in order to stimulate foreign investment in domestic processing industries, since it lacks the capital to build out these industries itself.[63] More broadly, part of the Indonesian Mining Law's implementation includes a new ministerial regulation to encourage the development of high-value industries in Indonesia by phasing out exports of raw materials, including coal, by 2014.[64] Although this has clearly not happened in full, Indonesia means to send a strong signal to trading partners that they need to invest in Indonesia's processing and manufacturing industries in order to have a share of raw materials. The policy has yet to yield the desired results, and for some commodities it is being relaxed. The unsuccessful result of this policy demonstrates Indonesia's quandary in being unable to generate the needed capital to advance its resource sector from raw material export alone. It also reveals Indonesia's preference for an investment-based relationship around coal over the export model that previously defined its resource relationship with China.

Despite these diplomatic and policy efforts to attract investment, Indonesia has sent mixed signals to investors. While the 2009 Mining Law simplified licensing processes for foreign investors, the political volatility and inconsistency in Indonesia's investment and export policies has led many potential coal investors to be reluctant. Overseas demand has historically played a large role in Indonesia's coal production industry. If the export market was cut off in favor of domestic priorities, foreign investors would be hit hard.[65] Foreign investors in mines must now reduce their ownership stake to 49 per cent after a mine has been operating for ten years.[66] In addition, the Indonesian government is considering raising royalty rates for some mining operations.[67] These coal-related regulations have been negotiated in a complex and unpredictable network of power players, from major international mining companies to the Indonesian Supreme Court and the Ministry of Energy and Mineral Resources.[68] Yet, additional infrastructural investment in both mines and plants is necessary for Indonesia to make its coal industry more efficient.

Fortunately for Indonesia, China is not known for shying away from investment in politically volatile countries. Empirical evidence indicates that high political risk in the recipient country does not discourage Chinese investors as it does for Western investors.[69] Indeed, Chinese investment in Indonesia's coal sector is booming. China is investing in coal power generation projects that are critical for bringing electricity to more parts of Indonesia at low cost. Between 2008 and 2013, China held at least $4.34 billion worth of agreements to provide public finance to Indonesian coal power plants and mine-mouth plants (a coal-fired power plant supplied by an adjacent coal mine). Additional projects since 2013 are listed in Table 4.1.

The investments profiled are largely from Chinese state-owned enterprises (SOEs) with financial backing from either the China Exim Bank or the China Development Bank. Some of these SOEs have joint venture companies with Indonesian firms that receive construction contracts and finances. In other cases, Chinese firms win Engineering, Procurement, and Construction (EPC) contracts in Indonesia. The form of investment and actors in China–Indonesia coal partnerships are often complex. For example, the Bukit Asam plant in South Sumatra is being jointly built and managed by a joint venture between China Huadian Corporation; a state-owned electricity utility, which holds a majority stake in the plant; and Bukit Asam, an Indonesian state-owned mining company. The China Exim Bank is providing the $1.2 billion loan. Meanwhile, Bukit Asam has already arranged a 25-year power purchase agreement with the Indonesian state-owned electricity utility, using coal from its own mines. Finally, Bukit Asam granted a multi-billion dollar contract to the China Railway Group to build and operate a railway line from Bukit Asam's coal mine (adjacent to the power plant) to Lampung, a coastal province near Jakarta.

Many of these projects are the direct results of political efforts around investment. The Bukit Asam deal was brokered during the Indonesia–China Economic Cooperation Forum, which also coincided with President Jokowi's visit to China in early 2015.[70] China Huadian, the Chinese partner in the Bukit Asam venture, has been involved in a number of Indonesian coal projects in recent years, including a 760 MW plant agreement brokered by the governor of Bali during a visit to China in 2010.[71] A Huadian subsidiary in Indonesia was the

TABLE 4.1
Selected Recent Chinese Investments in Indonesian Coal Plants

Year	Investor	Type	Location	Cost (Millions)	Partner	Capacity (MW)
2013	CDB	Power Plant	Central Java	$700	PT SSP	660
2013	Sinomach	Power Plant	S. Sumatra	Unknown	Sinar Mas	300
2013	Gezhouba, Exim	Power Plant	W. Kalimantan	$240	PT PLN	210
2014	Shenhua	Mine-Mouth	E. Kalimantan	Unknown	Adaro Energy	600
2014	Sinomach	Power Plant	Riau	$675	PT BTN	450
2015	Huadian, Exim	Mine-Mouth	S. Sumatra	Unknown	Bukit Asam	1240

Sources: Listed in footnote.[72] Data limitations are significant, as many investments are not publicly listed. Therefore, the numbers referenced should be considered the low end of possible investment amounts.

first foreign-funded company licensed to operate a power plant in Indonesia. The Shenhua-Adaro deal also arose during a China–Indonesia event at the recent Asia-Pacific Economic Cooperation (APEC) talks in Beijing in November 2014.[73]

In addition to coal-fired power plants, China is also investing in mining operations. Some mine-mouth plants sell their excess coal to international markets. Coal projects in Kalimantan tend to produce coal for the international market, given their proximity to international shipping routes. Chinese investment in mine-mouth plants in this area could indicate an interest in sending this coal to China. In contrast, Chinese involvement in Sumatran mining could reflect an interest that goes beyond resource acquisition, since most Sumatran coal will likely be used for the Indonesian market. The Indonesian government is focusing on coal resource development in Sumatra, where energy demand is growing rapidly and which could also supply Java, Indonesia's most populous island. Chinese-funded coal projects in Sumatra include the Sinomach-Sinar Mas plant and the Bukit Asam plant (see Table 4.1).

China's involvement in Indonesian coal mining goes beyond project-based finance, given that mines tend to be less centralized and capital-intensive than massive power plants. China has invested directly in major Indonesian coal mining companies. In 2009, China Huaneng Group offered several bids for a majority stake in PT Berau Coal, Indonesia's fifth largest coal producer. In 2013, the China Investment Corporation invested $1.9 billion in Bumi Resources, Indonesia's largest coal company,[74] widely seen as a bailout for the controversy-ridden company. China's investments in Indonesian mining further entangle the two countries through coal, as many Chinese investors now have direct financial interests in continuing to import coal from Indonesia.

Discussion

An incident with a Chinese ship in March 2016 provoked discussion on Indonesia's sovereignty over the Natuna Islands, mirroring disputes throughout the South China Sea between China and other coastal countries in the region.[75] Yet in the face of China's expanding maritime power in the region, China–Indonesia relations remain strong due to increasing long-standing economic ties through resources and infrastructure, as with the coal industry.

The vagaries in the China–Indonesia coal relationship reflect divergences and convergences of economic and political interests within each country. While Chinese imports of Indonesian coal have declined in recent years, their coal relationship is likely to continue in a significant manner through investment. Although Indonesia has ample coal resources, the country's poor infrastructure and a volatile regulatory environment have led to a major need for foreign investment in domestic coal production to meet growing domestic energy consumption. China is an ideal prospect for filling this funding gap, given its own overseas investment priorities and practices. The Indonesian government faces the challenge of opening coal mining and coal power plants to investment from China, while also trying to increase control over the mining process. This gradual process is forming the foundation of a deeper, more long-term economic relationship in which China is a financier as well as a consumer of Indonesian coal.

The decisions of both countries with respect to coal will have a major impact on the environmental and energy future of Asia and the world. As Indonesia builds more coal-fired power plants to meet domestic electricity needs, enabled by increasing finance from China, potential environmental implications also arise. The coal reserves that Indonesia seeks to develop for domestic use are lower quality than the export-oriented developments in Kalimantan. In addition, Indonesia has little incentive to spend additional capital on efficiency technologies for these power plants, as domestic coal is so ample and cheap. Previous public international financiers of coal plants in Indonesia had overarching climate and clean energy goals that led them to focus specifically on high efficiency coal plants.

In contrast, Chinese financial institutions have a different set of priorities for their investments, preferring large centralized projects rather than technological retrofits and favouring economic development goals more than or on par with environmental needs.[76] Chinese SOEs and the Chinese government are suspicious of — and less beholden to — international civil society and media entities that informally regulate social and environmental impacts overseas. In addition, smaller private Chinese companies operating overseas are poorly regulated by the central government. Through these notional mechanisms, Chinese-managed coal plants in Indonesia could have a different pattern of social and environmental outcomes than plants with other sources of finance.

The proliferation of low-efficiency coal plants in Indonesia could lead to local air pollution as well as carbon emissions detrimental to the global climate. On a more macroscopic level, decreased exports and increased finance from China in Indonesia's coal's sector could be an indicator of coal "leakage", as health and environmental concerns in one country push its polluting industries abroad. Chinese investment adds another layer to this phenomenon, as both countries can capitalize on the technological know-how and capital accumulation that China gained as a result of its coal economy.

This study, drawing from multiple disciplinary frameworks, is an imperfect case study of the linkages between geopolitical relationships and coal trade, production, and investment patterns. Further qualitative analysis is essential to verify these linkages and deepen the understanding of Chinese and Indonesian actors and their relationships. In addition, future research would benefit from a more robust quantitative dataset. If trade and investment data improve, more powerful economic tools can be used to understand the outcomes of the trade-to-investment shift. Demonstrating the importance of improved quantitative data for understanding the China–Indonesia coal relationship, in November 2015, China revealed that it had underreported its annual coal consumption by up to 17 per cent.[77] China's appetite for coal (including imported coal) remains voracious, and even recent trade slumps do not yet necessarily indicate a long-term transition away from Indonesian coal for China.

Despite these complexities, the China–Indonesia coal relationship is a valuable way to view the priorities of the two countries in a highly interconnected system of trade and finance. Coal can demonstrate the complex geography of commodity relationships between countries. Recent trends in the China–Indonesia coal trade reveal resource security priorities in both countries, as well as the broader economic benefits of remaining entangled through investment.

NOTES

1. International Energy Agency (IEA), *Southeast Asia Energy Outlook 2015* (Paris: IEA, 2015).
2. Ibid.
3. Moritz Paulus, Johannes Trueby, and Christian Growitsch, "Nations as Strategic Players in Global Commodity Markets: Evidence from World

Coal Trade", EWI Working Paper (Energiewirtschaftliches Institut an der Universitaet zu Koeln, 2011), available at <https://ideas.repec.org/p/ris/ewikln/2011_004.html>.

4. International Energy Agency, *Key World Energy Statistics 2012* (Paris: Organisation for Economic Co-operation and Development, 2013), available at <http://www.oecd-ilibrary.org/content/book/key_energ_stat-2012-en>.

5. Energy Information Administration, "Analysis of the Impacts of the Clean Power Plan", 2015.

6. World Wildlife Fund, "Global Coal: The Acceleration of Market Decline", 2015, available at <http://d2ouvy59p0dg6k.cloudfront.net/downwards/global_coal_the_acceleration_of_market_decline_report.pdf>.

7. Council on Foreign Relations, "Boston Review: Living with Coal: Climate Policy's Most Inconvenient Truth", 18 September 2009, available at <http://www.cfr.org/coal/boston-review-living-coal-climate-policys-most-inconvenient-truth/p20230> (accessed 28 October 2015).

8. International Energy Agency (IEA), *World Energy Outlook 2014* (Paris: IEA, 2014).

9. Tim Wright, *The Political Economy of the Chinese Coal Industry: Black Gold and Blood-Stained Coal* (Routledge, 2012).

10. Ibid.

11. Greenpeace International, "Point of No Return", January 2013, available at <http://www.greenpeace.org/international/en/publications/Campaign-reports/Climate-Reports/Point-of-No-Return/> (accessed 3 May 2015).

12. Luthfi Fatah, "The Impacts of Coal Mining on the Economy and Environment of South Kalimantan Province, Indonesia", *ASEAN Economic Bulletin* 25, no. 1 (1 April 2008): 85–98.

13. IEA, *World Energy Outlook 2014*.

14. Wright, *The Political Economy of the Chinese Coal Industry*.

15. Council on Foreign Relations, "Boston Review: Living with Coal: Climate Policy's Most Inconvenient Truth".

16. Bart Lucarelli, "The History and Future of Indonesia's Coal Industry: Impact of Politics and Regulatory Framework on Industry Structure and Performance", Program on Energy and Sustainable Development, Freeman Spogli Institute for International Studies, Stanford University, Stanford, California, US. Retrieved 10 May 2010.

17. Bernadetta Devi and Dody Prayogo, "Mining and Development in Indonesia: An Overview of the Regulatory Framework and Policies", International Mining for Development Centre, March 2013, available at <http://im4dc.org/wp-content/uploads/2013/09/Mining-and-Development-in-Indonesia.pdf>.

18. Ibid.

19. Moritz Paulus and Johannes Trueby, "Coal Lumps vs. Electrons: How Do Chinese Bulk Energy Transport Decisions Affect the Global Steam Coal Market?", *Energy Economics* 33, no. 6 (November 2011): 1127–37.

20. Gang He and Richard Morse, "China's Coal Import Behavior and Its Impacts to Global Energy Market", in *Globalization, Development and Security in Asia* (World Scientific, 2013), pp. 69–85, available at <http://www.worldscientific.com/doi/abs/10.1142/9789814566582_0032>.

21. Kevin Jianjun Tu and Sabine Johnson-Reiser, "Understanding China's Rising Coal Imports", Carneige Endowment for International Peace, 16 February 2012, available at <http://carnegieendowment.org/files/china_coal.pdf> (accessed 29 April 2015).

22. Ibid.

23. Paulus and Trueby, "Coal Lumps vs. Electrons".

24. Oxford Institute for Energy Studies, "China's Coal Market — Can Beijing Tame 'King Coal'?", December 2014, available at <https://www.oxfordenergy.org/wpcms/wp-content/uploads/2014/12/CL-1.pdf> (accessed 29 April 2015).

25. Jianjun Tu, "Industrial Organization of the Chinese Coal Industry", Stanford University Program on Energy and Sustainable Development, 2011.

26. Tu and Johnson-Reiser, "Understanding China's Rising Coal Imports".

27. Dennis Hickey and Baogang Guo, *Dancing with the Dragon: China's Emergence in the Developing World* (Lexington Books, 2010).

28. He and Morse, "China's Coal Import Behavior and Its Impacts to Global Energy Market".

29. Tu and Johnson-Reiser, "Understanding China's Rising Coal Imports".

30. He and Morse, "China's Coal Import Behavior and Its Impacts to Global Energy Market".

31. Ibid.

32. Yuyu Chen, Avraham Ebenstein, Michael Greenstone and Hongbin Li, "Evidence on the Impact of Sustained Exposure to Air Pollution on Life Expectancy from China's Huai River Policy", *Proceedings of the National Academy of Sciences* 110, no. 32 (6 August 2013): 12936–41.

33. *BBC News*, "China Decries Shenyang Pollution Called 'Worst Ever' by Activists", 10 November 2015, available at <http://www.bbc.com/news/world-asia-china-34773556> (accessed 11 November 2015).

34. "China's Coal Market — Can Beijing Tame 'King Coal'?".

35. IEA, *Southeast Asia Energy Outlook 2015*.

36. *Reuters*, "China's Coal Use Falling Faster than Expected", 26 March 2015, available at <http://www.reuters.com/article/2015/03/26/china-coal-idUSL3N0WL32720150326>.

37. World Coal, "China Coal Imports from Australia and Indonesia Slip in February", 23 March 2015, available at <http://www.worldcoal.com/

coal/23032015/China-coal-Australia-Indonesia-February-2093/> (accessed 19 April 2015).

38. IEA, *Southeast Asia Energy Outlook 2015*.
39. Devi and Prayogo, "Mining and Development in Indonesia: An Overview of the Regulatory Framework and Policies".
40. Energy Information Administration, "Indonesia", 5 March 2014, available at <http://www.eia.gov/countries/analysisbriefs/Indonesia/indonesia.pdf>.
41. Ibid.
42. IEA, *Southeast Asia Energy Outlook 2015*.
43. Ibid.
44. Energy Information Administration, "Indonesia".
45. IEA, *Southeast Asia Energy Outlook 2015*.
46. Ibid.
47. Energy Information Administration, "Indonesia".
48. Neil Gunningham, "Managing the Energy Trilemma: The Case of Indonesia", *Energy Policy*, Decades of Diesel, 54 (March 2013): 184–93.
49. IEA, *Southeast Asia Energy Outlook 2015*.
50. International Energy Agency, "Energy Supply Security: Indonesia", 2014, available at <http://www.iea.org/publications/freepublications/publication/ESS_Indonesia_2014.pdf>.
51. Energy Information Administration, "Indonesia".
52. Ibid.
53. International Energy Agency (IEA), *Medium-Term Coal Market Report 2014* (Paris: IEA, 2014).
54. IEA, *Southeast Asia Energy Outlook 2015*.
55. Ching Kwan Lee, "The Spectre of Global China", *New Left Review* 89 (October 2014): 28–65.
56. Charles Wolf, Xiao Wang, and Eric Warner, "China's Foreign Aid and Government-Sponsored Investment Activities: Scale, Content, Destinations, and Implications" (Santa Monica, CA: RAND Corporation, 2013), available at <http://www.rand.org/pubs/research_reports/RR118.html>.
57. Climate and Finance Policy Centre, "China's Outward Foreign Direct Investment in 2013", available at <http://www.ghub.org/cfc_en/?p=591> (accessed 4 May 2015).
58. IEA, *Southeast Asia Energy Outlook 2015*.
59. Takahiro Ueno, Miki Yanagi, and Jane Nakano, "Quantifying Chinese Public Financing for Foreign Coal Power Plants", 2014, available at <http://www.pp.u-tokyo.ac.jp/research/dp/documents/GraSPP-DP-E-14-003.pdf>.
60. Lucarelli, "The History and Future of Indonesia's Coal Industry".
61. International Energy Agency, "Development Prospects of the ASEAN Power Sector", 2015.

62. Zhao Hong, "China–Indonesia Economic Relations: Challenges and Prospects", *ISEAS Perspective* 2013 #42 (Singapore: Institute of Southeast Asian Studies, 4 July 2013).
63. Energy Information Administration, "Indonesia".
64. Devi and Prayogo, "Mining and Development in Indonesia: An Overview of the Regulatory Framework and Policies".
65. Council on Foreign Relations, "Boston Review: Living with Coal: Climate Policy's Most Inconvenient Truth".
66. IEA, *Southeast Asia Energy Outlook 2015*.
67. Energy Information Administration, "Indonesia".
68. Julia Puspadewi Tijaja, "The Proliferation of Global Value Chains: Trade Policy Considerations for Indonesia", *TKN Report*, January 2013, available at <http://www.iheal.univ-paris3.fr/sites/www.iheal.univ-paris3.fr/files/global_value_chains_indonesia.pdf>.
69. Diego Quer, Enrique Claver, and Laura Rienda, "Political Risk, Cultural Distance, and Outward Foreign Direct Investment: Empirical Evidence from Large Chinese Firms", *Asia Pacific Journal of Management* 29, no. 4 (13 January 2011): 1089–104.
70. *Nikkei Asian Review*, "Tambang Batubara Bukit Asam: Indonesian Coal Producer Gets $1.2B Loan from China", 28 March 2015, available at <http://asia.nikkei.com/Business/AC/Indonesian-coal-producer-gets-1.2B-loan-from-China> (accessed 23 November 2015).
71. *Bali Update News*, "Bali News: Old King Coal to Soon Power Bali", 1 November 2010, available at <http://www.balidiscovery.com/messages/message.asp?Id=6457> (accessed 23 November 2015).
72. SourceWatch, "Cilacap Sumber Power Station", available at <http://www.sourcewatch.org/index.php/Cilacap_Sumber_power_station>; Wataru Suzuki, "Indonesian Coal Producer Gets $1.2B Loan from China", *Nikkei Asian Review*, 28 March 2015, available at <http://asia.nikkei.com/Business/AC/Indonesian-coal-producer-gets-1.2Bloan-from-China>.
73. *Adaro - Positive Energy*, "Adaro Energy and Shenhua Signed MOU to Develop Initially a 2x300 MW Mine Mouth Coal Fired Power Plant in East Kalimantan", 24 November 2014, available at <http://www.adaro.com/news/read/55/Adaro_Energy_dan_Shenhua_Menandatangani_Nota_Kesepahaman_Untuk_Mulai_Mengembangkan_Pembangkit_Listrik_Mulut_Tambang_Bertenaga_Batubara_dengan_Kapasitas_2_300_MW_di_Kalimantan_Timur> (accessed 23 November 2015).
74. Chengjin Wang and César Ducruet, "Transport Corridors and Regional Balance in China: The Case of Coal Trade and Logistics", *Journal of Transport Geography*, Changing Landscapes of Transport and Logistics in China, 40 (October 2014): 3–16.

75. *Reuters*, "China, Indonesia to Boost Security Ties despite South China Sea Spat", 26 April 2016, available at <http://www.reuters.com/article/us-china-indonesia-idUSKCN0XN20R>.

76. Deborah Bräutigam and Tang Xiaoyang, "Economic Statecraft in China's New Overseas Special Economic Zones: Soft Power, Business or Resource Security?", *International Affairs* 88, no. 4 (1 July 2012): 799–816.

77. Chris Buckley, "China Burns Much More Coal Than Reported, Complicating Climate Talks", *New York Times*, 3 November 2015, available at <http://www.nytimes.com/2015/11/04/world/asia/china-burns-much-more-coal-than-reported-complicating-climate-talks.html>; PT Coalindo Energy Indonesian Coal Index (ICI), available at <http://coalindoenergy.com/adaros-capacity-of-power-generation-is-potential-to-rise/>; Sinomach, "CNEEC Indonesia SUMSEL-5 (2 x 150 MW) Pithead Coal-fired Power Plant Project Commenced Construction", 25 September 2013, available at <http://www.sinomach.com.cn/en/MediaCenter/News/201412/t20141209_22074.html>; SourceWatch, "BTN Dumai Power Station", available at <http://www.sourcewatch.org/index.php/BTN_Dumai_power_station>; Wu Chongbo, "Forging Closer Sino–Indonesia Economic Relations and Policy Suggestions", Ritsumeikan International Affairs 10 (2011): 119–42, available at <http://r-cube.ritsumei.ac.jp/bitstream/10367/3402/1/asia10_wu.pdf>; AidData, "Exim Bank Loans $240m to Indonesia for Takalar Steam Coal Power Plant", available at <http://china.aiddata.org/projects/39373?iframe=y>.

5

INDONESIA–CHINA ENERGY AND MINERAL TIES
The Rise and Fall of Resource Nationalism?

Zhao Hong and Maxensius Tri Sambodo

Energy relations between China and Southeast Asian countries have extended from energy trade cooperation to equity investment and infrastructure construction cooperation in recent years. However, several factors are pushing some Southeast Asian countries toward resource nationalism and protectionism. Local politicians and general public reportedly are concerned of China's ambitious plans for energy resources exploitation in Southeast Asia. This chapter examines different concerns of and responses to China's energy resource related investments in Indonesia, and demonstrates that the actual impact of Chinese energy investment depends not only on China, but also on the recipient country's domestic politics, regulatory system and state capacity.

Introduction

Indonesia is rich in energy and mineral resources and it has allowed foreign companies to explore and exploit its oil and gas reserves since the early 1960s. Chinese national oil companies (NOCs) have long demonstrated interest in Indonesia's energy resources and they have developed many oil and gas exploration projects there. After the global financial crisis in 2008, China accelerated its foreign direct investment (FDI) to Indonesia and, for the first time, China was among the top five countries for levels of FDI to Indonesia in the last quarter of 2014.[1] This may elevate energy cooperation to a new level.

However, although Chinese government has planned to stake a long-term strategic energy investment in Indonesia through a range of policy incentives and Chinese capital has poured into resource and energy-related infrastructure sectors, emerging factors have been pushing the two countries' energy ties toward difficulties and competition. Concerns in Indonesia that an increasing trade deficit with China will affect national economic security have stirred debates over how to protect Indonesian resource and mineral industries while maintaining trade ties with Beijing. Fearful of falling into a relationship of "dependency development" with China or other foreign mineral-investing country, Jakarta implemented a new law banning the export of unprocessed ore in January 2014. Although the law aims to increase added value for mineral resources prior to export, the new regulations will affect Sino–Indonesian energy resource cooperation. Within this context of fraught political and economic relations, this chapter addresses the following questions: In what direction is the China–Indonesia energy tie going — towards cooperation or conflict? And can it provide a basis for a broader bilateral relationship?

Overview of the Indonesian Energy Sector

Historically, Indonesia has been an attractive country for mineral investors because of its rich mineral wealth. Indonesia ranks among the top ten countries in the world for proven reserves of copper, nickel, tin, bauxite and coal. It produces more than 15 per cent of the global nickel supply and 3 per cent of the global copper supply. It is also the world's largest coal exporter.[2] The role of energy and mineral resources to the national economy can be assessed with these three key indicators: their contributions to gross domestic product (GDP), exports, and state revenue. Based on these indicators, we can argue that the energy sector has moderately contributed to the Indonesian economy, although coal has become one of its engines for growth. The energy sector has also created a buffer for national exports and contributed significantly to state revenue from tax and non-tax sources.

As seen from Table 5.1, the share of three main energy sectors as a proportion of GDP was about 11.5 per cent between 2010 and 2014. The sector including crude oil, gas and geothermal had the highest share. In 2011, the contribution of the energy sector to GDP reached its highest level because of a surge in coal production. Thereafter, the contribution of energy decreased gradually. Between 2010 and 2014, average economic growth was about 5.7 per cent. During this time, the coal sector contributed positively to economic growth while the two other main energy sectors experienced negative growth.

For exports, as shown in Table 5.2, the share of energy-related products as a proportion of total exports was on average 27 per cent between 2005 and 2014. More recently, however, coal has become a major source of export revenue for Indonesia. In 2009, export revenue from coal surpassed that of natural gas. Sumatera and Kalimantan Islands have the largest deposits and production of coal. Indonesia exports 67 per cent of its coal to Asian destinations, especially Japan, Taiwan, China and India. Natural gas reserves are spread out across the country, but most deposits are found on Natuna Island. Natural gas is exported through gas pipelines and mostly as liquefied natural gas (LNG).[3]

Energy and mineral resources contribute to state revenue from two main sources, namely tax and non-tax revenue.[4] In 2015, tax and non-tax revenue from oil decreased due to the decline of Indonesia's crude oil price (ICP). However, the contribution from minerals and coal increased. As shown in Table 5.3, the contribution of energy to state revenue declined from about 33 per cent in 2006 to 10 per cent in 2015. This is also because the contribution of non-energy resources to tax and non-tax revenue has been rising.

The role of the energy sector in supporting the economy has been changing due to a rapid increase in domestic energy consumption. Oil consumption has grown much faster than oil production. Since the early 1990s, oil production has been declining. As a result, between 1991 and 2013 growth in oil production was negative. Indonesia reached peak oil production in the early 1990s. Afterwards it became increasingly necessary to increase exploration activities for new reserves. In contrast, gas production grew by 9.4 per cent between 1970 and 2013, peaking

TABLE 5.1
Contribution of Energy Sector to GDP
(in billions of Rupiah)

Energy Sector	2010	2011	2012	2013	2014	Average Annual Growth (%)*
Crude Petroleum, Natural Gas, and Geothermal	336,170	444,068	492,894	519,210	506,445	–2.4
Coal and Lignite Mining	160,733	253,026	270,519	275,988	251,303	11.3
Manufacture of Coal and Refined Petroleum Products	233,822	284,099	298,403	310,863	331,743	–1.6
GDP	6,864,133	7,831,726	8,615,705	9,524,737	10,542,694	5.7
Share of energy related products to GDP (%)	10.6	12.5	12.3	11.6	10.3	

* At 2010 constant prices
Source: Central Bank of Indonesia.

TABLE 5.2
Contribution of the Energy Sector to Exports
(in millions of US$)

	2005	2006	2007	2008	2009	2010	2011	2012	2013	2014
Coal	4,179	6,190	6,977	10,305	13,765	17,801	26,924	26,248	24,359	20,814
Crude oil	7,259	7,911	9,380	11,442	8,008	11,219	14,166	12,723	12,188	8,840
Natural gas	10,243	11,863	12,165	16,254	9,778	12,968	18,196	17,671	15,689	14,942
Liquefied natural gas	8,734	9,953	9,722	12,785	7,188	9,432	12,961	11,943	10,568	10,294
Total export	86,995	103,528	118,014	139,607	119,645	158,074	200,787	188,496	182,089	175,290
Share of energy-related products to total export (%)	24.9	25.1	24.2	27.2	26.4	26.6	29.5	30.0	28.7	25.4

Source: Central Bank of Indonesia.

Zhao Hong and Maxensius Tri Sambodo

TABLE 5.3
Contribution of Energy and Mineral Resources to State Revenue
(in billions of Rupiah)

Year	Tax Revenue from Oil and Gas	Non-tax Revenue from Oil	Non-tax Revenue from Gas	Non-tax Revenue from Mineral and Coal	Non-tax Revenue from Geothermal	Total State Revenue from Energy-related Sector	Domestic Revenue	Share of Energy Revenue to the Total State Revenue (%)
2006	43,188	125,145	32,941	6,781	–	208,055	636,153	32.7
2007	44,001	93,605	31,179	5,878	–	174,662	706,108	24.7
2008	77,019	169,022	42,595	9,511	941	299,089	979,305	30.5
2009	50,044	90,056	35,696	10,369	400	186,566	847,096	22.0
2010	58,873	111,815	40,918	12,647	344	224,597	992,249	22.6
2011	73,096	141,304	52,187	16,370	563	283,519	1,205,346	23.5
2012	83,461	144,717	61,106	15,877	739	305,901	1,332,323	23.0
2013	88,747	135,329	68,300	18,621	867	311,864	1,432,059	21.8
2014	83,890	154,750	56,918	23,560	580	319,697	1,633,053	19.6
2015	50,919	72,999	22,638	31,679	584	178,819	1,765,662	10.1

Source: Government state budget (various issues).

in 2010. At the same time, gas consumption also increased rapidly, approaching levels of production after 2000. Only coal shows double-digit growth for both production and consumption. Indonesia is still the world's largest exporter of thermal coal, although its domestic demand has increased rapidly as well.

China and India are two of Indonesia's largest export markets, accounting for 31 per cent and 22 per cent of Indonesia's total coal exports in 2011, respectively.[5] Because the growth of gas and coal production was higher than that of consumption, in 2013, the share of export from LNG and coal production was about 88 per cent and 73 per cent, respectively.[6] However, growing demand on domestic energy consumption has compelled the Indonesian government to secure energy production for domestic consumption before export, although, for gas and coal, production is still higher than consumption (see Table 5.4). To secure domestic supply, the government implemented the domestic market obligation (DMO) policy for gas and coal. According to National Medium Term Development Plan, in 2019, the DMO for gas and coal is 64 per cent and 60 per cent respectively or it increases from 53 per cent and 24 per cent respectively. Thus, domestic allocation for coal increases about 36 per cent. A rapid increase of DMO on coal aims to secure a primary energy supply for steam coal power plant after government plan to add 25.8 gigawatt of coal power plant by 2019.

TABLE 5.4
Growth of Oil, Gas, Coal Production and Consumption in Indonesia

	Oil		Gas		Coal	
	Production	Consumption	Production	Consumption	Production	Consumption
Year	1991–2013	1965–2013	1970–2013		1981–2013	
Growth at corresponding year (%)	–3.2	5.8	9.4	8.2	20.7	16.3

Source: Calculated from BP Statistical Review of World Energy (June 2014).

China's Interests in the Indonesian Energy Sector

As Asia's energy demand is booming, it is also fuelling economic growth and rising standards of living. The *World Energy Outlook 2012* by the International Energy Agency (IEA) predicts that global energy demand will increase by a third from 2010 to 2035, with Asia accounting for nearly two thirds of that growth.[7] China and India alone will account for half of global demand growth. China, which only recently became the world's largest energy consumer, will account for nearly 40 per cent of world energy demand growth from 2011 to 2035.[8]

To fulfill its growing demand for oil, China depends on three NOCs for supply: China National Petroleum Corporation (CNPC), China Petroleum and Chemical Corporation (Sinopec Group) and China National Offshore Oil Corporation (CNOOC).[9] Currently, China's NOCs are international operators in more than forty countries, producing 2.5 million barrels of oil equivalent per day of oil and gas overseas (in 2013) and supplying 59 per cent of China's oil and gas demand.[10] IEA has suggested that five key motivations for NOCs to invest abroad are: (i) to expand oil and gas reserves and production, (ii) to diversify energy supplies to avoid risks, (iii) to become "international NOCs", (iv) to develop an integrated supply chain, and (v) to gain technical know-how and streamline managerial capacities.[11] The five motivations are pursued with five different strategies.[12] In order to become international NOCs, China's NOCs develop partnerships through mergers and acquisitions with other NOCs and IOCs (international oil companies). This strategy is well reflected in its energy development in Indonesia.

Bilateral energy cooperation between China and Indonesia is not new. It can be traced back to the 1980s. In February 1988, Petrochina signed an offshore production sharing contract (PSC) with Indonesia. The contract area was located in Tuban, East Java. In 1994, CNOOC obtained 2.8 per cent of the share of an Indonesian Malacca oilfield through capital mergers and acquisitions, starting its first forey into Indonesian energy exploration and development. In 2002, CNOOC bought a Spanish oil company's assets in Indonesian oil fields at a price of US$850 million. In the process, CNOOC became Indonesia's largest offshore oil producer. In April 2004, Sinopec purchased American Devon Energy's oil and gas assets in Indonesia as its first foray into the Indonesian energy exploration and development market.[13] In 2005, CNOOC obtained 16.9 per cent of the shares of a British Gas Corporation LNG project in Indonesia.[14] Such acquisitions continued in 2008 with CNOOC purchasing interests in Husky's Indonesia project, which was previously owned by Canada's largest energy company, to explore deep water blocks. Furthermore, Chinese NOCs have also been interested in constructing storage facilities in Indonesia. Total acquisitions by Chinese NOCs in Indonesia between 2002 and 2011 was worth about US$2.45 billion, or about 1.6 per cent of total Chinese overseas oil and gas upstream acquisitions during that same period.[15]

TABLE 5.5

Chinese Overseas Oil and Gas Upstream Acquisitions in Indonesia, January 2002–December 2011

Data	Company	Assets	Share (%)	Contract value (US$ billion)
December 2011	Sinopec	Purchased 18% stake from Chevron's deep-water gas fields (three blocks) in the Gendalo-gehem natural gas development in East Kalimantan of Indonesia. In addition to gas, oil is also produced in these blocks.	18	0.68
December 2010	Sinopec	Acquired 18% of Chevron's Gendalo-Gehem deep-water gas project in Indonesia	18	0.68
2008	CNOOC	Purchased 50% interest in Husky (Madura) Energy's assets in Indonesia	50	0.125
2004	CNOOC	Tangguh (BG)	–	0.105
2003	CNOOC	Tanguh (BP)	–	0.275
2002	CNOOC	Purchased Repsol's Yacimientos Petroliferos Fiscales upstream assets (Southeast Sumatra etc.) in Indonesia	–	0.585
2002	CNPC/PetroChina	Purchased Devon Energy Corporation for six blocks in Indonesia	100	0.585

Source: IEA, *Update on Overseas Investments by China's National Oil Companies: Achievement and Challenges since 2011* (Paris: IEA, 2014) and Marwan Batubara, Menggugat Pengelolaan Sumber Daya Alam: Menuju Negara Berdaulat [Defending Natural Resource Management: Towards a Sovereign Nation] (Jakarta: KPK-N, 2009).

Although the share of Chinese overseas oil and gas upstream acquisitions in Indonesia and the inflow of investment from China were minor, China's investment flow to Indonesia's mining sector has been increasing rapidly. The latest figures indicate that most Chinese FDI has flown to the mining sector. As seen from Table 5.6, in 2014 the share of China's FDI to the mining sector accounted for 99 per cent of China's total FDI to Indonesia (Share 4). In terms of total FDI flows to the mining sector, China's share increased from 20 per cent in 2005 to 39 per cent in 2014 (Share 3), although the share of its FDI has only increased from 3.6 per cent to 4.8 per cent (Share 2).

The main reason for the increase in China's FDI in the mining sector was China's increased demand for coal. When China became a net importer of coal in 2007, it shifted its focus to Indonesia. Coal from Indonesia has become increasingly attractive to the prosperous coastal regions of China, potentially displacing domestic Chinese production that must be transported by rail and shipped long distances from Shanxi and Mongolia. As part of a growing effort by Chinese companies to secure future coal supply, Shenhua — China's largest coal producer — announced a US$331 million coal project in Sumatera in July 2010 and, in October of the same year, China's sovereign wealth fund injected US$1.9 billion into Bumi Resources — Indonesia's largest coal producer.[16]

Energy cooperation between China and Indonesia through equity capital and investment has brought more energy trade between the countries. As Table 5.7 shows, the total export of energy-related products (HS-27) between 2006 and 2011 increased from US$27.6 billion to US$69 billion. After 2011, however, it began to decline.[17] We can see that as total exports increased, the proportion of Indonesia's mineral exports to Japan and Korea declined.[18] Meanwhile, the share of Indonesia's exports to China was relatively stable, as value increased steadily from US$3.1 billion in 2006 to US$8.3 billion in 2013, growing on average by 24 per cent per year. These changes show that China and India have become more important to the Indonesian energy export market. However, as the next section discusses, recent trends in mining sector laws have created new challenges in the Indonesian export market for resource based products.

TABLE 5.6
FDI Inflows to Indonesia
(in millions of US$)

Year	Total FDI Inflow (1)	Total FDI Inflows in Mining Sector (2)	Total FDI Inflows from China (3)	FDI inflows from China in Mining Sector (4)	Share of FDI Inflow in Mining Sector to Total FDI Inflow (2) : (1)	Share of Total FDI Inflow from China to Total FDI Inflow (3) : (1)	Share of Total FDI Inflow in Mining Sector from China to Total FDI Inflow Mining Sector (4) : (2)	Share of FDI Inflow from China Mining Sector to Total FDI Inflow from China (4) : (3)
2005	8,336.0	1,225.8	299.5	238.8	14.7	3.6	19.5	80
2006	4,914.0	322.1	124.0	123.0	6.6	2.5	38.2	99
2007	6,928.0	1,904.0	178.0	170.0	27.5	2.6	8.9	96
2008	9,318.0	3,609.5	531.0	534.0	38.7	5.7	14.8	101
2009	4,877.0	1,301.1	359.0	357.4	26.7	7.4	27.5	100
2010	13,771.0	1,896.0	354.0	354.0	13.8	2.6	18.7	100
2011	19,242.0	3,418.0	215.0	150.0	17.8	1.1	4.4	70
2012	19,138.0	1,822.0	335.0	285.0	9.5	1.8	15.6	85
2013	18,947.8	2,486.8	67.6	32.2	13.1	0.4	1.3	48
2014	22,276.3	2,710.7	1,066.9	1,065.9	12.2	4.8	39.3	100

Note: "Share" is in percentage.
Source: Calculated from the Bank Central Republic Indonesia.

TABLE 5.7
Share of Mineral Fuels, Oils and their Distillation Products (HS 27) Exports from Indonesia (%)[19]

Year	China	India	Japan	Rep. of Korea	Singapore	Total Countries*	Total Export HS 27 (Billion US$)
2006	11.2 (3.1)	2.5	39.4	18.2	4.5	75.8	27.62
2007	12.0 (3.5)	3.0	40.6	16.6	5.9	78.1	29.21
2008	11.5 (4.6)	3.4	40.3	14.6	7.0	76.8	39.78
2009	14.1 (4.7)	6.2	26.7	14.8	7.2	68.9	32.95
2010	12.9 (6.0)	5.3	25.8	17.9	9.0	70.9	46.77
2011	12.9 (8.9)	6.9	27.8	16.9	10.6	75.2	68.92
2012	12.8 (8.1)	7.9	26.0	17.3	10.4	74.5	63.39
2013	14.4 (8.3)	9.7	24.8	13.1	11.0	72.9	57.41

* Indicates sum of China, India, Japan, Republic of Korea and Singapore; figure in bracket indicates in billion US$.
Source: Calculated from the United Nations Commodity Trade Statistics Database.

Mineral Resource Nationalism on the Rise?

Many Chinese investors consider Indonesia to be one of the most promising investment destinations for mineral resources. Yet recent regulatory changes in sectors ranging from mining to oil have complicated foreign ownership and production, which has given Indonesia the reputation of a country where resource nationalism is on the rise.[20]

The 2009 Mining Law marked a key shift towards resource nationalist policies. Article 112 holds that after five years of production, foreign companies with a mining business license (IUP) and a Special Mining Business License (IUPK) need to divest portions of their ownership. Article 170 also states that five years after the enactment (i.e., in 2014), foreign companies must process a certain degree of minerals inside Indonesia.

The government has also linked policies on foreign ownership to investments in smelter facilities. According to Government Regulation No. 77 in 2014, divestment is conditioned by the presence of smelter facilities and whether mining is underground or open pit. As Table 5.8

TABLE 5.8
Export for Selected Mineral Products
(in millions of US$)

No.		2010	2011	2012	2013	2014
1	Mineral fuels & oils*	18,726	27,444	26,408	24,780	21,058
2	Cooper	3,306	3,811	1,886	1,738	1,967
3	Ores, slag, and ash	8,149	7,343	5,083	6,544	1,919
4	Tin	1,735	2,439	2,132	2,129	1,814
5	Iron and steel	1,102	1,353	875	652	1,148
6	Nickel	1,436	1,218	993	942	1,058
7	Aluminium	772	869	784	693	665
	Total export of 7 commodities (No. 1–7)	35,224	44,476	38,161	37,479	29,628
	Share in total exports	27.15	27.45	24.93	25.00	20.30

* Because non-oil and gas exports are calculated here, the figures in Table 5.8 are lower than in Table 5.7.
Source: Calculated from Economic Profile, Ministry of Trade, available at <http://www.kemendag.go.id/en/economic-profile/indonesia-export-import/growth-of-non-oil-and-gas-export-commodity> (accessed 16 March 2015).

shows, foreign companies that do not have smelter facilities can retain only up to 49 per cent ownership after ten years of production, while foreign companies that have smelter facilities may retain up to 70 per cent of ownership over the same period. However, foreign companies that have smelter facilities can have a maximum of only 60 per cent ownership after fifteen years of production. In the next fifteen years, these companies may then regain another 11 per cent of additional ownership. This policy effectively strips foreign companies of majority control over mining assets, unless they have smelter facilities.[21]

The Indonesian government has also placed stronger restrictions on the export of raw materials. It argues that by imposing export duties, it can control the trade of raw material or ore, increase added value inside Indonesia, and ensure availability of mineral resources for the domestic market.[22] It has imposed export duties on mineral extraction and stipulated that raw materials be processed domestically from 2014 onwards.[23] Although this policy has been challenged by countries that obtain raw materials from Indonesia, including Japan, India and China, the Indonesian government has argued that its actions are justified to prevent over exploitation of mineral resources, meet domestic demand, and promote downstream industries within Indonesia.

The government has also banned the export of certain raw minerals and required mining companies to build smelting facilities for domestic processing. The first regulation regarding export duties was implemented on May 2012. It imposed uniform export duties of about 20 per cent.[24] In the case of ores, it covered twenty commodities under the HS-26, which included ores, slag, and ash. Although the number of commodities covered under HS-26 was reduced from twenty to ten, export duties are being gradually increased to 60 per cent by July 2016. As shown in Table 5.8, the total export of metal products declined significantly after the government implemented export duties in 2012. The export of seven commodities reached a peak of US$44.5 billion in 2011 and then declined substantially to about US$29.6 billion in 2014. The share of minerals in total exports also decreased from about 27 per cent to 20 per cent between 2010 and 2014.[25]

There are several reasons why resource nationalism in Indonesia is on the rise. The main goal is to strengthen the role of national mining companies. Indonesia has five state mining and oil companies, including Aneka Tambang, Inalum, Pertamina, Bukit Asam (BA), and Timah.[26] The government aims to enhance the participation of these national companies and reduce their reliance on exporting resources as an engine of economic growth.

However, many observers believe that Indonesia's resource nationalism is driven by political motivations, reflecting an ongoing struggle between central, provincial and local governments for control over the issuing of mining permits.[27] The struggle dates back to 1998, when Suharto's fall set Indonesia on a path towards decentralization. Thirty-one years of highly centralized governance based in Jakarta under Suharto led to deep social and economic imbalances between Java and the outer islands. In response to calls for political and economic decentralization, President Bacharuddin Jusuf Habibie moved to limit the central government's authority to matters of military and policy security in 1999. Provincial governments were granted limited independence from Jakarta on social policies, while local and district governments gained control over economic policies, including control over the issuing of mining permits. In provinces such as East Kalimantan and Southeast Sulawesi, mining plays a large role in the local economy and conflicting claims over extractive projects became an on-going source of political tension between central and regional governments. In 2011, however, President Yudhoyono released the country's "Master Plan" for economic development through 2025, which called on Indonesia to transform itself from a natural resource exporter to an industrial manufacturing hub. In pursuing this goal, a degree of recentralization is needed to facilitate long-term strategic planning.

Resource nationalism is also a topic that political leaders may leverage for their own electoral ambitions. A strong agenda that privileges domestic industry over foreign investors appeals to popular nationalist sentiment and will garner votes, particularly where it concerns the ownership of natural resources. Notably, deliberations over the 2009 Mining Law occurred in the lead up to Indonesia's 2009 Presidential Elections. Similar arguments on resource exports and domestic industries were put forward again as campaigns heated up for the July 2014 Presidential Election. Indonesia announced an export ban on raw ore in January 2014.

Nevertheless, it is believed that without proper and transparent standards in operation and procedures, the divestment of foreign ownership cannot be successful. The experience of divestment for Newmont Nusa Tenggara (which began producing cooper and gold in 2002) indicates that a complex conflict of interest among central and local governments, state-owned companies and private companies led to long drawn out disputes on implementation. Thus in 2006, the divestment needed to be implemented but currently it is still left with 7 per cent of foreign shares that need to be sold. Most foreign investors also do not support the new divestment rules. Investment in the mineral sector is often a long-term proposition, so companies may not want to be involved in a project over which they will have little control in the future.[28]

Indonesia also faces dilemmas in implementing the minerals export ban. The Indonesian government recognizes the importance of shifting from an economic model heavily reliant on raw material exports to one in which Indonesia refines its own metals, minerals and ores, both for export and more importantly for domestic use in manufacturing industries. The underlying economic rationale for the ban is to stimulate domestic smelting and processing capacity, which is presumed to lead to significantly higher added value in mineral exports.[29] However, such interventions come with substantial risks as the industry may respond to the incentives differently from the policy intention. For example, foreign investors may reduce their investment in Indonesia. Historically, foreign investors have often chosen to export raw commodities from Indonesia because other countries already have well-developed processing capabilities.

Without adequate investment and capital inflows, Indonesia will find itself squeezed between slowing foreign demand for raw materials and the inability to shift to greater domestic consumption for key minerals. Overseas processors are less compelled to invest in Indonesia because they can secure ore supplies elsewhere. (Indonesia accounted for less than 2 per cent of global production in copper, lead and zinc in 2012 and it does not have a major share of reserves for any of these commodities.[30]) Investment in bauxite and iron ore are more likely to be viable if the raw ore can be accessed cheaply, placing Indonesia at a disadvantage compared to other countries, such as China and India. Thus "the export ban will inevitably lead to a dramatic decline of output in Indonesia's extractive industries, damaging foreign investment and economic growth, and disrupting global mineral markets".[31]

These new regulations, especially the export ban on raw materials, will certainly affect Sino–Indonesian energy cooperation. China is highly dependent on Indonesia for nickel, bauxite, copper, and coal. In 2012,

Indonesia produced 16 per cent of the world's nickel ore, and supplied 58 per cent of its nickel import demand and 48 per cent of its bauxite import demand. Most of Indonesia's exports of these metals go to China and Japan. In 2013, China sourced 66 per cent of its alumina (refined bauxite for smelting into aluminum) from Indonesia (up from 64 per cent in 2012) and 57 per cent of its bauxite (on par with 2012 levels). In 2012, about 6 per cent of China's copper ore imports came from Indonesia.[32] China is driving Indonesian coal export growth. However, the appetite for Indonesian coal in China is gradually reducing, as the Chinese government has discussed a ban on coal imports with low energy content as mostly found in Indonesia, while favouring higher quality Australian and South African coal.[33] In this sense, the export ban might have more impact on Indonesia itself than China.

Concerns over Cooperation with China

Indonesia–China energy cooperation is far from smooth. Public debate over the Indonesia–China energy trading agreement arose in 2009.[34] One of the debates was focused on a price agreement for shipping LNG from Tangguh to Fujian, China. In 2002, the price agreement was US$2.4/MMBtu (million British thermal units) and the maximum ceiling price for oil was about US$25/barrel. In 2006, the market price of LNG increased to about US$3.35/MMBtu and the price of oil to US$38/barrel. A first renegotiation was conducted by the government in 2006, which boosted the LNG price increase to US$3.3/MMBtu. A second renegotiation was attempted again in 2010, but it failed. After a meeting with Chinese President Hu Jintao in 2012, Indonesian President Susilo Bambang Yudhoyono asked the Minister of Energy and Mineral Resources to renegotiate the contract with CNOOC. Finally in June 2014, the negotiation was concluded and Indonesia obtained a better price on LNG of US$8.65/MMBtu.

The other concern has been an increasing trade deficit with China. Indonesian trade with China was slightly higher than with Vietnam, but less than with Singapore, Malaysia and Thailand. However, among ASEAN countries, Indonesia had the second highest trade deficit with China, after Vietnam. Jakarta believes that the growing non-minerals trade deficit with China is the main reason for the trade deficit. According to Indonesia's data, Indonesia had a trade surplus of US$820 million with China in 2005, but US$14 billion deficit in 2014 when the oil and gas sectors were excluded.[35] As Figure 5.1 shows, most of Indonesia's exports to China are resource-intensive products. For example, Indonesia's mineral exports to China increased from 26.2 per cent in 2000 to 56 per cent in 2013 as a proportion of total exports to China. Meanwhile, high value-added products, such as electrical machinery and transport equipment, accounted for over 50 per cent of Indonesia's imports from China.

This increasing trade deficit with China can be largely attributed to the low competitiveness of Indonesia's manufactured products. During the 1980s and 1990s, the Indonesian government largely failed to assist domestic firms in upgrading their technological capabilities. By the mid-1990s, Indonesia lagged behind its East Asian neighbours on most

FIGURE 5.1
Commodity Structures of Indonesia's Exports to China, 2013

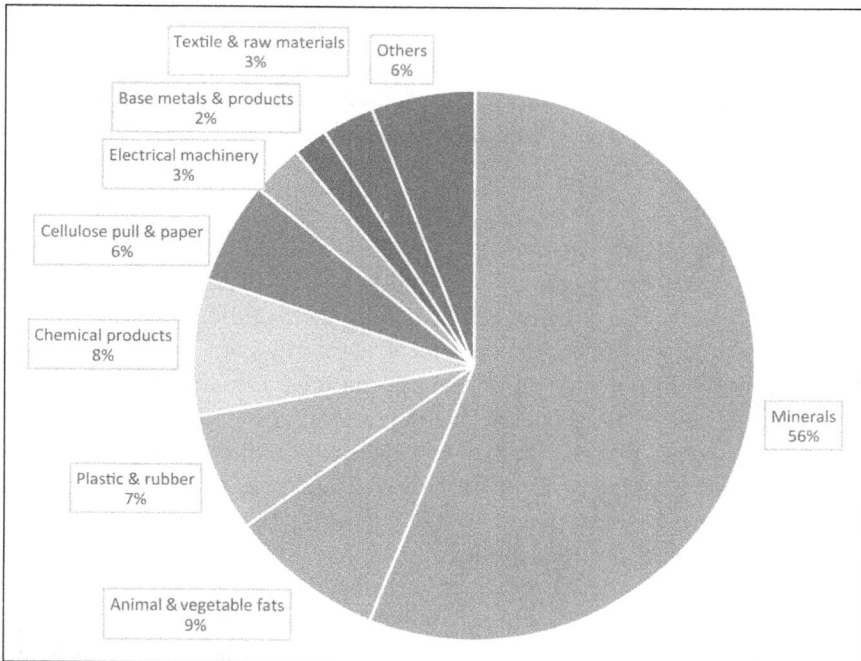

Source: Ministry of Commerce of China, *Country Report* (2014).

FIGURE 5.2
Commodity Structures of China's Exports to Indonesia, 2013

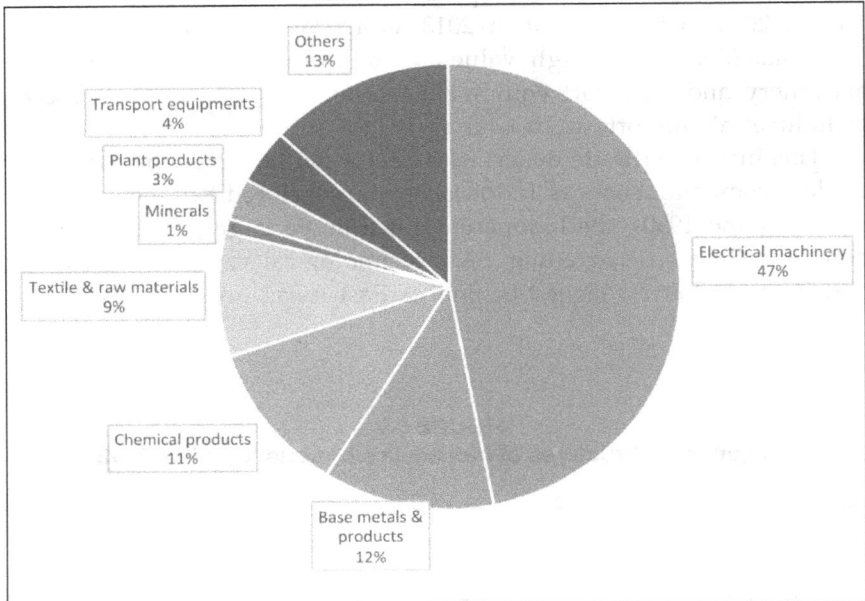

Others
13%

Transport equipments
4%

Plant products
3%

Minerals
1%

Textile & raw materials
9%

Electrical machinery
47%

Chemical products
11%

Base metals &
products
12%

Source: Ministry of Commerce of China, *Country Report* (2014).

technology indicators. According to Indonesian statistics, its spending on research and development was very low (0.2 per cent of GDP). It also had very few patent applications (twelve between 1981 and 1990), very few scientists and engineers were engaged in research and development (183 per million of the population), enrolments in tertiary education were low (10 per cent of the relevant age group in 1991), and few young adults had science or engineering degrees (0.4 per cent of 20–23-year-olds).[36] Catching up in the technological field also requires exposing domestic enterprises to the rigours of foreign competition. Because of this requirement, policymakers in China liberalized its trade and investment regime as much as, if not more than, their Indonesian counterparts. However, growing fears that this widening trade gap might affect Indonesia's national economic security have stirred debates over how domestic industries can remain competitive as the country seeks to improve trade ties with Beijing. In turn, this has also aroused domestic economic and resource nationalism.

Can Sino–Indonesian Energy Cooperation become the Basis for a Broader Bilateral Relationship?

As the largest country in Southeast Asia, Indonesia illustrates the diplomatic complexities that are involved in relations with China. At the bilateral level, Indonesia has increasingly become more comfortable with China. Although initially reluctant, Indonesia has forged a closer bilateral relationship with China, culminating in the signing of a strategic partnership in 2005, which was upgraded to a comprehensive strategic partnership during Chinese President Xi Jinping's visit to Jakarta in 2013. More and more Indonesians see China, compared with the United States, as an increasingly positive partner. For example, the Pew Research Global Attitudes Survey released in 2013 showed that China's favourability in the eyes of Indonesian respondents increased from 58 per cent in 2010 to 67 per cent in 2013, while the number of respondents with a favourable view of the United States rose slightly from 59 per cent to 61 per cent. Another 69 per cent of the respondents replied that China will have a great impact on Indonesia, especially in terms of the economy, increasing from 60 per cent in 2008.[37]

Moreover, the survey also showed that 54 per cent of respondents agreed that China considered Indonesia's interests when making international policy decisions, increasing from 50 per cent in 2008. In comparison, only 52 per cent indicated that the United States takes into account Indonesia's interests in making international policy decisions. Although Indonesian elites like the idea of United States engagement in the region and dislike the thought of a dominant Chinese role, they have far more confidence in the Chinese commitment to the region than they do in that of the United States.[38] Most Indonesians no longer see China as an ideologically threatening state, but as an economic opportunity and challenge.

The growing bilateral economic engagement can be gauged from the fact that despite the global financial meltdown, the two countries achieved the target of bilateral trade of US$30 billion in 2008. Bilateral trade increased from US$19 billion in 2006 to US$68 billion in 2013, registering more than 300 per cent growth in seven years. The countries have agreed to increase the volume of bilateral trade to US$80 billion by 2015.[39] Increasing bilateral trade has helped Indonesia reduce its dependence on Western markets. Due to expanding trade with China, Indonesia's over-reliance on particular export destination countries has decreased. For example, from 2000–12, the export market shares of United States, Japan and Europe decreased from 51 per cent to 37 per cent, while China's share increased from 3.6 per cent to 12 per cent.[40] It was the Asian emerging economies, mainly China, India and those in ASEAN, that subsequently compensated for Indonesia's decelerating exports to developed countries.

Based on mutual need and benefit, the relationship between Indonesia and China is likely to become stronger and grow further in the future. Viewed through China's lens, Indonesia's bountiful mineral wealth has elevated relations between Jakarta and Beijing to a position of strategic importance. Moreover, in Beijing's view, recent political reform and economic growth has made Indonesia re-emerge on both the international and regional stage with expanded prestige both in the East and West. As it bolsters its strength, Indonesia's weight and importance in the region's balance of power will only grow, particularly with respect to China and the United States. While welcoming United States rebalancing toward Asia, some in Indonesia have raised concerns that Washington has placed too much emphasis on the military dimension of this strategy. From Jakarta's perspective, the importance Washington attaches to Indonesia and ASEAN should not simply be derivative of China's rise but instead be based on the intrinsic value of the country and the subregion.[41]

In Indonesia's strategic calculations, China's importance lies primarily in it being a growing source of foreign investment that Indonesia desperately needs to develop its domestic natural resources and infrastructure. It needs a huge amount of investment in its energy sectors, including energy-related infrastructure like gas pipelines and seaports. According to the Indonesia Medium Term Development Plan (2015–19), the government has three top priority sectors to develop — food, energy, and maritime resources. Where energy infrastructure is concerned, if oil and coal production were to decline, gas will become the future of primary energy supply for Indonesia. In response to these targets, the government plans to develop gas infrastructure, such as pipelines, gas stations, and city gas networks. Connecting supply locus and market locus among the islands is one of the greatest challenges in optimizing gas utilization. Most of the gas is produced in the eastern part of Indonesia, and it needs to be shipped by sea to the western part. However, Indonesia's poor infrastructure has been the major problem and challenge.

For example, while its overall index has improved over the past few years, the country's infrastructure index remains very low: 76th for physical infrastructure, 103rd in terms of quality of ports, and 98th in electricity supply.[42] A World Bank study in 2010 found that the cost of shipping a 40-foot container from Padang to Jakarta is US$600 while the same container can be shipped from Jakarta to Singapore (three times the distance between Padang and Jakarta) for only US$185. The quality of port facilities remains alarmingly low and shows no sign of progress, and the electricity supply continues to be unreliable and scarce. China with its total outward FDI of US$101 billion in 2013 has potentially a big role to play in Indonesia's infrastructure sectors.

Conclusion

Although energy relations between China and Indonesia have thus far generally proved to be mutually beneficial, concerns and uneasiness among Indonesians about the nature and impact of the relations prevail. Thus, to what extent the expansion of energy cooperation between the two countries can grow will depend on whether local communities in Indonesia feel that their concerns are being addressed. Particular areas of concern are the continuing impact of Chinese investment and trade on energy supply, local jobs and the erosion of the competitiveness of Indonesian companies by the growing presence of Chinese companies, the unbalanced trade relations, and perceptions that expanding commercial relations have exerted a detrimental influence on Indonesian foreign policy.

However, compared with some other Southeast Asian countries, such as some Southeast Asian peninsular countries like Myanmar and Vietnam, the dynamics in the overall relations between Indonesia and China are rather different. Although initially reluctant to engage with China, Indonesia has forged a closer bilateral relationship with China. Improved relations culminated in the signing of a strategic partnership in 2005, which was upgraded to a comprehensive strategic partnership during Chinese President Xi Jinping's visit to Jakarta in 2013 and also encouraged Beijing's to develop closer relations with ASEAN. Both countries are keen to assert themselves on the international and regional stage, and they can position themselves as part of a new world order that is more representative of contemporary geopolitical realities. Both countries have visions of becoming maritime powers as well. Therefore, the strategic potential of China's investment in energy related-infrastructure and seaports is not limited to enlarging Sino–Indonesian energy trade, but also extends Indonesia–China relations more broadly and can help realize Indonesia's ambitions for becoming a maritime power.

NOTES

1. Okezone Finance "China Masuk Top Investor Indonesia" [China Became a Top Five Investor in Indonesia], 28 January 2015, available at <http://economy.okezone.com/read/2015/01/28/20/1098445/china-masuk-top-investor-indonesia> (accessed 27 May 2015).
2. According the World Coal Association, based on the purpose of use, there are two types of coal. First is steam coal, also known as thermal coal. Thermal coal is mainly used in power generation. Second is coking coal or also known as metallurgical coal. It is mainly used in steel production.
3. There are two gas refineries producing more than 96 per cent of gas in Indonesia, namely Bontang in East Kalimantan Province and Teluk Bintuni in West Papua Province.
4. According to Law No. 33 of 2004 on the financial budgets of central and local governments, 84.5 per cent of revenue from oil is owned by the central government and 15.5 per cent is allocated to the local governments, including 3 per cent for provincial government, 6 per cent for the district or city government where production is located, 6 per cent for distribution among other districts or cities within the province, and 0.5 per cent for primary education. Similarly, in the case of natural gas, the allocation between central and local government is 69.5 per cent and 30.5 per cent, respectively. The local 30.5 per cent allocation is structured as follows: 6 per cent for provincial government, 12 per cent for the district or city where the gas is exploited, 12 per cent for distribution to all districts or cities within the province, and 0.5 per cent for primary education. In 2014, the Minister of Finance allocated about Rp 36.6 trillion [US$2.9 billion] to provincial, district and city governments.

5. International Energy Agency (IEA), *Southeast Asia Energy Outlook 2013* (Paris: IEA, September 2013), p. 72.

6. This figure was calculated using information from the Indonesian Energy Handbook – Ministry of Energy and Mineral Resources.

7. International Energy Agency (IEA), *World Energy Outlook 2012* (Paris: IEA, 2012).

8. Ibid.

9. CNPC and Sinopec focus on onshore oil exploration while CNOOC focus on offshore oil exploration.

10. International Energy Agency (IEA), *Update on Overseas Investments by China's National Oil Companies: Achievement and Challenges since 2011* (Paris: IEA, 2014).

11. Ibid.

12. Ibid.

13. Li Tao, "*Qian xi zhongguo-dongmen de nengyuan hezuo*" [An Analysis of China–ASEAN Energy Cooperation], *Southeast Asian Studies*, no. 3 (2006).

14. Zhao Ping, "*Shiyou jingkou zhanglue da tishu*" [Speeding Up Oil Strategy], Chinese Foreign Investment, no. 8 (2005).

15. IEA, *Update on Overseas Investments by China's National Oil Companies: Achievement and Challenges since 2011*.

16. Anthony Deutsch, "Asia Giants' Scramble for Coal Reaches Indonesia", *Financial Times*, 9 September 2010.

17. Between 2006 and 2009, exports to Japan and Repubic of Korea declined. This was mainly due to the impact of the global financial crisis that mostly hit these countries.

18. In 2013, Indonesia exported crude oil to two major countries, Japan and the United States; coal was exported mainly to India and China; while LNG was exported mainly to Japan and South Korea.

19. HS 27 includes mineral fuels, mineral oils and products of their distillation; bituminous substances; mineral waxes.

20. Eve Warburton, "In Whose Interest? Debating Resource Nationalism in Indonesia", *Kyoto Review of Southeast Asia* 15 (March 2014).

21. Mateo Cabello, "Indonesia: Mining White Paper", *Oxford Policy Management*, November 2013.

22. Ministry of Finance Regulation, available at <http://www.jdih.kemenkeu.go.id/fullText/2012/75~PMK.011~2012Per.htm> (accessed 17 March 2015).

23. Paul J. Burke and Budy P. Resosudarmo, "Survey of Recent Developments", *Bulletin of Indonesian Economic Studies* 48, no. 3 (2012).

24. Ministry of Finance Regulation No. 75/PMK.011/2012. In the case of mining and quarrying, the export duties covered mineral-metal, mineral-non-metal, and precious stones.

25. Less than seven months after the export ban policy was implemented, the Indonesian government revised the policy on export duties. First, the government increased the number of HS-26 products covered from 10 to 11 commodities. Second, although the export duties scheme is similar to previous regulations, the government aims to ease export duties if the mining company can show serious commitment to building smelter facilities. The export duties are divided into three categories based on the progress in developing smelter facilities. For example, if the progress of constructing smelter facilities reached 7.5 per cent (stage one), the export duties are 7.5 per cent; if the progress reached between 7.5 and 30 per cent (stage two); if the progress is above 30 per cent, the export duties are 0. The export duties are flat up to January 2017. This implies that if one has smelter facilities constructed up to 30 per cent in 2015, and does nothing after that, one can enjoy zero export duties until January 2017.

26. Aneka Tambang produces ferronickel, nickel ore, gold, bauxite, and coal. Inalum's main products are aluminum (ingot) and hydropower (with capacity 426–513 MW). Pertamina has business in the oil, gas, and geothermal sectors. Tin is the main product of Timah. Bukit Asam has core business in coal mining, power generation, logistic, and methane gas.

27. Stratfor Global Intelligence, "Indonesia Struggles with an Export Ban", available at <http://worldview.stratfor.com/article/Indonesia_struggles_export_ban> (accessed 26 March 2015).

28. Jason Allford and Morkti P. Soejachmoen, "Survey of Recent Developments", *Bulletin of Indonesian Economic Studies* 49, no. 3 (2013).

29. World Bank, "Investment in Flux", *Indonesia Economic Quarterly* (March 2014): 21.

30. US Geological Survey of Metals and Minerals (2013).

31. John Kurtz and James Van Zorge, "The Myth of Indonesia's Resource Nationalism", *The Wall Street Journal*, 1 October 2013.

32. "Indonesia Struggles with an Export Ban".

33. IEA, *Southeast Asia Energy Outlook 2013*, p. 74.

34. This paragraph is a summary of "Renegosiasi Berhasil, Harga Jual Gas Tangguh Sesuai Harapan" [Renegotiation was Successful, the Price of Tangguh LNG as We Expected], available at <http://www.esdm.go.id/berita/migas/40-migas/6862-renegosiasi-berhasil-harga-jual-gas-tangguh-sesuai-harapan.html> (accessed 17 March 2015).

35. Indonesia Central Statistics Agency Figures.

36. Michael T. Rock, "What Can Indonesia Learn From China's Industrial Energy Saving Programs?", *Bulletin of Indonesian Economic Studies* 48, no. 1 (2012).

37. Global Indicators Database, available at <http://www.pewglobal.org/database/indicator/24/country/101/> (accessed 15 May 2015).

38. Bates Gill, Michael Green, Kiyoto Tsuji, and William Watts, *Strategic Views on Asian Regionalism: Survey Results and Analysis* (Washington, D.C.: Center for Strategic and International Studies, February 2009), p. 15.

39. Zhou Yan, "Indonesia Seeks More Chinese Investment", *China Daily*, 3 May 2011.

40. Based on IMF Direction of Trade Statistics Yearbook (2012).

41. Dewi Fortuna Anwar, "An Indonesian Perspective on the U.S. Rebalancing Effort Toward Asia", *NBR Commentary*, 26 February 2013, available at <http://nbr.org/downloads/pdfs/outreach/Anwar_commentary_02262013.pdf> (accessed 26 March 2015).

42. Makarim Wibisono, "Indonesia and Global Competitiveness", *Jakarta Post*, 10 October 2011.

6

THE DIRECTION, PATTERNS, AND PRACTICES OF CHINESE INVESTMENTS IN PHILIPPINE MINING

Alvin A. Camba

The literature on Chinese overseas foreign direct investment (FDI) in the Global South has generally pursued a global approach or state-to-state analysis. However, these overlook changes at the local level as Chinese FDI has rekindled the emergence of historical tensions among groups, local struggles for control, and anxieties toward globalization in the twenty-first century. Thus, this gap presents an opportunity to analyse China's engagements at multiple levels with a variety of regional actors across different scales, places, and contexts. This chapter shows how the patterns and practices of Chinese investments in Philippine mining differ from conventional multinational mining investments. First is the method of production. Multinational mining companies focus on large-scale mining (LSM), but Chinese investments tend to gravitate towards artisanal small-scale mining (ASM) to evade scrutiny from national authorities and hostile reactions associated with current territorial disputes between the two countries. Second is the method of accumulation. While multinationals use capital-intensive, ASM capitalizes on labour-intensive extraction, community-centred and house-driven support. And last are the host country linkages. While multinational mining companies need the support of national government agencies to pursue resource extraction, Chinese mining relies more on connections with overseas Chinese communities in the Philippines to access subnational political elites, such as regional politicians, governors, and local officials.

Introduction

This chapter shows how the patterns and practices of Chinese investments in Philippine mining differ from conventional multinational mining investments. While Ching Kwan Lee has previously argued that there are different practices between Chinese and multinational mining companies in Zambia,[1] the author suggests that Chinese investment in the Philippine mining sector presents an analogous case. First is the method of production. Multinational mining companies focus on large-scale mining (LSM),[2] but Chinese investments tend to gravitate towards artisanal small-scale mining (ASM) to evade scrutiny from national authorities and hostile reactions associated with current territorial disputes between the two countries. Second is the method of accumulation. While multinationals use capital-intensive infrastructure and formalized ways of pursuing extraction, ASM capitalizes on multiple methods and ways: labour-intensive extraction, community-centred and house-driven support, and flexible infrastructure. And last are the host country linkages. While multinational mining companies need the support of national government agencies to pursue resource extraction, Chinese mining relies more on connections with overseas Chinese communities in the Philippines to access subnational political elites, such as regional politicians, governors, and local officials. Multinational mining companies directly contribute to, and benefit from, the national mining sector, while Chinese investments focus on bolstering regional and local development through bypassing the national level.

These are not hard and fast distinctions between the People's Republic of China (PRC) and the rest, but rather they are variations in patterns and practices.[3] There is no single China, but instead a multiplicity of state agencies, national and regional state-owned enterprises (SOEs), provincial entrepreneurs, and private investors.[4] Similarly, there is no single Philippine state in relationship with China, but rather a variety of competing actors, including national political elites, regional bosses, local governments, non-governmental organizations, civil society, and many others.[5] But even as multiplicity and complexity characterize Filipino–China relations, Chinese investment in Philippine mining still exhibits patterns and practices that diverge from multinational mining companies.

This research builds on the existing literature on Chinese investments in the Global South focusing on development outcomes and political consequences.[6] The author uses the case of ASM in the Philippines to analyse Chinese outward direct investment from the global into the local.[7] With the limited share of Chinese investments in the formal mining sector, this research looks at two fundamental questions: (1) how does Chinese investment work in the ASM sector; and (2) is there a distinct logic to Chinese investment when compared with multinational mineral investments in the mineral sector? Stephen Bunker has written about two dominant modes of capital accumulation in the early 1970s,[8] namely Western vertically integrated companies and Japanese non-hierarchical production. While the former buys ownership all along the production chain, Japanese firms own the highest chain and then outsource the lower chains to companies in host countries. Extending Bunker's dichotomy to the twenty-first century, the author argues that the differences between Chinese mining firms and conventionally "western" multinational mining companies can be distinguished in a similar fashion.[9] Understanding these differences changes how we come to expect developmental outcomes from Chinese investments, strategize regulatory interventions, and understand other aspects of Philippine–Chinese relations.

The research for this chapter is based on fieldwork the author conducted while based in the Philippines in the summers of 2014 and 2015.[10] The author conducted participant observation at ASM sites in Zambales, Cagayan, and Compostela Valley, and thirty-five semi-structured interviews with politicians,[11] miners, officials, Chinese investors, and local representatives from these towns, based on snowball sampling.[12] Apart from that, the author acquired primary data on Chinese investments in the Philippines from the Philippine Board of Investments, Philippine Statistical Authority, and various Philippine Investment Promotion Agencies. These data were further triangulated with national and local media on Chinese-funded ASM in the Philippines, as well as with reflections from his own previous research on the wider Philippine mining sector.[13] This chapter presents preliminary findings from that research.[14]

Chinese Investments in the Philippine Economy and Mining Sector

While the Philippines has been a mineral exporting country since the American colonial period, the post-Marcos governments (1986 onwards) starting from Aquino consolidated the neoliberal model in the context of political instability and economic indebtedness, which was extended to complex industries like mining and petroleum.[15] In 1987, Corazon Aquino also laid down the initial foundations of the neoliberal political economy model by stressing the role of foreign companies in economic recovery, which was expressed in NEDA's Medium-Term Philippine Development Plan (1987–92), dismantling state monopolies, adopting the Executive Order (EO) 266 — an investment omnibus code — and strengthening administrative reforms.[16] EO 266 awarded generous tax holidays, duty-free importation and tax exemption for the first five years for any foreign investment.[17]

Fidel Ramos (1992–97) embraced sweeping liberal economic reforms as a way of catching up with the country's neighbouring Asian tigers. In particular, he aggressively secured bilateral treaties, promoted privatization of public services, and in the minerals sector secured an ally in the Senate to revitalize the extractive industries through a neoliberal mining framework. Gloria Macapagal Arroyo, who became Philippine president from 2001 to 2009, was the principal author of the bill that would be passed as the Philippine Mining Act of 1995 (RA 7942), which became the state's answer to foreign mining investors' demands to reduce uncertainties in the extractive industries.[18] The Act enabled the full foreign ownership of mining companies through the Financial Technical Assistance Agreement (FTAA). A streamlined version of up to 40 per cent equity ownership of mining companies, called the Mineral Production Sharing Agreement (MPSA), was also enacted in the law. Seen in Table 6.1, foreign mining companies have stimulated the Philippine sector in recent years.

As more foreign mining began to explore the Philippines, pressure from regional actors and local governments led to the People's Mining Act of 1991 (RA 7076). Under certain conditions, the Act empowers local governments to designate legitimate and protected small-scale mining operations within their locales. These areas cannot be sequestered by LSM operations. In some estimates, there are at least 3,000,000 ASM workers in perhaps close to a thousand small-scale mining operations in the Philippines, which was a large increase from around 50,000

TABLE 6.1
Economic Contribution of the Philippine Mining Industry
(in US$ million)

	2008	2009	2010	2011	2012
Mining Contribution to GDP	$1.205	$1.381	$1.955	$2.237	$1.724
	0.70%	0.80%	1.00%	1.00%	0.70%
Total Mining Investment	$604.2	$719.5	$1,053.1	$1,149.7	$791.7
Export Share (Metal Mining)	$2,498	$1,470	$1,929	$2,840	$2,265
	5.2%	3.9%	3.8%	6.0%	4.9%
Export Share (Non-metallic)	$211	$156	$162	$177	$145
	0.4%	0.4%	0.3%	0.4%	0.3%

Source: Mines and Geosciences Bureau (2013), adapted.

workers in a hundred ASM operations during the Marcos period. In the twenty-first century, ASM as a concept cannot truly capture the variety of small-scale mining operations and methods. While labour-intensive, flexible, and community-centred, some of these operations also import advanced machineries and sophisticated technologies. Some of them receive funding worth millions of dollars for years to conduct mineral extraction. While this was not the case before in the 1970s, ASM cannot really capture the variety of mineral operations in the twenty-first century. In the Philippine case, small-scale mining operations are those designated by the 1991 People's Mining Act. One key variation is the huge difference between the mining areas used by ASM and LSM.

While these changes are taking place in the Philippines, China's outward investment strategy in the form of overseas developmental assistance (ODA) and proposed investment projects started in the late 1990s. In rapid succession, the Arroyo administration (2001–9) announced several huge projects investments with China in agriculture, infrastructure, mining, and energy sectors.[19] However, many of these projects were delayed by bureaucratic holdups and corruption allegations. In 2004, the Philippines and China agreed on a US$503 million deal to build a 32-kilometre railway linking Metro Manila to seaports and resource production facilities in the northern provinces. In 2010, China National Machinery and Equipment Group (CNMEG), the company that won the bid, had only completed 15 per cent of the project.

Commentators on the issue reported that the project was abandoned due to a vague bidding process, corruption reports, and the inability of CNMEG to efficiently operate.[20] China was also tied to the Zhongxing Semiconductor Co. (ZTE) scandal, one of the biggest corruption scandals during Arroyo's tenure, which eroded her dwindling legitimacy. The ostensible link between Chinese investments and corruption scandals has strengthened the succeeding Benigno Aquino government's (2009–16) preference for investments from other countries.

As seen in Table 6.2, China's lackluster share in the biggest mining companies and mines in the Philippines, even despite the willingness of Chinese mining companies, can be explained by the broader politics of foreign ownership and changes in the Aquino administration.[21] First, the full foreign ownership of companies remains a controversial political issue to the Philippine public. Regional politicians interested in monopolizing the local economies and civil society intent on keeping Philippine resources for Filipinos easily unite against national moves to allow full foreign ownership. While the FTAA allows full foreign ownership in the mining sector, any Philippine administration risks significant political opposition unless there is some degree of public consensus of the benign motivations of the foreign investor. This perception is further highlighted in the mining sector given the scale of operations, leading to media exaggeration of social exploitation and environmental degradation. In most cases, foreign investors take the limited foreign ownership in the MPSA to evade the national controversy. Whether or not to allow foreign ownership rests on significant political ties and uncontroversial political relations. Canadian and Australian firms largely dominate the foreign ownership of the largest Philippine mining companies.

Second, Aquino's presidency rekindled territorial disputes in the South China Sea between the Philippines and China. Since the current territorial disputes in the South China Sea popularized China's bullying in the Philippine media, it became increasingly difficult for politicians to endorse Chinese investments in the country, much less lobby for full foreign ownership. In Aquino's time, Philippine media and politicians ceaselessly hammered the South China Sea issue, casting negative perceptions on potential Chinese investment or other forms of engagement.[22] There were joint natural resource development agreements and positive signs during Arroyo's time, but Aquino rescinded them during his tenure. Along with these changing perceptions, the limited space of foreign ownership in mining companies led to endorsing

TABLE 6.2
Seven Major Mining Companies in the Philippines

Mining Company	Broad Ownership Structure	Major Mines Extracted	Locations or Names of Some Major Mining Projects in the Philippines
Apex Mining	90% Filipino, 10% Foreign	Silver, Copper, Lead, Gold	Compostela Valley, Cagayan Region
Atlas Consolidated	80% Filipino, 20% Canadian	Gold, Copper, Nickel	Baguio, Abra, Trinidad
Manila Mining Corporation	80% Filipino, 10% Foreign	Gold, Copper, Silver	Suriago Del Norte, Masbate Area
Phil Ex Mining Corporation	80% Filipino, 20% Chinese (Pacific First Corporation Indonesian Company dDomiciled in Hong Kong)	Gold, Copper, Black Sand	CAR, Santo Tomas Region, Bumulo Project
Oceania Gold	45% Canada, 35% Canada, 20% New Zealand	Gold, Nickel, Lead	Nueva Vizcaya and Quirino
Sagittarius Mines	55%, Canadian (Indophil Resources), 45% Australian (Glencore Xtrata)	Gold, Copper, Silver	Tampakan Copper-Gold Project
TVI Resources	30% Canada, Filipino owned but based in Canada	Gold, Silver, Magnetite	Mindoro, Agusan Del Sur

Source: Various sources collected by the author in 2015.

partnerships with Western multinationals instead.[23] The United States, Japan and other countries from Europe, the Middle East, and Southeast Asia are popularly seen as more beneficial to the Philippines. Companies from these states benefit from a perception of benevolence built on the historical hegemony of the United States and Japan in post-war Philippines. The national media, education institutions, internal regional rivalries, and liberal civil society organizations reinforce these positive and often misleading perceptions of the West.

The decision to allow full foreign ownership became more difficult for Chinese investors because of the associated territorial disputes in the South China Sea. As a result, Chinese investments funnelled into ASM operations across the country. Though many Chinese mining firms have invested in large-scale mining operations, their operations and mineral areas were much smaller than those of the multinational and Filipino companies. Alternatively, Chinese companies became major buyers of Philippine copper and nickel to supplement their supplies from Indonesia. Comparatively, as seen in Figure 6.1, Chinese investments in Philippine mining pale in comparison to Indonesian mining. At the regional level, Chinese investments in the Philippines remain very small at US$4.68 billion in comparison with other Southeast Asian states (see Figure 6.2).

FIGURE 6.1
Chinese FDI in Indonesian and Philippine Mining
(in millions of US$)

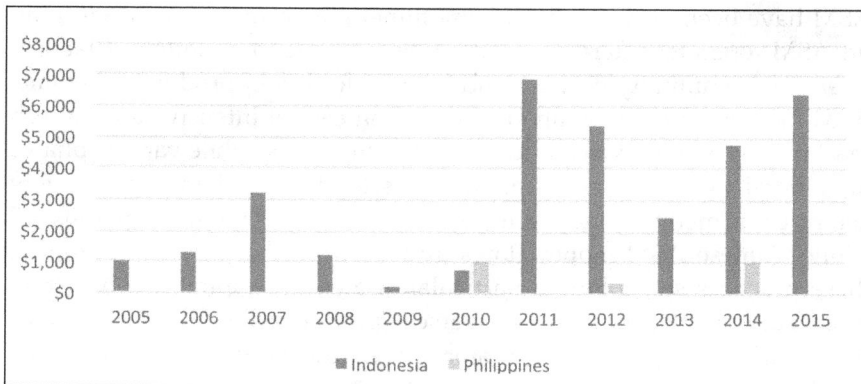

Source: Data from the American Enterprise Institute, Chinese Overseas Investment Tracker.

FIGURE 6.2
Chinese FDI Stock in ASEAN States, 2015
(in millions of US$)

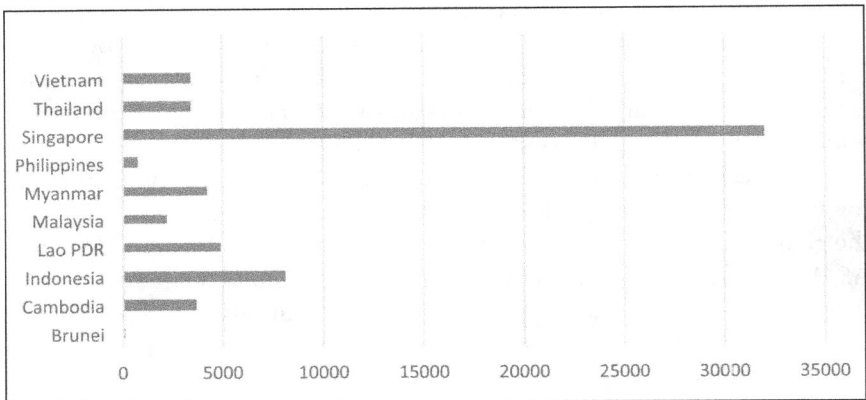

Source: Data from the 2015 Statistical Bullet of China's Outward Foreign Direct Investment.

Chinese Tendencies for Small-scale Mining

As multinational mining dominated LSM in the Philippines, Chinese mining companies started to increasingly fund ASM operations. This section discusses the differences in their methods of capital accumulation. Multinational mining capital tends to concentrate on LSM, notably by building long-term fixed-capital mineral infrastructure, auxiliary roads and shipping routes, and cultivating ties with national government officials.[24] As multinationals dominate LSM, their mineral investments in ASM have been very small. Chinese mineral investments, concentrating on ASM, operate differently. Their operations tend to follow a labour-intensive, community-centred, and household-driven production regime. ASM focuses on manual labour rather than capital-intensive and hi-tech machines. Workers are recruited from nearby towns, while various phases of mineral extraction occur at the household-level. Some civil society actors confirm the tendency for ASM to rely on labour from towns.[25] While Chinese ASM money flows directly from the global to the local, these practices still follow a particular strategy of capital accumulation to cheaply extract minerals for overseas production, which would have been difficult for Chinese investors given the political climate in the South China Sea.

While no current data can confirm the overall scale of Chinese ASM operations in the Philippines, their activities are well-known within the Philippine mining sector.[26] In Figure 6.3, Australian and Canadian transnational mining companies dominate the Philippine LSM.[27] Chinese investment in ASM mirrors Stephen Bunker's description of Japanese non-hierarchical production, in which Japanese companies own the highest node on the production chain and outsource the lower nodes to companies in host countries. In particular, it capitalizes on the skilled workers left behind by the fall of the state-led mineral regime in the early 1980s.[28] The Chinese-funded middleman, engineers, and key local officials in the province decide on employment decisions.[29] Hiring workers in the area fosters loyalty in the locale, gaining allies and preventing intervention from the national government. The extracted minerals are eventually transferred to the middlemen who will deliver the goods to relatively unregulated ships and ports in the coastal areas. The numerous areas for transshipment smuggling allow minerals to be extracted and exported beyond the conventional quota. It is a distinct strategy of accumulation since it not only allows mineral extraction to keep up with production and market demand in China, but also avoids intervention by the Philippine state at the national-level or interruptions of interstate politics.

In the Philippine provinces of Masinloc and Appari, Chinese ASM uses the community's local knowledge and unemployed residents to explore potential mining areas.[30] During exploratory phases, Chinese ASM sends consultants to conduct geological assessments to ascertain the potential profitability of different sites. Community members who would like to be employed as labourers contribute their customary and "local" understanding of geography to make the project more feasible. In Cagayan, a member of the mining team told the author that "they had ancestral knowledge of the magnetite source's location, but they never had the capital to begin mining themselves."[31] In the conversation, he said he feels rewarded by their direct involvement and their employment in the mining operation.[32] A combination of local knowledge from communities, expertise from skilled engineers, and political maneuvering from local politicians form the conditions for successful Chinese ASM, despite scrutiny and vigilance from the national government. LSM, conversely, uses consulting firms from the cities and engineers from the National Association to conduct their scientific assessments. Approval of their operations goes through the

FIGURE 6.3
Locations of Ongoing and Developing LSM Projects, 2010

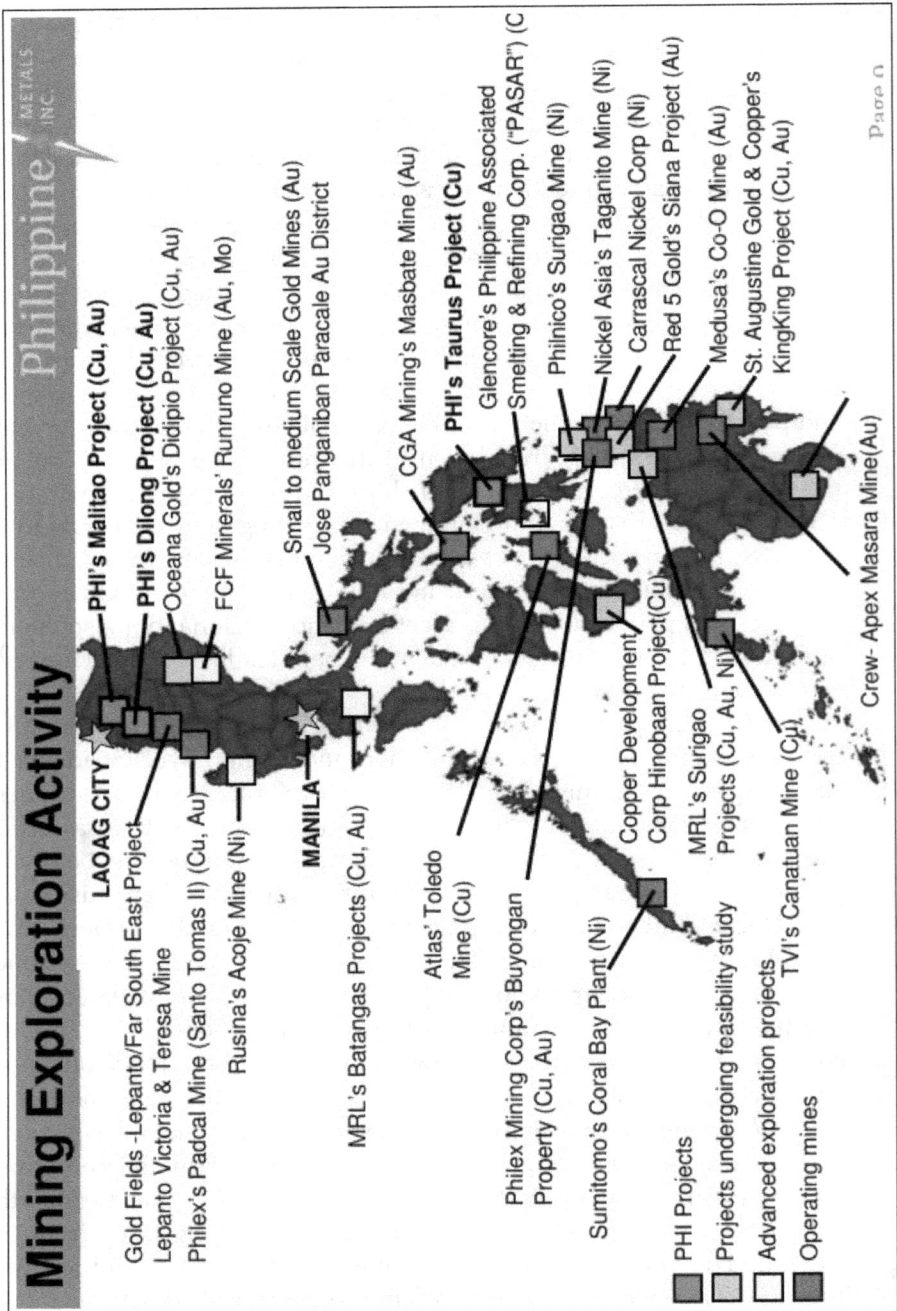

Mining Exploration Activity

Philippine METALS INC

LAOAG CITY

Gold Fields -Lepanto/Far South East Project
Lepanto Victoria & Teresa Mine
Philex's Padcal Mine (Santo Tomas II) (Cu, Au)

Rusina's Acoje Mine (Ni)

MANILA

MRL's Batangas Projects (Cu, Au)

Atlas' Toledo Mine (Cu)

Philex Mining Corp's Buyongan Property (Cu, Au)

Sumitomo's Coral Bay Plant (Ni)

PHI's Malitao Project (Cu, Au)

PHI's Dilong Project (Cu, Au)
Oceana Gold's Didipio Project (Cu, Au)

FCF Minerals' Runruno Mine (Au, Mo)

Small to medium Scale Gold Mines (Au)
Jose Panganiban Paracale Au District

CGA Mining's Masbate Mine (Au)

PHI's Taurus Project (Cu)

Glencore's Philippine Associated
Smelting & Refining Corp. ("PASAR") (C

Philnico's Surigao Mine (Ni)

Nickel Asia's Taganito Mine (Ni)

Carrascal Nickel Corp (Ni)

Red 5 Gold's Siana Project (Au)

Medusa's Co-O Mine (Au)

St. Augustine Gold & Copper's
KingKing Project (Cu, Au)

Crew- Apex Masara Mine(Au)

Copper Development
Corp Hinobaan Project(Cu)

MRL's Surigao
Projects (Cu, Au, Ni)

TVI's Canatuan Mine (Cu)

PHI Projects
Projects undergoing feasibility study
Advanced exploration projects
Operating mines

Page 0

Source: Global Metal News (2010)

FIGURE 6.4
Locations of ASM Gold Mining Operations, 2015

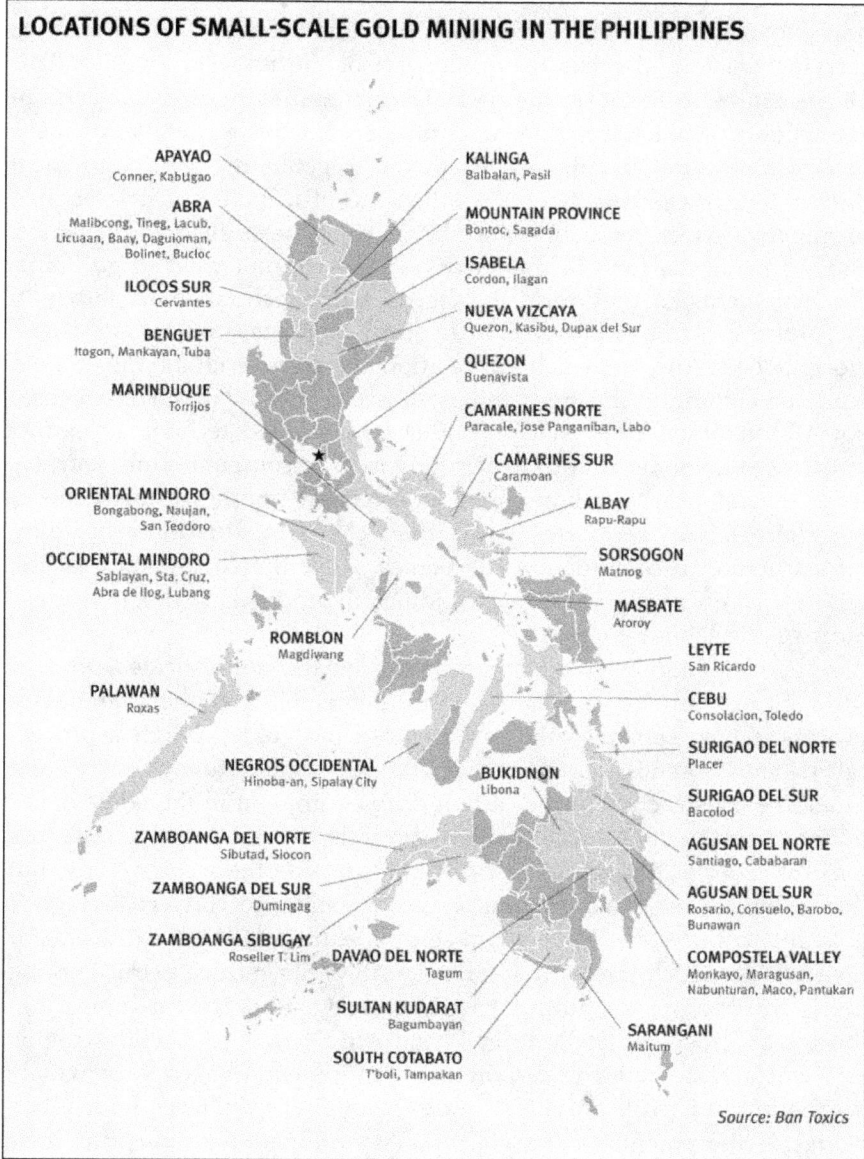

LOCATIONS OF SMALL-SCALE GOLD MINING IN THE PHILIPPINES

APAYAO
Conner, Kabugao

ABRA
Malibcong, Tineg, Lacub,
Licuaan, Baay, Daguioman,
Bolinet, Bucloc

ILOCOS SUR
Cervantes

BENGUET
Itogon, Mankayan, Tuba

MARINDUQUE
Torrijos

ORIENTAL MINDORO
Bongabong, Naujan,
San Teodoro

OCCIDENTAL MINDORO
Sablayan, Sta. Cruz,
Abra de Ilog, Lubang

ROMBLON
Magdiwang

PALAWAN
Roxas

NEGROS OCCIDENTAL
Hinoba-an, Sipalay City

ZAMBOANGA DEL NORTE
Sibutad, Siocon

ZAMBOANGA DEL SUR
Dumingag

ZAMBOANGA SIBUGAY
Roseller T. Lim

KALINGA
Balbalan, Pasil

MOUNTAIN PROVINCE
Bontoc, Sagada

ISABELA
Cordon, Ilagan

NUEVA VIZCAYA
Quezon, Kasibu, Dupax del Sur

QUEZON
Buenavista

CAMARINES NORTE
Paracale, Jose Panganiban, Labo

CAMARINES SUR
Caramoan

ALBAY
Rapu-Rapu

SORSOGON
Matnog

MASBATE
Aroroy

LEYTE
San Ricardo

CEBU
Consolacion, Toledo

SURIGAO DEL NORTE
Placer

SURIGAO DEL SUR
Bacolod

AGUSAN DEL NORTE
Santiago, Cababaran

AGUSAN DEL SUR
Rosario, Consuelo, Barobo,
Bunawan

BUKIDNON
Libona

DAVAO DEL NORTE
Tagum

SULTAN KUDARAT
Bagumbayan

SOUTH COTABATO
T'boli, Tampakan

COMPOSTELA VALLEY
Monkayo, Maragusan,
Nabunturan, Maco, Pantukan

SARANGANI
Maitum

Source: Ban Toxics

Source: Human Rights Watch (2015).

provincial and national environmental boards with little input from the local population.[33]

Chinese ASM firms also utilize the different modes of work in the household. Children of miners often provide auxiliary labour, such as carrying waste and rocks from one part of the mine to another. This allows adults to focus on the more labour- and skill-intensive parts of extraction.[34] Women provide food to the men onsite, delivering bags of rice alongside lunchboxes of cooked vegetables, fruits, and meat during longer working hours. As the wife of a male miner described to me, "we wake up in the morning to cook meals for our husbands, and we bring the food to their work so they do not need to go home. Our son helps in the work, but heeds the direction of his father in the process."[35] In a framework of feminist Marxists, Chinese ASM uses work outside the wage labour relation by appropriating the unpaid work of communities and families in economically depraved areas.[36] For multinational LSM, machines and sophisticated technology extract minerals at a very efficient rate. Usually mining companies hire workers from several nearby communities to construct mining infrastructures, but there have been cases when they chose to outsource building infrastructure to construction companies. After the construction phase, LSM relies on capital-intensive, automated processes and relinquishes most of the labour force.[37]

The extraction of gold remains a particularly sophisticated process. ASM miners rely on the carbon-in-pulp method and the cyanidation process, which separates the carbon from the gold through a process called "elution" and releases cyanide at a high temperature. Chinese ASM funds the construction of the carbon-pulp equipment in the home of the miners, but the household members provide free labour by converting the ASM's gold. Miners also extract gold outside the working day, but they often do not have the equipment or capital to convert the "raw" gold into a pure one. Miners can sell the extra gold they extracted for themselves, which means that Chinese ASM can induce extra working hours by creating an output based incentive to work on mines.[38] In these cases, men go to the tunnels and extract raw gold while women and children operate the carbon-pulp equipment. Wages vary widely for the kind of work provided, but everyone receives basic hourly pay alongside the potential extra gold.[39] LSM multinationals construct their own facilities and machines to convert the gold.

A former environmental official in Nueva Vizcaya told the author that it is very difficult to catch small-scale miners. Potential minerals could be found anywhere in the mountains and fields, spanning hundreds of hectares hidden in the edges of the land. The official said, "We would only hear about mining in those areas weeks after they already conducted the exploration and extraction. And when we send people there, they're already gone and moved on."[40] The national state's weak capacity to conduct data collection, incomplete geographical information, and limited manpower explain the inability to guard potential mineral areas from unsanctioned ASM activities. Additionally, the rollback of state employment because of the processes of privatization, deregulation, and liberalization hinder the national government from employing more people. The loyalty of Filipino officials in the areas, such as regional politicians and local governments, depends on political bargains and compromises to share power, delegate mandate, and disburse development rents.[41] It is in this context that the Philippine state has been largely unable to police local governments, private sector funding, and ASM.

The difference between the two "modes" of extraction also reflects the kinds of actors involved in negotiating capital accumulation. LSM operates within the framework of global mineral extraction and intrastate agreements. Chinese ASM, as a product of China's own capital export and the multiplicity of actors, do not operate similarly. Chinese extraction activities differ substantially from LSM operations that can employ the local populace for the construction of mineral sites. LSM operations follow the regulations stated by the Philippine government. A Social Development Mineral Plan between the mining companies and the communities is needed for every LSM operation in the country. Consultation procedures to acquire social acceptability also take place. Multinationals and companies agree on the returns of extraction quickly with the national government facilitating the negotiations. Beyond the period of consultation, multinationals do not need to update or consult with the communities in the other phases. Mining companies construct fences that hinder community members from venturing into their former communal lands.[42] Large-scale companies, furthermore, hire security companies and military guards to patrol their mineral lands at all times. These moves not only alienate the communities, but also reinforce preferences for Chinese ASM.[43]

In sum, Chinese ASM follows a distinct practice, a strategy of capital accumulation, to cheaply extract minerals below the radar of geopolitical tensions and the politics of foreign ownership. On the one hand, it would be a mistake to see Chinese ASM firms as necessarily more benign actors that enable local development. Since firms operate informally with little rules and regulations, the accounting of minerals, the payment to communities, and accountability to stakeholders cannot be fully taken into account. The smuggling of Philippine minerals to the global market diminishes tax revenues and social services takes place through these decentralized, multiple, and fluid actors. On the other hand, it would be a mistake to simply paint Chinese investments as exploiters of the weak Philippine state. Local communities have often suffered from the opportunism and manipulation of multinational mining companies.[44] The "success" of ASM operations has depended on strong regional linkages and consensus among the Filipinos locals and Chinese investors. A more nuanced interpretation, which goes back to one of this volume's central themes, is that Chinese investments rekindled historical tensions and struggles among different groups and regional actors for social position and economic gain. In this sense, Chinese investments provided the opportunity for local actors to reassert themselves nationally underneath the ongoing transformation of the global economy.

Bypassing the Nation-state through Chinese Diaspora Networks and Local Political Relations

Conventional transnational multinational companies need either national support or local intermediaries to facilitate investments. In an interview with a CEO of a major mining company, he expressed that "formally establishing ties with the Philippine government would lead to the equitable outcome of sharing profits ... while investors provide technology, the Philippines provides the resources". The belief rests on a state-to-state framework between the Philippine and Western governments, reinforcing the centrality of the national state within global economic shifts. Chinese ASM, however, outsources their operations by partnering with regional actors that often bypass national-level actors. While China's economic rise has enabled a plethora of China-based capitalists to engage in the developing world, these actors face the difficulties of adapting to postcolonial politics, often being blamed for their government's political activities.

Given these hurdles, Chinese investors tap into the network of Chinese communities in the host countries. Many Chinese in the developing world deal with tensions of group identification, such as in the Philippines and Thailand, or encounter state discrimination from the local population, such as in Malaysia and the Indonesia. Unlike Singapore, Hong Kong, and Taiwan, Chinese-majority countries that allowed the overseas population to establish a distinct cultural identity separate from the PRC and the Southeast Asian states, the overseas Chinese communities constantly deal with finding a place in their respective societies. A shared cultural artifact between the Chinese investors and the overseas Chinese communities, the widespread of use of Chinese mandarin or the local dialect, allows the ease of communication to foster common agendas. These historical structures enable some Chinese communities in host countries to easily collude with Chinese investors, capitalizing on the former's network with the local population's elites to reduce transaction costs and enable capital accumulation. These elites use their political leverage and economic influence over communities to facilitate and enable Chinese investors to fund informal and/or illegal ASM.

In the Philippines, there are two historical bases for this. First, migration from China to the Philippines changed population dynamics and ethnic relations in significant ways. The Philippines has historically been a recipient of vast numbers of Fukkien and Hakka migrants since the fifteenth century.[45] Though the Spanish regime initially institutionalized discrimination based on Catholicism and race, the place of the Chinese in the Philippines changed in the nineteenth century. The Spanish needed to infuse new sources of capital into the colony and the Chinese population provided that window of opportunity. The decision allowed Filipinos, who were partial owners of land based on the customary socio-economic structures, and Chinese, who had capital from their migration, to intermarry and spur the Philippine economy. In the twentieth and twenty-first centuries, the Chinese population hovered at one per cent of the Philippine population or around 1.5 million people.[46] An estimated 5–10 million people who have partial Chinese heritage are now considered to be Chinese mestizos.[47] While the Chinese mestizos are more widespread across social and class lines, most of the Chinese belong to the professional, business, and retail sectors.

Second, Chinese migration also changed the ethnic composition of Philippine political and economic elites. The Cojuancos and Lopezes, some of the most well established political and economic elites, trace their lineage back to Chinese migrants. The Chinese mestizos' emergence goes back to Spain's decision to relax racial and ethnic lines to draw in Chinese money after Latin America's independence.[48] While the Chinese mestizos have always been part of the Philippine elite classes since the nineteenth century, new elite Chinese groups became prominent in the late twentieth and twenty-first centuries.[49] These new Chinese elites, known as the *Taipans*, started to control a significant chunk of the Philippine economy. Their rise rivalled and to some degree eclipsed the Chinese mestizos and the land-based "old rich" Filipino elites.[50]

The historical migration and relative strength of the Chinese mestizos and *Taipans* in Philippines help mediate Chinese investments in ASM. From Metro Manila, the Chinese *Taipans* may dominate the highest echelons of the economy in the country, but the focus on them neglects the role of Chinese in the provinces and the Chinese mestizos who became political and regional elites across the country. Many of these Chinese-related political dynasties emerged and eventually controlled the politics and economics of different provinces. These groups have competed, colluded, and collided with different kinds of Filipino elites. Many of these families, or local elites with Chinese heritage, remain firmly linked to their East and Southeast Asian counterparts.[51] The Philippine government's weak state capacity makes it not only possible, but also logical for Chinese investors to bypass the national state and go directly to regional actors. Chinese mining firms tap into these political and economic Chinese mestizo networks to conduct ASM operations.

Different kinds of Filipino actors bring up different narratives of exploitation between multinational and Chinese mining companies. For transnational capital, social movements bring up the narrative of exploitation more conventionally tied to core-periphery relations.[52] American mining companies dominated the mineral industry for much of the early twentieth century, but in the past twenty years Australian and Canadian mining companies have been the most active not only in the Philippines but also all over the world. Some interpret Australia's and Canada's recent domination and concentration in the extractive sector as their rise in the global economy.[53] Leon Dulce, a spokesperson of *Kalikasan,* one of the major anti-mining organizations in the country, said, "Australians and Canadians are the most active players in the mining sector."[54] Developmental initiatives by the Philippine government give global mining firms a huge advantage. Philippine national government sees itself as an equal partner in developing a vibrant, competitive, and profitable industry. Investments in LSM affirm the Philippine government's gatekeeping or vanguardist powers of national patrimony into these potentially resource-rich areas.[55] An important point is that the official power of the Philippine national state to hold the ultimate decision over resource domains remains untrue for most of the populace in the provinces, despite policy initiatives from Manila. While this may legally be the case, the tenuous state formation, weak party-mass links, regional elite powers, and the multiplicity of cultural identities across the country suggest otherwise.

Chinese and multinational companies also elicit different racial narratives and justifications. While it may be hard to extrapolate the views of Chinese-funded ASM, the author interviewed a Chinese capitalist in Masinloc in the province of Zambales. From the perspective of the Chinese capitalist, "the unreliability, corruption, and foreign policy issues with the Philippine national government make doing business very shaky."[56] Investment co-ownership, in the view of Chinese firms, "can be reversed due to whim, fear, or preference for Western investors"[57] by the Philippine government. However, the Chinese investors view regional and local elites in the provinces with some degree of healthy optimism. Chinese firms see them as "partners that cannot be influenced by Western powers and looking after their own [development]".[58] In the author's fieldwork in Mindanao, a former provincial politician in one of the provinces in the South said that "rather than let the national government get the development money, it is good that the Chinese are heading straight to the provinces."[59] Regional Chinese elites mestizos are seen as trustworthy partners and racialized as "Chinese". Multinational companies partner with, and rely on, the national state.

Some Chinese firms in Africa seem to be willing to lose out on short-term profit in exchange for long-term resource security.[60] However, it appears to be different in the Philippines. The Chinese perception of preferential treatment to the Western firms comes up saliently and frequently in the interviews. These perceptions, sustained within the broader context of the South China Sea, pressure the Chinese investors to bypass the national state and deal with regional elites to conduct operations. These actors, in the view of the Chinese, are more reliable not only because of their lack of involvement in foreign policy, but also their "Chinese background". Racialized frames around what is Chinese comes up in these discussions despite similar Chinese mestizo figures in the national government. In sum, the historical regional linkages between the Chinese in the PRC and the overseas Chinese in the Philippines emerge in the context of Chinese investments in Philippine ASM.

In sum, the concentration of Chinese investments in ASM has not taken place in a vacuum, but was enabled by historical ethnic constructions. The migration of the Chinese population in the Philippines before and during Spanish colonization, the formation of the Chinese-mestizo class, and the economic power of both groups cemented strong foundations for the successful linkages between Chinese investors and overseas Chinese. The Philippines presents an interesting case of overseas Chinese cooperation despite amiable ethnic relations with the local population. Unlike Indonesia and Malaysia, the Chinese in the Philippines never encountered persistent anti-Chinese sentiments from the Filipinos. Chinese and Chinese mestizos became highly accepted and esteemed members of the Philippine society. Those cooperating with Chinese investors became strong actors in the localities. Venturing in ASM signifies their desire to partake of accumulation and an economic opportunity against the Philippine state's strategy to approach globalization. The Chinese investors identify the "Chinese identity" of their overseas Chinese business partners even if the Chinese heritage has passed on centuries ago. As the "fear" of China's geopolitical and economic ambitions hovers in the background, the strategy reflects the PRC's position within global politics. Fear of the PRC leads to multiple accumulation strategies. Conversely, multinational LSM affirm state-to-state relations, capitalizing on the dominance of their respective home states in the structure of the global political economy.

Conclusion

Through preliminary ethnographic research in three Philippine provinces, the author has shown how the patterns and practices of Chinese investments in Philippine mining differ from multinational mining investments in these three main ways: (1) method of production; (2) method of accumulation; and (3) host country linkages. Successful ASM operations between Chinese investors and the overseas Chinese population in the Philippines challenge the national state's developmental role. The successful linkage not only comes from the socially embedded Chinese mestizos, but also the historical inequalities among different provinces in the Philippines. Debates on Chinese investments have informed scholarly understandings of Chinese political economy, geopolitical trajectories and developmental trajectories of many economies in the Global South. This chapter analyses how investments, territorial disputes, and historical elite formation can distinct patterns and practices. By discussing the movement of Chinese investments towards ASM, the author was able to locate theoretically the place of foreign policy, ASM, and elite class formation in the discussion of political economy.

This chapter's contribution lies in empirically grounding Chinese investments in an often overlooked region in the discussion of "Global China". the author provides a grounded understanding of Chinese investments in a particular economic sector. Most works on Chinese investment in the Global South focus on the impact at the national or sectoral level, but miss the dynamics or relations among competing actors at subnational scales and contexts. This work sheds light on these dynamics by discussing their respective perspectives, sentiments, and hopes. The research provides not only an empirical study to begin further research of Chinese activities in the Philippines, but also intra- and inter-regional comparisons across the world. Comparative ethnographic studies of Chinese capital across the world could also be pursued to further understand the micro-level transformations of class and ethnic differences within China's "new" context of development.

Acknowledgements:
The author would like to thank the numerous small-scale mining firms, communities, and local government officials for the interviews.

NOTES

1. Ching Kwan Lee, "Raw Encounters: Chinese Managers, African Workers and the Politics of Casualization in Africa's Chinese Enclaves", *The China Quarterly* 199 (2009): 647–66. See also Ching Kwan Lee, "The Spectre of Global China", *New Left Review* 89 (2014): 29–65.
2. While the author defines multinational corporations as Australian and Canadian companies, the definition could also encompass firms from the region. Furthermore, there is literature on the distinctiveness of Japanese firms from Chinese and Western firms. The literature in international relations argues that there is a human rights baggage attach to Western investments, while none exist for the Chinese companies.
3. Deborah Bräutigam, *The Dragon's Gift: The Real Story of China in Africa* (Oxford: Oxford University Press, 2009); Elizabeth Economy and Michael Levi, *By All Means Necessary: How China's Resource Quest is Changing the World* (Oxford: Oxford University Press, 2014); Hung Ho-fung, *The China Boom: Why China Will Not Rule the World* (Columbia University Press, 2015); Lee (2009; 2014).
4. Hung Ho-fung, "America's Head Servant?", *New Left Review* 60 (2009): 23.
5. Alvin A. Camba, "Philippine Mining Capitalism: The Changing Terrains of Struggle in the Neoliberal Mining Regime", *Austrian Journal of South-East Asian Studies* 9, no. 1 (2016): 69–81. See also Jewellord T. Nem Singh and Alvin A. Camba, "Neoliberalism, Resource Governance and the Everyday Politics of Protests in the Philippines", in *The Everyday Political Economy of Southeast Asia*, edited by Juanita Elias and Lena Rethel (UK: Cambridge University Press, 2016).
6. Deborah Bräutigam and Tang Xiaoyang, "African Shenzhen: China's Special Economic Zones in Africa", *The Journal of Modern African Studies* 49, no. 1 (2011): 27–54. Deborah Bräutigam and Haisen Zhang, "Green Dreams: Myth and Reality in China's Agricultural Investment in Africa", *Third World Quarterly* 34, no. 9 (2013): 1676–96. Marek Hanusch, "African Perspectives on China–Africa: Modelling Popular Perceptions and Their Economic and Political Determinants", *Oxford Development Studies* 40, no. 4 (2012): 492–516. Kevin Gallagher and Roberto Porzecanski, *The Dragon in the Room: China and the Future of Latin American Industrialization* (Stanford: Stanford University Press, 2010). Vivien Foster, William Butterfield, Chuan Chen, and Nataliya Pushak, *Building Bridges: China's Growing Role as Infrastructure Financier for Sub-Saharan Africa*, Trends and Policy Options no. 5 (Washington, D.C.: World Bank, 2008).
7. ASM in this sense does not denote a particular manual activity. Rather, it describes a series of activities, such as digging, marking, panning

and shoveling, cleaning, and transporting that lead to the extraction of minerals. Meanwhile, the term "informal" denotes mining by individuals, groups and cooperatives that is carried out without formal restraints and sometimes even illegally. In this way, "informal mining" (also known as "artisanal" or "small-scale" mining) is distinguished from "formal", capital-intensive mining carried out by state or transnational mining companies. Informal mining has the following characteristics: (1) reliance on physical labour for all types of operations, making minimal use of technology; (2) lack of legal mining-licenses, titles, leases and claims to the mineral areas for exploratory and extractive activities; (3) low levels of productivity per mining operation, resulting from relatively small geographical areas and water resources; (4) absence of economic, health and environmental security for miners, workers and local communities; and (5) the transient character of employment due to the seasonal dependence of mining. Figure 6.4 shows the on-going small-scale mining operations in the country.

8. Stephen G. Bunker and Paul S. Ciccantell, *Globalization and the Race for Resources* (Baltimore, Maryland: John Hopkins University Press, 2005).
9. Bunker wrote about Japanese firms during the Cold War. The similarities between Chinese and Japanese firms may apply more during that time. The similarities today depends on empirical research on both topics.
10. Research on the subject was conducted as an independent researcher prior to the author's matriculation at Johns Hopkins University.
11. The author anonymized the name of the town in Mindanao due to the present concerns in the Philippine mining.
12. Michael Burawoy, *The Politics of Production: Factory Regimes Under Capitalism and Socialism* (London: Verso Books, 1985). Gillian Hart, "Denaturalizing Dispossession: Critical Ethnography in the Age of Resurgent Imperialism", *Antipode* 38, no. 5 (2006): 977–1004. See also Lee (2009).
13. The author has previously worked on the Philippine mining industry. See Alvin A. Camba, "From Colonialism to Neoliberalism: Critical Reflections on Philippine Mining in the 'Long Twentieth Century'", *The Extractive Industries and Society* 2, no. 2 (2015): 287–301. Also see Camba (2015); Nem Singh and Camba (2016). The author conducted fieldwork in 2009, summer 2013 and 2014, and December 2015. He spoke with various representatives from the big mining companies in the country — Benguet Corporation, Indophil Resources, and Rio Trinto — civil society representatives, local officials, community members, and indigenous group leaders.
14. With the new Philippine administration under Rodrigo Duterte in June 2016, the chapter may not capture new processes emerging from his tenure.

15. Camba (2015), pp. 287–301.
16. Ibid.
17. Camba (2016).
18. Nem Singh and Camba (2016).
19. Roel Landingin, "Chinese Foreign Aid Goes Offtrack in the Philippines", *The Reality of Aid, South-South Cooperation: A Challenge to the Aid System* (2010): 87–94.
20. Landingin (2010).
21. Alunan (2013) lists some Chinese mining companies in the Philippines since 2001: China Metallurgical Construction Corp., Echeng Iron and Steel Group Co,. Ltd., Epochina Mining Corporation, Guo Long Mining Corporation, Jiangxi Rare Earth and Metals Tungsten Group, Jiangxi Rare Earth and Rare Metals Tungsten Group, Jinchuan Nonferrous Metals, Konka Fulim Mining and Development Corp., Lian Xing Song Carving Co,. Corporation, Macao Quanta Mining Co,. Ltd., Nicua Mining Corporation, Oriental Synergy Mining Corp., Peng Cheng Metallic Resources Corp., Prime Rock Mining Company, Rock Check Steel Group Company Ltd., Shanghai Baosteel Group, Shenzhen Zhao Heng Industrial Co. Ltd., Shenzhou Mining Corporation, Singtech Mining and Trading Co. Ltd., Inc., Sinian International Corporation (Bohol and Cebu operations), Sinophil Mining and Trading Corp., Wei-Wei Group, Yinlu Bicol Mining Corporation, Zhongli Mining Corporation, Zijin Mining Group Company Ltd. For reference, see Rafael Alunan III, "What's Yours is Mine", *Business World*, 30 July 2013.
22. Paterno Esmaquel II, "Binay: 'China Has Money, We Need Capital'", Rappler.com, 14 April 2015, available at <http://www.rappler.com/nation/89880-binay-china-philippines-south-china-sea>; Ayee Macaraig, "Binay Joint Venture with China Must Be Corruption-Free", Rappler.com, 10 July 2015, available at <http://www.rappler.com/nation/98940-binay-joint-venture-china-corruption>.
23. Renato Cruz De Castro, "The Obama Administration's Strategic Pivot to Asia: From a Diplomatic to a Strategic Constrainment of an Emergent China?", *The Korean Journal of Defense Analysis* 25, no. 3 (2013): 331–49.
24. Camba (2016).
25. Interview with an Executive Director from a Non-Governmental Organization in Quezon City, 6 June 2014.
26. The author's fieldwork in 2010, 2014, and 2015 revealed the perceptions.
27. Camba (2015).
28. Boris Verbrugge, "The Economic Logic of Persistent Informality: Artisanal and Small-Scale Mining in the Southern Philippines", *Development and Change* 46, no. 5 (2015): 1023–46. See also Camba (2015).

29. Interview with Chinese small-mining firm representative, Zambales, 8 July 2014.

30. Interview with former provincial politician, Compostela Valley, 29 July 2015.

31. Interview with Chinese small-mining firm representative, Zambales, 8 July 2014.

32. Interview with local representative, Cagayan, 21 July 2014.

33. Nem Singh and Camba (2016).

34. Interview with a miner, Cagayan, 21 July 2014.

35. Interview with a miner, Cagayan, 22 July 2014.

36. Jason W. Moore, *Capitalism in the Web of Life: Ecology and the Accumulation of Capital* (London: Verso Books, 2015).

37. Camba (2015).

38. Interview with a miner, Cagayan, 21 July 2014.

39. Ibid.

40. Interview with former local government unit official, Nueva Vizcaya, 26 June 2014.

41. Interview with Congressional Staff, Committee on National Communities, Quezon City, Philippines, 19 October 2013.

42. Nem Singh and Camba (2016).

43. Interview with Congressional Staff, Committee on National Communities, Quezon City, Philippines, 19 October 2013.

44. Camba (2016); Verbrugge (2015).

45. Nick Cullather, *Illusions of Influence: The Political Economy of United States–Philippines Relations, 1942–1960* (Stanford: Stanford University Press, 1994).

46. Philippine Statistical Authority, "Provincial Summary: Number of Provinces, Cities, Municipalities and Barangays, by Region", 2015; Antonio S. Tan, *The Chinese in the Philippines, 1898–1935: A Study of Their National Awakening* (Quezon, Philippines: Printed by R.P. Garcia Pub. Co., 1972); Edgar Wickberg, *The Chinese in Philippine Life, 1850–1898* (Ann Arbor, Michigan: University Microfilms, 1965).

47. Philippine Statistical Authority (2015).

48. Cullather (1994); Wickberg (1965).

49. Victor Purcell and Alice Li, *The Chinese in Southeast Asia* (London: Oxford University Press, 1965). For the recent emergence of Chinese elites in the Philippines, see Walden Bello, Kenneth Cardenas, Jerome P. Cruz, Alinaya Fabros, Mary A. Manahan, Clarissa Militante, Joseph Purugganan and Jenina J. Chavez, "State of Fragmentation: The Philippines in Transition", *Focus on the Global South and Friedrich Ebert Siftung* (2014).

50. Bello et al. (2014).

51. Interview with former provincial politician, Compostela Valley, 29 July 2015.
52. Camba (2016); Nem Singh and Camba (2016).
53. Stuart Kirsch, *Mining Capitalism: The Relationship between Corporations and Their Critics* (Oakland, California: University of California Press, 2014).
54. Interview with an organisation member, Kalikasan-PNE, Quezon City, 13 June 2014.
55. Interview with Kayzer Llada, Commercial Specialist, Philippine Associated Smelting and Refining Corporation (PASAR), Makati City, 11 June 2014.
56. Interview with Chinese small-mining firm representative, Zambales, 8 July 2014.
57. Ibid.
58. Interview with former provincial politician, Compostela Valley, 29 July 2015.
59. Ibid.
60. Economy and Levi (2014), see also Lee (2014).

7

DEVELOPMENT COOPERATION WITH CHINESE CHARACTERISTICS
Opium Replacement and Chinese Rubber Investments in Northern Laos

Juliet Lu

As China's role in the global economy has grown, it has become increasingly involved in developing economies through aid and overseas development assistance. China promotes a range of projects under the umbrella of "development cooperation" and has, on occasion, proclaimed an alternative development model to that of its Western counterparts. Such statements draw attention to how development operates as both a rhetoric and a practice for channelling foreign investment abroad. This chapter examines how narratives of a "Chinese model" has shaped the rationalization and practices of Chinese rubber companies in Laos through the Opium Replacement Program (ORP). The ORP is a Chinese state project active since 2004, which aims to eradicate opium cultivation by providing alternative agricultural livelihoods. This project catalyzed an influx of Chinese capital into northern Laos that has drastically transformed local agricultural systems, livelihoods and land uses. The ORP is an important case for analysing the challenges and contradictions that arise when Chinese concepts and China's unique historical experience of development are transplanted into other contexts.

FIGURE 7.1
Yunnan Alternative Development Association Poster
Promoting the Opium Replacement Programme which reads,
"Alternative Crop Cultivation - Serve the Ban on Drugs"

Source: Yunnan Alternative Development Association.

Introduction

In the late 1990s, the global price for natural rubber soared and Xishuangbanna — a major rubber producing area in remote southwest China — grew rich. Across the border in northern Laos, farmers and state officials looked on in envy at this miracle cash crop. Some began experimenting with it. When the Chinese government established the Opium Replacement Program (ORP) in 2004, it was welcomed in Laos as further support for the growing rubber sector. As a result, Lao rubber cultivation expanded significantly. The ORP was established to incentivize large-scale agribusiness projects to provide alternative livelihoods schemes as substitutes for opium cultivation in Laos and Myanmar. Chinese companies rushed to take advantage of the ORP and its financial supports. Though a range of cash crops were allowed, the vast majority of ORP funding went to establishing rubber plantations. Amid the initial rush, Chinese investors, Lao farmers and Lao state officials recounted stories of rubber's success in Xishuangbanna as justification. Rubber was a silver bullet that contributed to development, poverty alleviation, and modernization. But in 2011, as global rubber prices plummeted, dreams of new motorbikes, paved roads, corporate profits and hefty government revenues were suddenly dashed.

The drop in rubber prices also dashed claims that China's development success was wholly replicable in other contexts. The ORP was based on an emerging discourse of Chinese development in remote borderland areas, for which Xishuangbanna's historical experience with rubber was a paragon. The discourse reinforced Chinese state visions of development as a mutually beneficial engagement between China and recipient countries like Laos. Yet, at the same time, it created preferential conditions for large Chinese investors, such that the benefits to recipient country stakeholders appeared as dwarfed in comparison. This chapter argues that state-backed narratives of a "Chinese model" for development cooperation reflect Beijing's aspirations for international recognition as a development success story and, hence, a historical model for the developing world. This, however, creates a selective blindness to differences in local context and the socio-historical conditions that made rubber a miracle crop in Xishuangbanna but not in Laos.

The analysis for this chapter is based on fieldwork conducted in China and Laos between 2012 and 2015 to study four ORP companies operating in northern Laos. It includes analysis of policy and legal documents, interviews with state officials, local land users and company managers, and a review of Xishuangbanna's rubber sector history. The chapter begins with an overview discussion of Chinese overseas development cooperation to suggest that the ORP was established in a moment of alignment between Lao and Chinese economic interests. It then examines the history of rubber promotion in China to demonstrate how a range of actors in both countries came to see rubber as an ideal crop for bringing development to northern Laos. Finally, the chapter analyses how discourses of Chinese development cooperation shaped investment practices of individual Chinese rubber companies in Laos and the logic by which the managers of these companies justified and promoted rubber's expansion.

China's Development Cooperation Narratives

The ORP was established at a moment when Chinese and Lao economic interests aligned. Over the last two decades, China has gone from being heavily focused on domestic development to increasing its foreign aid and foreign direct investments. Just as China was seeking new markets for investment abroad and new sources of raw materials to fuel its own expansive economic growth, Laos was also redefining its own development approach. It had begun shifting towards an emphasis on attracting more foreign direct investment — primarily in land and natural resources — as a means to kick-start its own stuttering economy. The Chinese and Lao governments, which maintained a considerable diplomatic distance as a legacy of Cold War era tensions, began in the 1990s to embrace bilateral cooperation on multiple fronts, particularly in economic development. China went from being an insignificant contributor to the Lao economy in the 1990s to its primary investing country by 2007. By 2011, Chinese investors had gained concessions for over an estimated 200,000 ha of state-leased land, plus vast areas of land through contract farming agreements with local land users.

The expansion of Sino–Lao economic cooperation parallels a global rise in China's development cooperation efforts, including various forms of aid and development initiatives throughout the Global South.[1] State-sponsored development schemes support large-scale infrastructure projects abroad, but they also include large-scale land acquisitions by

Chinese commercial actors.[2] In many ways, China, as the leader among emerging market donors (e.g. Brazil, Russia, Saudi Arabia), departs from traditional approaches to foreign aid modelled by the OECD countries (also referred to as traditional donors). China typically opts for laissez-faire approaches to aid granting support without political demands on recipient country governments, which contrasts with traditional models of "tied aid". Furthermore, Chinese and emerging donor activities are not overseen by the OECD's Development Assistance Committee, meaning they can set their own standards of implementation, notably in such areas as transparency and accountability. The emergence of this alternative regime for development has raised concerns that "emerging donors might support 'rogue states,' increase levels of indebtedness, ignore environmental protections, focus on extracting resources, and undermine the improvements that have been made over the past several decades".[3] Emerging donors, however, respond that what they offer is South–South cooperation, free from the neo-colonialism and neo-imperialism that defined the Bretton Woods institutions and North–South development cooperation in the post-World War II era. In contrast, they argue, South–South cooperation is founded on a common development experience.

China also views foreign investment as essential to supporting development, particularly in postcolonial contexts characterized by unequal or dependent economic relations.[4] China itself benefitted greatly from economic aid and investment first from the Soviet Union and then through foreign direct investment, much of which came from members of the Chinese diaspora in Hong Kong, Taiwan, and Singapore. Reliance on external support has also defined how Beijing approached development for the hinterland western regions, which depended on influxes of capital from the more developed coastal regions and foreign sources. In 1999, President Jiang Zemin enacted the Western Development Strategy (西部大开发) to mobilize state funding for infrastructure projects and draw foreign investment into the western provinces. Large-scale agriculture investments were seen as providing a range of development benefits, including wage labour opportunities, infrastructure, market access, and skills and technology transfer.[5] Thus, the driving assumption behind Chinese development cooperation is that China is able to export its development success abroad, particularly as defined by its transition to modern, commercial agricultural production systems.

There also exists a deep divide between traditional and emerging market donors in thinking about development cooperation as either compatible with or counter to commercial interests. Traditional donors have preferred trade and aid to be separate since development aid was established during the postcolonial era. At that time, colonial powers were criticized for exploiting colonies for their own economic gains and the development project was meant to be a departure from that model. Thus, prevalent attitudes towards aid at that time categorized any development intervention that primarily benefitted the donor country as highly problematic. Furthermore, the drive for profit is what many critics have pointed to as the reason corporations cannot be relied upon to deliver development benefits like livelihood opportunities.[6, 7]

China, in contrast, embraces commercial projects and trade as necessary channels for capital investment. The Chinese government has funded extensive investments in infrastructure and key industrial sectors as a form of development cooperation. Even as China has attracted international criticism for funding major hydroelectric dams, road and infrastructure construction, and large-scale land acquisitions, Beijing has rarely questioned whether the project of funding development while gleaning a profit for Chinese companies is inherently contradictory.

This difference in thinking about the overlap of profit and development aid extends to the involvement of private capital in development interventions, specifically independent corporations that are expected to prioritize profit over other objectives. The Chinese state has historically had a very different relationship with companies than governments in the Global North. China's economic rise was negotiated through heavy state-led macroeconomic planning under which corporations were vehicles for both economic organization and the provision of social services. Until the early 1980s, state-owned enterprises provided a range of state services (e.g. education, health care) and production in all sectors was organized through "work units" (*danwei*) within those enterprises. In the early 1980s, China began its transition to a market economy in a process characterized by sudden and rapid growth through economic liberalization and the influx of foreign direct investment. While state-owned enterprises have been gradually privatized, they remain pillars of the country's economy. The Chinese state also continues to exercise great power over economic planning, market functions in key sectors, and state-owned enterprise operations. Thus, companies' capacity to operate profitably while providing development benefits to the local economy is largely taken for granted in China, to the point that many Chinese plantation managers in Laos argue that when the company profits the community benefits.

In particular, China considers its companies' extensive investments in large-scale land acquisitions (what some have referred to as "land grabs") as development cooperation projects. Intense debate has arisen over the benefits of these land deals to recipient countries. The global rise of large-scale foreign land acquisitions has been associated with environmentally destructive resource extraction, minimal benefits to the local economy, threats to land tenure security and risks to local state sovereignty — though outcomes of such acquisitions are context specific. Some observers have derided the Chinese model of development cooperation as a hollow justification for unfettered resource extraction and capital accumulation in underdeveloped countries. For example, China's rubber investments in Laos have been cited as dispossessing local land users,[8,9] putting a dangerously large amount of land in the hands of foreign companies,[10] and increasing rural poverty while investors make a quick gain.[11] These negative impacts of Chinese investments mirror the risks and destructive outcomes recorded across the globe as a result of the rise in global land grabs. However, Chinese actors tend to view them in a different light from many of their critics.

A number of scholars have also pointed to the importance of the convergence between investor interests and recipient states. A country like Laos, which has experienced a chronic lack of capital to fund rural development, has welcomed Chinese aid and foreign land investments openly. Situated on China's southwest border, Laos is designated by the United Nations as a "Least Developed Country". It receives significant foreign aid from traditional OECD country donors. Yet looking beyond aid, the Lao government seeks to attract foreign capital as a driver of development. Laos has a history of using land regulations (for lack of capital revenues to invest in infrastructure) as an approach to encouraging economic development.[12] In the 2000s, the government declared a strategy of "turning land into capital",[13,14] which involved granting land to foreign investors to spur economic development. The logic underlying the call to turn land into capital resonates with Chinese investors who contrast Laos as "land rich, labour scarce" with China as "labour and capital rich, land scarce". Indeed, the Lao government emphasizes its land abundance in many international investment promotion materials. By 2011, the Lao government had granted land concession contracts to Chinese corporations covering roughly 200,000 ha of state land,[15] making China the top investor in Laos.

With these debates and trade-offs in mind, Chinese development cooperation serves as a powerful political narrative for framing Chinese large-scale rubber investments as ideal tools for eradicating opium by bringing development to rural northern Laos. It is promulgated by the Chinese state and enacted in practice by various Chinese companies implementing development projects abroad. The Chinese development cooperation concept, the author argued, is more than mere political justification for China's commercial interests in rubber, but rather it is based on China's own historical experience of development. China's development cooperation discourse suggests that its own economic success can be translated to other developing countries, reasoning that this can be done with the help of Chinese capital and technical guidance. Following the observed success of rubber in Xishuangbanna, the vast majority of Chinese investments in Lao agriculture are in rubber plantations. Rubber is promoted not only as a valuable cash crop but also as a tool for modernizing the hinterlands. As such, the promotion of rubber by both Chinese and Lao companies and government officials reflects deeper convictions and logics about what development looks like, how it is achieved, and who is affected on the ground. The ORP is an example of how China's own development experience characterizes a range of activities, including large-scale agricultural land investments for rubber, as a mutually beneficial and economically necessary form of development.

The Xishuangbanna Story: Rubber as a Model for Development

If China considers itself as a role model for the developing world, then Xishuangbanna would be an apt model for the upland regions of Southeast Asia — especially for the bordering region of northern Laos. Prior to the arrival of rubber, Yunnan Province played a peripheral role in China's development. However, when the US blocked international rubber imports to China in 1951, which were needed for military jet and truck tire production, China found itself with no rubber supply to speak of.[16] Consequently, it designated rubber as a "strategic national crop".[17] This sparked a drive among policymakers for China to establish its own rubber sector in the few places warm enough to grow rubber, invest state resources in rubber research and development, and support cultivators through hefty market protections, state subsidies, and other policy supports. More than just an income-generating cash crop, rubber's significance to national security also shaped Yunnan's path to development and Xishuangbanna's strategic importance to the rest of the country.

Before the nineteenth century, rubber was primarily cultivated at latitudes far south of Xishuangbanna. Having been brought over from Brazil to Asia, it had initially been planted mostly in Indonesia and Malaysia. Only three areas in China were suitable for natural rubber cultivation, namely Hainan, Yunnan and Guangxi Provinces. Yet even these areas have cooler climates and higher elevations than Malaysia or Indonesia.[18] As a result, rubber cultivation techniques had to be adapted to the Xishuangbanna context and maximizing productivity per land unit has been a central focus in China's rubber research and development efforts. Extensive research has been carried out by the Jinghong Tropical Crops Research Institute, originally a branch of Yunnan State Farms (云南农垦基团), the province's state-owned enterprise for rubber. When China's rubber plantations began to succeed in the 1960s, they were described as a "miracle of science".[19] This marked the start of rubber's status as a symbol of progress and modernity in China. And because research on rubber in Yunnan has largely operated in concert with Yunnan State Farms — even after the Tropical Crops Research Institute was separated from the company's operations — it performs nearly all of its rubber research in company plantation areas. This link between technical research and development for rubber production and commercial plantation operations continues to define the company's production model and set rubber sector standards for cultivation management, processing, and quality assessment.[20] Sturgeon (2013) therefore observes that, even today, state rubber plantations under the umbrella of the Yunnan State Farms enterprise "have a mythic cachet as emblems of socialist science and revolutionary zeal".[21]

Once the technology of rubber cultivation was adapted to the Xishuangbanna context, labour proved to be the second bottleneck to rubber's development. Initially, plantation managers considered local Dai minority smallholders (who share linguistic and cultural ties with communities in northern Laos and Thailand) too uncivilized for rubber production, which was believed (based on the Malaysian and Indonesian models) to require large-scale and factory-like production.[22] Instead, youth members of the People's Liberation Army came in waves to clear vast areas of land and plant rubber trees, and they were supplemented with Han Chinese who migrated into the province.[23] Only in the 1980s when China's industrial sector began driving up the demand for rubber did companies turn to local smallholders to expand production.[24] The Chinese state charged agricultural collectives with reallocating land from the collectives to households, allowing individual smallholders to convert their own land for rubber cultivation. Today, Chinese rubber companies have decades of experience collaborating with smallholder rubber farmers, which has resulted in widespread development benefits in Xishuangbanna.[25] But this collaboration emerged from a confluence between increased demand for new plantation lands in the face of ageing company plantations and the new system of household land allocation, which forced companies to engage smallholders to be able to access their lands.

Chinese rubber production has also enjoyed extensive protection and support from the state. Initially, its significance to national security, not capital accumulation, was the state's main justification for subsidizing rubber.[26] In its early stages, research and development for the rubber sector was heavily state-funded. As a crop of strategic importance, the quantity, quality and stability of supply was emphasized, not necessarily the cost-effectiveness of production. During the integration of smallholders in the 1980s, agricultural extension support for planting and tapping was supplied by state extension agents and through Yunnan State Farms.[27] Furthermore, Chinese SOEs and smallholders alike enjoyed protectionist trade policies that shielded them from global price fluctuations.[28] Currently, rubber has a new strategic role in the production of tires to feed China's automobile industry, which has grown exponentially. Despite cutting most tariffs and price controls after its accession to the WTO in 2001, China continues to claim rubber as a crop of significant national security importance and thus retains a 20 per cent import tax per metric ton of latex.[29] This reflects an effort by the Chinese state to dissuade domestic producers from converting to other crops or livelihoods.[30]

But the few suitable areas for rubber production in China do not come near to meeting national demand. High global prices and rising domestic demand through the early 2000s drove a planting frenzy by smallholders in Xishuangbanna, while little land is now available for further expansion.[31] Meanwhile, Yunnan's SOEs for rubber have been gradually privatized,[32] which has forced them to compete with other suppliers just when the productivity of their plantations has begun to wane as existing plantations near the end of their production cycle. Still, large state rubber companies retain notable political influence despite some moves towards privatization in recent years. In a context of these recent constraints on land and the drive for greater competitive efficiency, the ORP provided a much-needed boost (along with more state-based financial support and justification) for the Yunnan rubber sector to expand.

The story of rubber in Xishuangbanna is a compelling example of development, but it is important to register the complex historical conditions through which it occurred. The economic transformation of Xishuangbanna appears to prove rubber's broad instrumentality as a tool for modernization and development progress. Just over the border in northern Laos, communities practise the same shifting cultivation agriculture that many Dai in Xishuangbanna did, they share linguistic and cultural connections, and the mountainous subtropical landscapes appear quite similar. But with its strategic military and now economic importance, rubber is far more than a lucrative cash crop to China and it has been promoted, developed, and protected as such. It is also credited with transforming Xishuangbanna from a peripheral border region to a development success story, and its "backwards", uncivilized minority population into a civilized, disciplined workforce.[33] But this was as much a function of strong land tenure rights for smallholders as it was of effective company- and state-led development. As any political narrative may illustrate, the history of rubber's rise in Yunnan has been simplified, with examples of inequality, exclusion, and dispossession subsumed under what Sturgeon calls the "mythic cache" of state rubber farms. This myth of rubber translates into the Lao context in peculiar ways with both Lao and Chinese actors often glorifying its development benefits based on the Xishuangbanna example.

The ORP: Replacing Opium, Spurring Development

In conceptualization, the ORP is based on a similar reasoning on the drivers of opium cultivation as other alternative development initiatives.

But in implementation, it has been heavily shaped by narratives of the Chinese model for development cooperation and by the political, economic, and geographic factors linking opium cultivating areas and opium markets in the region. As many comparable programmes have attempted, the ORP aims to promote cash crop cultivation and agribusiness projects as an alternative to opium cultivation livelihoods. But it does so by establishing broad policy supports and financial incentives for Chinese agribusiness companies instead of targeting opium cultivation areas or funnelling support through the local state, civil society or directly to vulnerable households. This has resulted in a focus on rubber and a preference for large-scale plantation models of rubber cultivation. This results from a combination of the Chinese state's view of opium as a national security issue, the strategic role and specific interests of the Yunnan provincial government in the ORP's implementation, and the way Xishuangbanna's experience with rubber has resonated with larger Chinese discourses of development.

In the early 2000s, a rise in injection drug use in Southwest China was viewed by Beijing as a critical threat to China's national security and social stability.[34] The Chinese government established the ORP in 2004 to respond to this rise. The government aimed to reduce opium production and trade into China by supporting alternative agricultural activities, primarily in commercial agriculture. Qualifying companies receive reimbursements for up to 80 per cent of their initial exploration and project establishment costs, eases in customs requirements such as import/export tariffs and quotas, and exemption from interest on loans.[35]

Related legislation was initiated by the central government and in 2004, funding and administrative responsibilities were allocated directly to Yunnan Province, which borders the "Golden Triangle" countries (the border areas between Thailand, Myanmar and Laos) from which 40 per cent of the world supply of opiates and most of that found in China originates.[36] To finance ORP implementation, the Chinese government established a Special Fund to be managed by the Yunnan Provincial Government, which provided financial and regulatory support for Chinese agribusiness companies investing in northern Myanmar and Laos where most of Southeast Asia's opium cultivation occurs. As a result, Chinese companies poured into both countries to invest in large-scale rubber plantations.[37] While critics chalk the ORP up to political dressing for an economically motivated expansion of rubber into Laos,[38] proponents claim it has stimulated technology and capital transfer and facilitated Lao producers' access to Chinese rubber markets.[39]

According to state officials interviewed, Yunnan is the obvious staging ground for the ORP for a number of reasons. Starting in the 1980s, under China's period of economic liberalization, Yunnan was gradually reopened to cross-border trade, which allowed the resurgence of the drug trade with Southeast Asia. Civil unrest in Myanmar and the region's general disconnect from traditional markets also enabled and fuelled a thriving illicit opium cultivation economy. This has heightened pre-existing Chinese government concerns over economic and political stability in Yunnan, a remote Western border province. Thus, Chinese state concerns underpinning the ORP were based on the perceived connection between opium use, social instability and border politics — all of which were seen as rooted in the low economic development of the region surrounding Yunnan.

Because of this understanding of opium cultivation as a product of economic underdevelopment, rubber held appeal as an alternative to a wide range of actors. China's history with opium began in the Qing Dynasty when opium addiction crippled the country and opium eradication was a foundational effort after the PRC was established in 1949. It has come to represent backwardness, an era of China's humiliation and subjugation to foreign powers, and a threat to political stability and the legitimacy of the Chinese Communist Party.[40] Rubber, as it has come to represent scientific progress, modernity, and the economic transformation and political integration of the borderlands, is an ideal alternative to opium in ideological terms. But rubber grows at lower altitudes than opium and has a markedly different political economy, so how has rubber come to be seen as a replacement or an alternative to opium?

The ORP supports investment in the provinces in northern Laos, but investments are not required to overlap or coincide geographically with opium cultivation. Instead, large-scale commercial agriculture is seen by Chinese managers as the vehicle for opium replacement and Sino–Lao development cooperation. As the manager of Jinrun Rubber Company explained, the company's aim was to provide alternatives to opium not by physically replacing opium fields but through generating wage labour opportunities with large-scale plantations, even as he acknowledged that opium fields were far smaller and operate on a different economic scale from rubber plantations.[41] Other managers echoed this, emphasizing the generation of wage labour and company investments in roads, schools, clinics and other services that would attract opium cultivators into other sectors.

This has led to the reasoning that the more rubber is planted, the greater the development benefits for Laos and the greater the incentive for cultivators to abandon opium. The Program has set minimums for required capital invested and ORP benefits are calculated based on the area of land developed.[42] Minimum area and capital requirements encourage investors to seek expansive areas of land but also squeeze smaller companies out of participation. This logic that the more rubber planted the more development benefits will ensue can also be traced to the rubber sector in Xishuangbanna, where large, politically connected companies have been the bastions of government service and technical expertise provision in the rubber sector. Large-scale plantations are also expected to accompany greater infrastructure investments by companies, including road and factory construction.

Thus, the ORP reflects the perception among Chinese policymakers at national and provincial levels that opium cultivation is symptomatic of larger development issues, and that it should be discouraged not necessarily through targeted local level interventions, but instead by introducing commercial agriculture on a regional scale. For them, opium seemed more a symptom than a cause of underdevelopment, and large-scale agriculture was its remedy.

Chinese Development Cooperation through Rubber Plantations in Northern Laos

Many aspects of how China envisions development cooperation also resonate in Laos. Opium cultivation has similarly been seen by the Lao state as emblematic of cultural backwardness in the country's most remote areas and as cyclically linked to rural poverty. The story of rubber in Xishuangbanna has generated, especially in northern Laos where people close to the border have seen with their own eyes, the wealth generated by rubber. But uprooting the Chinese rubber experience and translating it into a Lao context has been a fraught process. The global drop in rubber prices placed rubber operations of all sizes and backgrounds under market stress and exposed a range of differences between Xishuangbanna and Laos, from their contrasting systems of land regulation to the lack of state oversight and support for the rubber sector. These differences demonstrate the problems inherent in a discourse that idealizes China's path to development to legitimize the expansion of Chinese capital on a new frontier.

Opium cultivation has a long history in Laos, but the Lao government has come to associate it, along with other politically undesirable land uses like shifting cultivation, with poverty, backwardness, and rural social insecurity.[43] Lao's own opium eradication efforts, driven by the United Nations Office of Drugs and Crime (UNODC) funding and oversight in the 1990s and early 2000s, featured intensive campaigns to destroy opium fields but provided few alternative livelihood solutions. By 2005, these efforts had drastically reduced opium production but in the process had exacerbated the economic difficulties faced by opium cultivating communities.[44] As Lao state officials were searching for alternative livelihoods to stem a backslide into opium,[45, 46] foreign land investments were increasingly seen as effective instruments for development.[47] Rubber was particularly appealing to Lao state officials because tree plantations (no matter what the species) are categorized as forest cover under national law, thus rubber plantations are considered to contribute to the government's ambitious reforestation goals.[48] Lao smallholders in areas bordering China also witnessed Xishuangbanna's rise to prosperity[49] and some villages in Luang Namtha had already adopted the cultivation techniques of their neighbours across the border years before the ORP.[50, 51] Finally, rubber prices in the early 2000s were at a historic high, driving a rubber planting frenzy in Xishuangbanna and buoying Chinese and Lao state actors' optimism that rubber could combat poverty and opium cultivation.

This optimism was high when ORP companies began pursuing land deals in Laos and their investments catalyzed the expansion of rubber. The majority of ORP rubber projects in Laos are located in Luang Namtha (47), Bokeo (19), Oudomxay (14) and Phongsaly (20), the four northern provinces bordering China.[52] ORP rubber companies constitute between one and two thirds of all rubber companies registered in each province,[53] representing a significant portion of the rubber sector in the region.

Narratives of Development at the Company Level

While both the Beijing and Yunnan governments have clear stakes and interests in the ORP, statutes and legislation regarding policy implementation and company oversight are vague. This gives ORP companies significant latitude when operating in Laos. All of the Chinese company managers interviewed for this study espoused the officially stated aims of opium eradication through alternative development. Specifically, they championed their rubber plantations as vehicles for technology transfer, generating wage-labour opportunities and enhancing infrastructure development. But in actually establishing and running plantations, their interpretation of larger narratives on the intersection between rubber, opium, and development in Laos were articulated through the management decisions they make and the way they compare themselves to other stakeholders in Laos.

Yunnan provincial officials and plantation managers frequently compared Xishuangbanna to northern Laos in interviews. They discussed comparable climatic, geophysical, sociocultural and agricultural systems between the two regions. They used this comparison to justify rubber production as the dominant crop among ORP companies. Yunnan State Farms placed special emphasis on the former prevalence of shifting cultivation in Xishuangbanna,[54] which remains a dominant land use in Laos. The State Farms also emphasized the linguistic and religious similarities between the Dai in China and many northern Lao ethnic groups. Due to these perceived similarities, Chinese companies believe their decades of experience with rubber in Xishuangbanna will smoothly translate into Lao contexts. Existing studies criticize the tone of Chinese exceptionalism in the ORP's policy level discourses, taking issue with China's "broader civilizing mission that envisages the transference of idealized Chinese qualities (such as scientific rationality, technical competence and entrepreneurial spirit) to 'backward' peoples" in Laos.[55] Yet through the author's interviews it seems that Chinese managers are inclined to agree with this charge of having a civilizing mission, but they hardly see this as a criticism. Presenting their own development model as a path to be emulated or imposed in Laos and referring to Laos as needing Chinese assistance to establish the foundations (physical and in character) necessary to progress economically is, in their eyes, a vision of optimism and comradery.

Furthermore, many company managers criticized other countries' development interventions. They were particularly disparaging of grassroots level or small-scale projects (e.g. water well maintenance and agro-biodiversity initiatives), which they portrayed as having failed the Lao people. Chinese managers referred to these development

170   Juliet Lu

projects as low-impact and short-term compared to their own. They
also proclaimed their own companies' dual motives of development and
profit as the strength of their approach, not as a contradiction or reason
for suspicion. The fact that rubber is a tree crop ties companies to the
land for a long-term period (20–30 years). Chinese plantation companies
typically sign land concessions or contract farming agreements with
government for the duration of that time. One manager interviewed
in Luang Namtha stated:

> Many others, Japanese and Western companies, come here and give a
> bit of money. That just solves surface problems ... Our company will
> be here for many decades. Of course our main purpose is to make
> money! But as such, we need to take care of our relationships, so we
> invest in the villages around here. We have a long term outlook.[56]

Managers see themselves as bringing lasting and sustainable development
impacts to Laos, just by virtue of the long-term nature of rubber
production. They legitimize their presence by the fact that rubber is a
significant investment, which ties them to the surrounding community
and the country of Laos. Cultivating strong village relations is central
to their business plans (e.g. the recruitment of labour and avoiding
the cost of conflict).

Not only was such reasoning promoted by Chinese managers, it
was also evident among Lao villagers and state agents. For example, the
Lao government established certain tax holidays for rubber concessions
as an incentive to encourage such long-term investments. Government
officials and villagers echoed this logic when they explained why
certain areas had been selected for Chinese rubber investments. They
said that the area was poor, rubber was a promising cash crop, and
it was expected that the company would provide a range of services
once rubber tapping had begun. This logic also champions the long-
term benefits that Chinese rubber might bring and ignores the local
level dynamics of opium replacement. In particular, it glosses over the
importance of opium for generating immediate income and relieving
financial distress for the most vulnerable households.

Scaling up Rubber Production without Local Rubber Producers

By the early 2000s, before ORP companies arrived, smallholders and
other Chinese companies had already initiated rubber cultivation. For
a number of reasons, however, the ORP eventually shaped the entire

rubber sector across northern Laos. This is in part due to the sheer scale of land ORP companies have sought for rubber plantation establishment, the influx of capital they brought, and the promise of Chinese market access they provided to the sector. But these companies have also come to dominate certain activities — especially processing — which allow them disproportionate power over the entire rubber sector in Laos.

Smallholder rubber production was first introduced to Laos in small villages located along the Chinese border by households with family ties to rubber producers in Xishuangbanna in 1999. It was then further promoted, though without consistent funding, by provincial governments.[57] Under contract farming, companies or individual investors provide inputs and training to villagers willing to provide labour and devote a portion of their own land to growing rubber. Contract farming ranges in the level of formality of production and benefits sharing arrangements, but companies typically agree to buy rubber at market value or set minimum prices for purchase once trees reach tapping stage. They then split earnings, minus the cost of inputs, with participating cultivators. Concessions, which involve a long-term lease (35–50 years) of state land to investors, are often preferred by companies because they afford greater control over operations and typically higher profit margins. However, they also require greater oversight by companies. Regardless of the cultivation model, nearly all inputs (seedlings, fertilizers, etc.) to the rubber sector are sourced from China and China is the final market for all rubber produced in northern Laos.

The ORP implementation framework does not indicate priorities in terms of whether companies enter into contract farming or concession agreements. That said, it does favour large-scale monoculture plantations as evidenced by minimum area requirements, the structure of financial incentives, and statements of managers interviewed. This is perhaps because in China, large state companies were the first movers in the rubber sector and have long acted as instruments for technology and capital delivery. ORP support is based on plantation area. As one company representative explained, "the more you invest, the more money you get from the government".[58] This has had the effect of encouraging companies to seek as much land as they can conceivably be granted, regardless of their calculated management capacity or local land availability. Chinese company representatives and Lao state officials alike also explained that large plantation projects were expected to require the company to engage in infrastructure construction (e.g. roads, bridges, irrigation systems), whereas smaller scale investments and extension services to individual smallholders did not necessarily require such major investments. As a result, ORP rubber projects are typically reported as being over 1,000 ha in area.

Whereas Yunnan State Farms in Xishuangbanna worked closely with the government to provide vast agricultural extension services and training, the relationship between ORP companies and Lao farmers has been inconsistent with the China case and across companies within Laos. Nor have the Lao or Chinese governments played a role in brokering or regulating that relationship. Some villages reported periodic trainings in rubber planning, maintenance, and tapping held by companies. Small infrastructure projects, including bridge construction, road improvement and electricity and water services, have also been funded and built by companies, although such investments occur regardless of whether companies are ORP participants. Furthermore, due to rubber's extended maturation period of three to five years, many households who have experienced unexpected hardships (e.g. low rice yields or unforeseen expenses) or lack alternative income sources, have been unable to spare land and labour in the years before rubber can be tapped. These households typically sell their rubber trees or the land itself to the company for cash, thus losing that land and the potential profits from rubber. Thus, since the establishment of plantations, benefits may have accrued to certain Lao actors, but the most vulnerable have often been negatively impacted.

The potential for rubber plantations to benefit local communities by providing jobs as tappers or plantation workers is still unclear because most areas have not yet reached tapping age. Already, however, recruiting and training labour in Laos is proving far more difficult for companies than in China, where rubber production is already much more established. Because of these labour constraints, companies expressed a limited interest in obtaining new land for plantation development. "We can't handle more, and this year we're realizing there is a labour problem", admitted one manager.[59] Another echoed this saying, "labour is the main bottleneck for rubber development here".[60] In China, the massive movement of migrant workers and students during the Cultural Revolution established the first wave of rubber plantations in Xishuangbanna. In comparison, Laos has an extremely limited labour supply. Companies would prefer for the Lao government to recruit workers, thus avoiding language barriers and their own poor knowledge of Lao labour markets. Many hire Lao villagers who speak Chinese as middlemen or Chinese wage labourers within the company, who organize groups of Lao labourers but take cuts from their wages. Companies must also invest in substantial training for Lao tappers, unlike in Xishuangbanna where a base of trained labour already exists.

Finally, ORP support has generated very different operating capacities between Chinese rubber companies in terms of processing and access to Chinese markets. ORP companies enjoy some financial and administrative support (e.g., low interest loans) but this matters little in the overall processes of accumulation because land prices and operating fees in Laos were already astonishingly low during the plantation establishment boom in the mid-2000s. Instead, processing and export afford the greatest profits and it is at these valuable nodes in the rubber commodity chain that ORP companies have come to control. Whereas basic processing of cup lump rubber is performed in Laos, secondary processing from cup lumps into sheets and blocks occurs in Xishuangbanna, where long-established facilities are widely available and their capacity for rubber processing exceeds the domestic sector's output.

Because it is a higher value-added activity, many more companies would like to establish rubber processing factories. However, ORP companies enjoy a pivotal competitive advantage in transportation and export to China because of their access to quotas which exempt them from China's steep rubber import taxes. Through 2011, when quantities of produced latex were still modest, smallholders evaded these import taxes through illicit trade. Non-ORP companies also profitted despite these taxes due to soaring prices. But a drop in global rubber prices — caused by a range of factors related both to global commodity price slumps and shifts in global rubber supply — has turned import taxes into a key bottleneck for the Lao rubber sector. This has extended the competitive advantage and market control of ORP companies.

No Protection from Boom and Bust Rubber Prices

As most rubber plantations in Laos are reaching tapping age, now is a critical time to consider whether ORP companies are poised to deliver on the development benefits they have promised. However, a drop in global rubber prices in 2011 has exposed some of the Program's deeper weaknesses, as well as key differences from what made rubber successful in Yunnan. Three troubling situations have resulted from this.

First, most large companies that planted in the early to mid-2000s have stopped tapping, choosing to wait out the price drop. In concessions, companies are sending their seasonal wage labourers (both Lao and Chinese) home and both ORP and non-ORP companies are waiting to move forward with processing factory construction. China already has significant stockpiles of rubber, and aside from a few strategic purchases, China's demand is minimal. With either ORP support or simply the low land prices initially paid, Chinese companies have invested minimal capital into plantations thus far. As a result, they also feel little urgency to move into tapping at such low profit margins.

Second, many households are far more vulnerable to global price fluctuations than ORP companies, whose initial costs were subsidized. Many cannot afford to wait out the price downturn. Rubber has replaced a range of former land uses, including shifting cultivation and other areas key to local livelihoods and subsistence. Typically, rice can be grown amongst rubber for the first three to five years, when labour demands are relatively low and even though production declines significantly after the second year as rubber canopies expand.[61] Come the fifth year, tapping is expected to provide necessary income, either through wage-labour on concession plantations or the sale of latex by smallholders and contract farmers.

The global price drop, however, has meant that Lao farmers whose land had been converted to rubber, either under contract farming or concession, must wait even longer before gaining access to the benefits of wage labour or sale. Those with other income sources (e.g. other cash crops or market activities) have also limited tapping. The most vulnerable households, however, have no choice. As one villager remarked, "all my fields are now rubber, we have to tap to eat".[62] In more extreme cases, some households have sold their rubber holdings back to the companies with which they had signed up for contract farming. Or they converted rubber to other cash crops (e.g. bananas) (Friis and Nelson 2016), which represents a huge loss of initial investment. A few villagers, particularly those residing near Yunnan Rubber's concession, have been hired as tappers, whereas those with limited land, food harvests, or alternative incomes have been driven to more desperate adaptation strategies, such as opening new, distant or low productivity shifting cultivation areas.

Finally, latex processing activities have been further consolidated in the hands of ORP companies. Most rubber companies throughout Laos have included plans to build processing factories, a necessary infrastructure to prepare rubber latex for transport. With the downturn in global prices, however, most have halted their factory construction plans and are waiting for higher prices before tapping their plantations. In Luang Namtha, Yunnan Natural Rubber (a subsidiary of Yunnan State Farms) now holds the only factory in operation and conducts all latex processing for the entire province.[63] It is simply not cost effective for other companies to process rubber. Thus the opening of Laos to foreign investment and the promotion of rubber through the ORP resulted in the rapid spread of rubber throughout northern Laos, only for the price drop to prevent expected windfalls for rural development. It has also resulted in the consolidation of market and accumulation power (through control at the processing level) in the hands of a few Chinese companies.

As a result of this consolidation, Yunnan Natural Rubber has an influence over the margin of profit for all rubber production in Luang Namtha. As the only processing factory in operation in the province, Yunnan is the only purchaser of raw latex. It therefore has the power to set prices and determine quality standards for all the latex it processes. Should rubber prices rise enough to encourage other ORP companies to finish building processing factories, Lao producers may benefit from an increase in competition among processors but until prices rise significantly, only ORP companies will be able to profit from exporting.

The control over rubber processing and export by ORP companies is more than a bottleneck for the rubber sector in northern Laos. Through the competitive edge provided by export quotas and the recent price downturn, ORP companies have consolidated control over the mechanisms through which rubber producers access Chinese markets. As a result of their control over them, ORP companies are able to outcompete other actors along the rubber commodity chain. This highlights the tension between corporate profit and local livelihoods development, and casts the suggestion that companies may act as agents of development in a critical light.

Conclusions

Through this examination of the multilevel processes driving China's ORP, the author has demonstrated how narratives of Chinese development cooperation translate on the ground. Because development cooperation is heavily informed by China's own development experience, examples of economic success are taken by Chinese actors as justification to promote similar approaches abroad. But in the case of rubber in Laos, key conditions prevent the replication of Xishuangbanna's path to development. In Xishuangbanna, the expansion of rubber was shaped by a strong state, trade protection for the rubber sector, smallholder land use rights and access to market services, and heavy state oversight of and collaboration with agribusiness companies. In northern Laos, in contrast, the rubber sector lacks protection from volatile global price fluctuations, individual farmers exercise much weaker land tenure claims vis-a-vis Chinese companies, and Chinese companies operate with far less direct state oversight from either the Chinese or Lao governments.

What is interesting about the ORP when examined at the local level is the dissonance between the lofty ideals promoted by Chinese actors at a policy narrative level and their applicability on the ground in Laos. Northern Laos is discursively framed as comparable to China's own rubber producing areas, while China's success in rubber is assumed to be reliant on a simple set of replicable factors. These two simplifications along with the overarching narrative of Chinese development cooperation are bundled together as a range of assumptions, which have shaped how Chinese actors perceive and promote corporate land investments as win-win endeavours. And while these assumptions frame companies' approaches to rubber investments in Laos, they do not translate on the ground into the same development path. It is too early and beyond the scope of this chapter to analyse the ORP's full impacts on rural development in Laos, in part because the rubber price drop has halted tapping. Rather, the author's objective has been to analyse the ORP as a window into the Chinese model of development cooperation, trace its impact through corporate practices in Laos, and identify the power, appeal, and contradictions contained within that model. Understanding the ORP in this way provides a lens, based on Chinese actors' own views of development, through which China's approach to developing countries as new markets, suppliers, and partners in development in the future can be examined.

These findings contribute both to region-specific policy questions and to the literature on China's economic development and investments abroad. In terms of policy, understanding the logic driving Chinese investments and the role companies play in development cooperation can inform Lao policies for managing Chinese investments to its own benefit. Chinese policymakers could also redesign future development cooperation initiatives to better monitor investors' activities or define expectations for development outcomes. Meanwhile, in the context of development cooperation, Chinese firm behaviour may depart from conventional corporate behaviour since they are expected to balance profit maximization with development outcomes. That said, the companies studied take a diverse set of approaches, thus examining the attitudes and beliefs of plantation managers operating in the country of investment is key to predicting how firms balance profit and development.

This chapter also aims to compliment other pieces in this collection which address different ways Chinese capital articulates in country contexts and how it is perceived and facilitated by host-country elements. Laos represents an interesting case because it has long been closed off from China's influence, but it has now opened up to a rapid influx of Chinese actors. Through the ORP and Chinese interests in a new frontier for investment, Laos is now becoming a context in which politics in Yunnan Province have great sway over how Chinese capital articulates with existing economic structures, land and resource systems and political dynamics. Because of surface similarities, such as climatic suitability for rubber and common ethnolinguistic ties, Laos represents an ideal context for examining the logics of Chinese capital, the impacts of development cooperation in practice, and the role of Chinese corporations in shaping market transition and development throughout the Global South. As such, the mixed outcomes of the ORP highlight the complexities and contradictions that arise in this approach to development cooperation and point to the need to study Chinese perspectives to uncover the logics driving Chinese investment as it continues to grow.

Too often analyses of Chinese investment treat Chinese investors as a monolithic type, whereas the author has observed a diverse range of Chinese actors become increasingly engaged in economic interventions abroad. The diversity of Chinese investors, their priorities, assumptions, and strategies for resource development must be better studied. It flattens the complex historical factors that inform Chinese actors' perspectives on development, and therefore what they intend in projects labelled development cooperation and thus how they design and implement them. Hence instead of documenting the risks and potential rewards of engaging Chinese companies in Lao land markets, the author suggests that a deeper analysis of Chinese development cooperation as a discourse shows what factors and development logics drive land deals. Such analysis can, in reflecting deeper logics of development among Chinese actors, also contribute to the formulation of better regulations for Chinese aid and investment in Laos.

NOTES

1. Ian Scoones, Lídia Cabral, and Henry Tugendhat, "New Development Encounters: China and Brazil in African Agriculture", *IDS Bulletin* 44, no. 4 (2013): 1–19.
2. Marcus Power, Giles Mohan, and May Tan-Mullins, *China's Resource Diplomacy in Africa: Powering Development?* (UK: Palgrave Macmillan, 2012).
3. Julie Walz and Vijaya Ramachandran, "Brave New World: A Literature Review of Emerging Donors and the Changing Nature of Foreign Assistance", *Center for Global Development Working Paper* 273 (2011).
4. Xiaoyun Li, Dan Banik, Lixia Tang, and Jin Wu, "Difference or Indifference: China's Development Assistance Unpacked", *IDS Bulletin* 45, no. 4 (2014): 26.
5. Lila Buckley, "Chinese Agriculture Development Cooperation in Africa: Narratives and Politics", *IDS Bulletin* 44, no. 4 (2013): 42–52.
6. Tania Murray Li, "Centering Labor in the Land Grab Debate", *The Journal of Peasant Studies* 38, no. 2 (2011): 281–98.
7. Pal Nyiri, "The Yellow Man's Burden: Chinese Migrants on a Civilizing Mission", *The China Journal* 56 (2006): 84–85.
8. Karen E. McAllister, "Rubber, Rights and Resistance: The Evolution of Local Struggles Against a Chinese Rubber Concession in Northern Laos", *The Journal of Peasant Studies* 42, nos. 3–4 (2015): 1–21.
9. Miles Kenney-Lazar, "Dispossession, Semi-Proletarianization, and Enclosure: Primitive Accumulation and the Land Grab in Laos", paper presented at the International Conference on Global Land Grabbing, 6–8 April 2011.
10. *The Economist*, "A Bleak Landscape: A Secretive Ruling Clique and Murky Land-Grabs Spell Trouble for a Poor Country", 26 October 2013, available at <https://www.economist.com/news/asia/21588421-secretive-ruling-clique-and-murky-land-grabs-spell-trouble-poor-country-bleak-landscape>.
11. Charlotte Hicks, Saykham Voladeth, Weiyi Shi, Zhong Guifeng, Sun Lei, Pham Quang Tu, and Marc Kalina, "Rubber Investments and Market

Linkages in Lao PDR: Approaches for Sustainability?", Sustainable Mekong Research Network, Bangkok, 2009.

12. Olivier Ducourtieux, Jean-Richard Laffort, and Silinthone Sacklokham, "Land Policy and Farming Practices in Laos", *Development and Change* 36, no. 3 (2005): 499–526.
13. Mike Dwyer. "Turning Land into Capital: A Review of Recent Research on Land Concessions for Investment in Lao PDR", Land Issues Working Group, Vientiane, Laos, 2007.
14. Ian G. Baird, "Turning Land into Capital, Turning People into Labor: Primitive Accumulation and the Arrival of Large-Scale Economic Land Concessions in the Lao People's Democratic Republic", *New Proposals: Journal of Marxism and Interdisciplinary Inquiry* 5, no. 1 (2011): 10–26.
15. Oliver Schoenweger, Andreas Heinimann, Michael Epprecht, Juliet Lu, and Palikone Thalongsengchanh, *Concessions and Leases in the Lao PDR: Taking Stock of Land Investments* (Bern, Vientiane: Geographica Bernensis, 2012).
16. C.C. Mann, "Addicted to Rubber", *Science* 325, no. 5940 (2009): 564–66.
17. Jefferson Fox and Jean-Christophe Castella, "Expansion of Rubber (Hevea Brasiliensis) in Mainland Southeast Asia: What are the Prospects for Smallholders?", *The Journal of Peasant Studies* 40, no. 1 (2013): 155–70.
18. Ibid.
19. Janet C. Sturgeon, "Cross-Border Rubber Cultivation between China and Laos: Regionalization by Akha and Tai Rubber Farmers", *Singapore Journal of Tropical Geography* 34, no. 1 (2013): 76.
20. Interview in Mandarin with TCRI researcher, 9 August 2014.
21. Sturgeon, "Cross-Border Rubber Cultivation between China and Laos", p. 76.
22. Janet C. Sturgeon, Nicholas K. Menzies, Yayoi Fujita Lagerqvist, David Thomas, Benchaphun Ekasingh, Louis Lebel, Khamla Phanvilay, and Sithong Thongmanivong, "Enclosing Ethnic Minorities and Forests in the Golden Economic Quadrangle", *Development and Change* 44, no. 1 (2013): 53–79.
23. Mann, "Addicted to Rubber", pp. 564–66.
24. Jianchu Xu, "The Political, Social, and Ecological Transformation of a Landscape: The Case of Rubber in Xishuangbanna, China", *Mountain Research and Development* 26, no. 3 (2006): 254–62.
25. Jianchu Xu, Jefferson Fox, John B. Vogler, Zhang Peifang, Fu Yongshou, Yang Lixin, Qian Jie, and Stephen Leisz, "Land-Use and Land-Cover Change and Farmer Vulnerability in Xishuangbanna Prefecture in Southwestern China", *Environmental Management* 36, no. 3 (2005): 404–13.
26. Weiyi Shi, "Rubber Boom in Luang Namtha: A Transnational Perspective", GTZ RDMA, 2008.
27. Sturgeon, "Cross-Border Rubber Cultivation between China and Laos".
28. Ibid., p. 55.

29. Jennifer Poole, "China to Cut Natural Rubber Import to Increase Local Supply", *Rubber Journal Asia* (2012).

30. Ibid.

31. Alex Smajgl, Jianchu Xu, Stephen Egan, Zhuang-Fang Yi, John Ward, and Yufang Su, "Assessing the Effectiveness of Payments for Ecosystem Services for Diversifying Rubber in Yunnan, China", *Environmental Modelling & Software* 69 (2015): 187–95.

32. Yunnan Provincial Government, "关于加快国有农垦企业改革与发展的通知", 2000.

33. Sturgeon, "Cross-Border Rubber Cultivation between China and Laos".

34. Tom Kramer and Kevin Woods, "Financing Dispossession: China's Opium Substitution Programme in Northern Burma", Transnational Institute, Amsterdam, 2012.

35. Mann, "Addicted to Rubber", pp. 564–66.

36. Paul Cohen, "Resettlement, Opium and Labour Dependence: Akha–Tai Relations in Northern Laos", *Development and Change* 31, no. 1 (2000): 179–200.

37. Weiyi Shi, "Rubber Boom in Luang Namtha".

38. Paul Cohen, "The Post-Opium Scenario and Rubber in Northern Laos: Alternative Western and Chinese Models of Development", *International Journal of Drug Policy* 20, no. 5 (2009): 424–30.

39. Cao Yin and Guo Anfei, "Alternate Crops Replacing Opium Poppies", *China Daily*, 26 June 2012.

40. Lin Lu, Yuxia Fang, and Xi Wang, "Drug Abuse in China: Past, Present and Future", *Cellular and Molecular Neurobiology* 28, no. 4 (2008): 479–90.

41. Interview with Jinrun manager, 14 November 2012.

42. 云南省禁毒网, "云南省开展境外罂粟替代种植项目管理办法(试行)", Xinhua News Agency, 2004.

43. Miles Kenney-Lazar, "Shifting Cultivation in Laos: Transitions in Policy and Perspective", Sector Working Group-Agriculture and Rural Development (SWG-ARD), 2013, p. 49.

44. United Nations Office on Drugs and Crime (UNODC), "Laos: Opium Survey 2005", 2005.

45. Weiyi Shi, "Rubber Boom in Luang Namtha".

46. Cohen, "The Post-Opium Scenario and Rubber in Northern Laos".

47. Dwyer, "Turning Land into Capital".

48. Yayoi Fujita and Kaisone Phengsopha, "The Gap between Policy and Practice in Lao PDR", in *Lessons from Forest Decentralization: Money, Justice and the Quest for Good Governance in Asia-Pacific*, edited by Carol J. Pierce Colfer, Ganga Ram Dahal and Doris Capistrano (London: Earthscan/CIFOR, 2008).

49. Mann, "Addicted to Rubber", pp. 564–66.

50. Weiyi Shi, "Rubber Boom in Luang Namtha".

51. Hicks et al., "Rubber Investments and Market Linkages in Lao PDR: Approaches for Sustainability?".

52. Some companies have more than one project; data compiled from <www. xinlaowo.com> ("云南省在老挝投资企业名录"), a Chinese language website on investing in Laos; cross-checked with public information from multiple online sources citing Yunnan Department of Commerce project approval information.

53. GoL, "State Land Leases and Concessions Inventory", edited by Deutsche Gesellchaft fur Internationiale Zusammenarbeit (GIZ), Ministry of Natural Resources and Environment (GoL) (Vientiane: Centre for Development and Environment, 2011).

54. Interview with company manager, 17 August 2012.

55. Cohen, "The Post-Opium Scenario and Rubber in Northern Laos", p. 6.

56. Interview with company manager, 23 December 2012.

57. Interview with M. Canet, 16 July 2013.

58. Interview with company manager, 9 November 2012.

59. Interview with company manager, 17 December 2012.

60. Interview with company manager, 13 December 2012.

61. Interview with a villager, 30 August 2014.

62. Interview with a villager, 28 August 2014.

63. Interview with M. Canet, 14 July 2014.

8

THE HIGH COST OF EFFECTIVE SOVEREIGNTY
Chinese Resource Access in Cambodia[1]

Siem Pichnorak

Foreign aid and economic investment have been effective tools used by China to leverage influence in many developing countries, including Cambodia. China is now the largest aid donor and investor to Cambodia. Despite positive aspects of China's presence, which include an increase of FDI inflow, infrastructure development, and economic growth, China's substantial investment in natural resource sectors also comes with significant costs. As one of the largest holder of economic land concessions, mining licenses and hydro dam construction projects, Chinese companies, often partnering with local companies and elites, have been involved with illegal land grabbing, deforestation, and human rights abuses. These issues have prompted countless outcries and protests from local peasants, civil societies and international community. In addition, opposition parties in Cambodia have leveraged these issues to accuse the ruling Cambodia People's Party (CPP) of promoting a pro-China policy. This chapter explores the complex issues that have emerged around Chinese investment in Cambodian natural resources. First, the chapter analyses structural factors in the Sino–Cambodian relations that have generated a favourable investment climate for Chinese resource companies. Second, it investigates various recurring

*problems emerging around resource sector projects of Chinese companies in
Cambodia. Lastly, the chapter evaluates the benefits and costs brought along
by Chinese investment in natural resource extraction in Cambodia with a
view to preventing the fallouts of such investment in the future.*

Introduction

If China has been "going global" since the early 2000s, it has gone
regional for far longer. China's Cold War-era influence in Southeast
Asia prior to the 1980s is well documented,[2] after which both foreign
and domestic difficulties led to a period of relative withdrawal. Since
the end of the Cold War in the early 1990s, however, China's foreign
policy toward its southern neighbours has increasingly re-embraced
its outwardly engaged version of "peaceful coexistence", as first
articulated in the 1950s. Seeking both regional political stability and,
perhaps more pointedly, access to natural resources for its own ongoing
economic development,[3] China's leaders are widely understood to see
Southeast Asia as a strategic sphere of influence.[4] In such a context, they
have looked to development cooperation as a core mode of regional
engagement for much of the last two decades.

As a key locale of this strategy, Cambodia receives close attention
from Beijing, given its pivotal position in the region.[5] Since 1996, the
relationship between Phnom Penh and Beijing has deepened. The two
countries signed a Comprehensive Strategic Partnership Agreement in
2010, extending their cooperation across sectors ranging from trade to
security. China is now a dominant foreign player in Cambodia due
to its high levels of investment, trade volume, and aid disbursement.
Between 1992 and 2013, Chinese aid to Cambodia amounted up to
US$2.7 billion in loans and grants.[6] In the same period, trade between
the two countries expanded to US$3.6 billion in 2012, up from US$57.3
million in 1995 and US$732.8 million in 2006.[7] From 1994 to 2013,
cumulative Chinese investment in Cambodia was worth an estimated
US$9.6 billion, making China the country's largest foreign investor.[8]

Natural resources have figured centrally in this partnership.
Among other strategic interests, China and Chinese investors see in
Cambodia a wealth of under-exploited timber, gas, oil, water, rubber,
fertile farmland, and mineral ores, including gold, silver and iron ore,
which have all been essential inputs to China's ongoing industrial
development. According to China's 10th Five-Year Plan, which saw

the launch of the country's "going global" policy, Beijing foresaw the need to ensure uninterrupted supplies of natural resources and committed to utilizing overseas supplies across a range of sectors. This has had major implications on key economic sectors, from agriculture and forestry to energy, minerals and infrastructure development.[9] It has also precipitated significant state support for Chinese business ventures abroad.[10]

This chapter examines the mix of strategic benefits and costs of this reinvigorated partnership for Cambodia. These costs and benefits overlap in some key areas of politics and the economy, and many are incommensurable. This means that while trade-offs exist, it is impossible to evaluate whether they are "worth it" or not — this chapter seeks no conclusions along these lines. Instead, this chapter focuses on describing the difficult nature of Cambodian land and resource governance — and, ultimately, of sovereignty itself — given that Sino–Cambodian collaboration brings certain benefits to the Cambodian sovereign state, but also exacerbates a number of already existing problems in the exercise of sovereign territorial control.

The following sections develop this claim using the concept of effective sovereignty, which refers to states' practical abilities to manage, govern and develop a country, its territory and its population(s). The concept contrasts with *formal* sovereignty, which refers to the legal dimensions of sovereignty, typically articulated in terms of rights — specifically the formal right to govern a particular territory and the population that lies within it.[11] The author argues, in short, that Sino–Cambodian relations are oriented toward enhancing the effective sovereignty of the Cambodian state, but that they also exacerbate the weak governance that already exists, and in doing so impose significant social and environmental costs. These costs are the high price of sovereignty referred to in the chapter's title.

This argument is elaborated sequentially. The next section describes four ways in which China's partnership with Cambodia seeks to reinforce the effective sovereignty of the Cambodian state over its territory and population. These include greater capacity to invest in "development"; the consolidation and distribution of economic resources; the state's monopoly on violence; and leverage vis-à-vis other donors. The third section looks at the other side of this analysis by examining the costs that attend these putative, if sometimes contradictory and socially uneven, benefits. It focuses on four issues that illustrate the conflicted and problematic nature of Cambodian sovereignty-in-practice. These issues — regulatory capture, land grabbing, negative social and environmental impacts, and the repression of popular protest — are not unique to Sino–Cambodian cooperation. In Cambodia, they accompany other foreign endeavors as well. Globally, they appear in contexts across the Global South. But given China's significant role in Cambodia — China is by far the largest foreign investor in Cambodia, representing 24.44 per cent in 2014 alone[12] — looking at the case of China helps to illustrate key features of a complex landscape.

This chapter contributes to the ongoing scholarly and popular efforts to understand the "soft power" dimensions of China's rise, as American domination is increasingly contested, both globally and in Southeast Asia. Coupled with rapid military modernization, the country's growing economic prowess has helped China gain significant political influence globally, and made China an attractive development partner for many smaller states, especially those with authoritarian regimes.[13] As illustrated below, this new type of partnership provides certain advantages for countries like Cambodia whose governments are dissatisfied with the intrusive foreign policy of traditional donors like the United States and international institutions. However, as this chapter also illustrates, these advantages also come at a certain price, whose full dimensions are still being realized.

Sino–Cambodian Relations: Efforts to Enhance Effective Sovereignty

Sino–Cambodian relations have been constructed around the so-called Five Principles of Peaceful Coexistence, first articulated as the basis of China's foreign policy in 1954.[14] Centred on ideals of mutual benefit and non-interference, the idea of "peaceful coexistence" plays well in Cambodia, where national sovereignty and territorial integrity were long subjected to foreign interference and, as a result, carry significant political sensitivity. China has figured centrally in modern efforts to build and maintain Cambodian national sovereignty. At the dawn of independence, King Norodom Sihanouk turned to Beijing for assistance and parlayed China's assurance and recognition of national sovereignty into significant political clout during the turmoil of the Cold War. In the years since, the sensitivity of Cambodian sovereignty has often been used for political ends.[15]

While some observers see the rhetoric of peaceful coexistence as papering over the implicit *quid pro quo* that guides many bilateral relationships, the principles of unconditional aid and respect for sovereignty provide the Cambodian government with important political cover. In April 2012, Prime Minister Hun Sen forcefully rejected the accusation of being a puppet or client state: "What I hate and am fed up with is talk about Cambodia working for China and must be under some kind of influence." Cambodia, he said, was "not going to be bought by anyone".[16] China, for its part, has proven adept in maintaining this image of remaining outside the fray of Cambodian politics. As the former Chinese Ambassador to Cambodia put it, "China supports Cambodia to develop its economy independently and with its [own] ownership."[17]

The remainder of this section examines four specific ways in which Chinese assistance has helped Cambodian authorities pursue "independent" development through bolstering their effective sovereignty over the territory, its resources, and key interest groups. From playing the developmental state to dealing with foreign donors, these dimensions of sovereignty-building cover the spectrum from the political to the economic, and from internal to foreign relations.

Economic Patronage and Political Stability

Perhaps most fundamentally, unconditional Chinese aid has enabled the ruling Cambodian People's Party (CPP) to stay in power by consolidating and maintaining political allies. The Cambodian state, according to Un Kheang, is structured by "interlocking pyramids of patron–clientelism networks" that were built during the 1980s and 1990s, and that extend throughout the economy.[18] Through various compromises and coalitions, these networks have become even more intertwined and complicated over time.[19] Today, the CCP's political power depends heavily on being able to maintain the flow of economic rent through these networks so as to maintain potential rivals as key allies. Beijing's willingness to accept and work with this arrangement has allowed Chinese aid and investment to play an important role in helping the CPP consolidate and maintain its power as a top political patron.

Given their existing capital and socio-cultural links, elite Chinese-Cambodians in particular play a prominent role in channelling foreign investment into this network. This class of businesspeople includes tycoons such as Ly Yong Phat and Lao Meng Khin,[20] who are active across a range of commercial and resource sectors, and who often hold government positions as well. Through the networks that surround this core, many other businesspeople have been able to build wealth and influence as well, such as some of the companies examined in the next section. The recent surge in Chinese FDI has strengthened this cementing of elite political-economic ties. Some Chinese business leaders have been given Cambodian nationality and hold positions in government, such as advisors or *Oknha*.[21] Over time, this pattern has become self-reinforcing: having a significant stake in the Cambodian economy, positions in state institutions, and the power generated from patronage networks, enables these groups to lobby Cambodian decision-makers to maintain close ties with Beijing and offer favourable treatment to Chinese investment in return.

Developmental State Policy

A second dimension of China's economic patronage extends into the realm of public infrastructure, where China's investment in projects such as hydropower dams meshes with the Cambodian government's efforts to promote and facilitate national development. After decades of civil war, peace and stability are prime social concerns. The CPP government understands these concerns clearly and has built its legitimacy by propagandizing its achievements in making peace, maintaining stability and, on the back of these, building development infrastructure such as roads, bridges, schools and irrigation systems. The symbolic importance of infrastructure helped the CPP win consecutive electoral victories from 1998 through 2008. China's role in financing infrastructure projects is palpable, especially in areas where traditional financers are hesitant to invest. At a groundbreaking ceremony for the China-funded expansion of a major national highway in February 2012, for example, Prime Minister Hun Sen praised Chinese investment publicly by stressing that China shows "respect" for recipient countries and that "China always responds to projects judged to be Cambodia's priority".[22]

Sovereign Power and Social Control

Popular legitimacy has its limits, however, and Chinese aid to Cambodia — consistent with China's support of poor authoritarian states elsewhere — also goes to the maintenance of police power and military force. Since the elections of 1993, Cambodia has been struggling on the bumpy road to democracy. This process is beset with challenges. Given ongoing challenges to the rule of law, rampant corruption, widespread patronage networks and executive usurpation of power, Cambodia is widely seen to be authoritarian or at least "electoral-authoritarian", given its formal electoral process.[23] Political deadlock and violent crackdowns on popular demonstrations further hinder democratization, as does increasing state pressure on civil and political rights. In this context, the un-conditionality of Chinese aid is only the beginning; police- and military-oriented support is another key piece of China's "implicit security guarantee"[24] to the CPP government. This is especially important when public infrastructure of the sort discussed above creates local opposition due to land and resource conflicts with local communities (see next section).

Sino–Cambodian security cooperation is relatively modest compared to the Cold War era, when the use of "hard" power and ideology were central to foreign policy. Today, China relies mainly on economic heft to build influence and enforce allegiance. However, in addition to the public infrastructure discussed above, Chinese aid also goes to build and maintain the state's means of coercion. China, with its own increasing military power, is a growing source of military support as other foreign partners appear hesitant in this realm. China is now a leader in providing military aid to Cambodia, offering finance, technology and equipment to modernize the Royal Cambodia Armed Forces. China provided 257 military vehicles as well as two MA60 airplanes to the Ministry of Defense of Cambodia in 2012. In 2013, China delivered 12 Z-9 military helicopters purchased under a US$195.5 million Chinese loan, signed two years prior. In 2014, China provided 26 military trucks and 30,000 sets of military uniforms.[25] The same year, Cambodian Minister of Defense Tea Banh revealed a plan to send 400 military personnel and civilians to study and train in China, and China has reportedly also funded a military institute known as the Combined Arms Officer School, located in Kompong Speu.[26]

Countering Donor Leverage

When Cambodian officials praise China for the unconditionality of its aid and its corresponding respect for Cambodian sovereignty, the implicit contrast is with other donors whose attempts to impose reforms are viewed as overly intrusive. These reform efforts focus on promoting respect for human rights, guaranteeing the rule of law, and governing development projects in accordance with transparency norms (for example, in the area of procurement), all of which threaten the CPP's heavy dependence on political-economic patronage. The explicit unconditionality of China's aid, in contrast, provides the Cambodian government with a safe haven and form of counter leverage against the criticisms and reform pressures of Western donors and multilateral institutions.

When criticisms and pressure for reforms come from traditional donors, China often voices its political support for the CPP's government and increases its aid to offset any losses.[27] Following the 1997 coup, for example, Prime Minister Hun Sen's government was severely pressured and condemned by Western powers. In contrast, China declined to join the condemnation and instead extended its economic and political support. Beijing provided an immediate $10 million loan, and six months later delivered US$2.8 million of military equipment.[28] Similarly, when Western donors and the UN pressed the CPP's government to draft and enact a long-delayed anti-corruption law in 2006, China stepped in by providing US$600 million in aid.[29] In 2009, the US suspended military aid in retaliation for the Cambodian government's controversial deportation of 20 ethnic Uighurs to China. The following year, China provided 257 new military trucks as well as 50,000 military uniforms to offset the lost American aid.[30]

Chinese support has contributed in a number of ways to the Cambodian government's capacities to deal with an array of domestic and foreign challenges. The dimensions discussed above gesture to the breadth and complexity of this assistance. However, these capacities also come with a significant cost, much of which relates to the resource-extractive dimensions of Chinese cooperation. Using examples from the agricultural, mining and energy sectors, the next section examines these challenges in more detail. It shows how Chinese projects have exacerbated Cambodia's already weak system of resource governance.

The Costs of China's Natural Resources Extraction

A number of factors facilitate Chinese investors' access in Cambodia, especially in projects related to natural resources. As in many other countries, Chinese State-Owned Enterprises (SOEs) play a significant role in the Cambodian natural resources sector. This helps to explain why investment projects are sometimes tied with foreign aid provided by the Chinese government. It could therefore be concluded that Chinese investors' favourable treatment from the Cambodia's government is at least in part due to close government-to-government relations.

Numerous social and environmental governance issues arise from natural resources exploitation in Cambodia. While Chinese companies are amongst the largest foreign concessionaires,[31] hydro dam,[32] and mining licenses, it is dangerous to generalize that these problems are all caused by "the Chinese". Nonetheless, the magnitude of Chinese investment in the country has virtually ensured that Chinese projects have become ensnared in governance issues over the last two decades. This section examines these challenges via four issues that figure centrally in land and natural resource-based investments. They are regulatory capture, land grabbing, negative social and environmental impacts, and repression of popular protest. The examples discussed below draw on Chinese projects from across Cambodia, but many of the issues are found in projects of other investor nationalities as well.

Regulatory Capture

Cambodia's legal system has numerous laws, regulations, policies, and decrees covering investment activities, the energy sector, environmental protection and natural resources management. These are oriented toward ensuring that the development of natural resources in the country will be sustainable.

Natural resource extraction projects often involve land concessions granted to investors for a long period of time. According to the Law on Concessions, there are two types of land concessions that could be granted for investment projects — economic land concessions, which enable beneficiaries to clear land for industrial and agricultural use; and the land concessions that allow beneficiaries to conduct mining, finishing, industrial development and port concessions. Economic land concessions are limited to a maximum of 10,000 hectares, and each concession area must be exploited within 12 months after the license is granted. Concession contracts could last up to 99 years.

Despite having what is often described as a fairly good set of laws on paper, however, many statutory procedures and requirements are not complied with, and in some cases are actively circumvented with the involvement of influential officials or bribes. A 2004 World Bank reported a high degree of informality as a "key factor contribut[ing] to low productivity", and found that "the share of sales revenue paid by Cambodian firms in the form of bribes is over twice that of Bangladesh and by far the highest among the [other countries studied]".[33] Despite incremental improvement in corruption index, Cambodia still ranked 156th out of 175 countries in 2014.[34] Cambodia's investment climate thus remains hampered by poor transparency and regulatory enforcement, as well as nepotism and corruption. All of these hamstring entrepreneurial activity and contribute to a range of social and environmental problems.

Land Grabbing

Primary among resource extraction-related problems is land grabbing, which is often considered as one of the most prevailing social problem in Cambodia. It frequently makes domestic newspaper headlines and provokes active public debate. Land grabbing is not unique to Cambodia. It happens in many developing countries where good governance is absent and laws are bypassed regularly by corrupt authorities. In Cambodia, unlawful confiscation of land property is precipitated by the fact that land division and entitlement have for so long been unclear and neglected by the government given the prolonged civil war. As key investors in resource sector in Cambodia, Chinese companies have been awarded extensive economic land concessions where forests are cut down and local residents evicted to pave the way for agribusiness plantations, primarily for sugarcane or rubber, or logging operations.

According to the Cambodian Center for Human Rights (CCHR), about half of Cambodia's total of 8 million hectares of land concessions granted since 1994 are currently in the hands of Chinese companies. From 1994 to 2012, 4,615,745 hectares of land concessions have been granted to 107 Chinese-owned firms — 3,374,328 hectares for forest concessions, 973,101 hectares for economic land concessions, and 268,316 hectares for mining concessions.[35] CCHR land reform project coordinator, Ouch Leng, has said that "Chinese companies control about a quarter of the 17 million hectares of agricultural land and forest available in Cambodia. Because of these concessions, many villagers have lost their homes and land." According to the International Federation for

Human Rights based in Paris, about 4 million hectares, or 22 per cent of Cambodia's total land area, have fallen under land grabbing, with indigenous communities as the predominant victims of confiscation. According to another study, roughly 770,000 people or 6 per cent of Cambodia's population have experienced land grabbing since 2000.[36]

Backed by local officials and tycoons, Chinese companies have gained access to vast land concessions and flagrantly bypassed the regulations. Land confiscations, often involving significant confrontations and violence, have been facilitated and backed by politicians, government officials, influential tycoons and members of the armed forces, who benefit from commissions or bribes provided by incumbent companies. Chinese firms are adept in dealing with local officials and receive extensive protection. Through connection with government officials and Okhna as well as support from Beijing to Cambodia's military, some Chinese companies appear to benefit from the protection of Royal Cambodian Armed Force (RCAF) who received donations in exchange for the protection of their business interests.[37]

The most noticeable company to be actively involved in land grabbing and illegal evictions is Pheapimex, owned by the powerful couple Choeng Sopheap and Cambodian senator Lao Meng Khin. The company is known to have linkages with different Chinese firms through joint ventures. Prior to recent reforms, Pheapimex controlled up to 7 per cent of Cambodia's land area through its concession holdings. In December 2000, Pheapimex signed an agreement with the Chinese State Farms Corporation to establish a US$70 million joint venture to grow eucalyptus trees and build a pulp and paper mill in Kompong Chhanang province. This project was funded by the EXIM Bank of China,[38] and drew on materials from the company's 315,028-hectare agro-industrial concession — Cambodia's largest — located in Pursat and Kampong Chhnang provinces.[39] It has been estimated that some 12,000 families have been affected by land disputes on these areas.[40] Pheapimex is also known to be involved in a 199,999-hectare, 99-year concession in Mondulkiri province — a joint venture with the Chinese company Wuzhisan[41] — as well as the proposed Stung Cheay Areng dam in Koh Kong province, discussed below.

Land grabbing and unlawful confiscation are a common challenge faced by local communities across Cambodia as regulatory system fails to protect their rights against extensive land concessions and confiscation. As major land concessionaires, Chinese investment in natural extraction in Cambodia has deteriorated the situation of land rights where rural people lack proper titles to the land they occupy. As limited by law, economic land concessions cannot exceed 10,000 hectares. However, Chinese companies have been able to acquire far larger concession areas, as illustrated above, through their ties to local elites and connections with government officials. These concession areas encroach on local communities' settlement areas, production areas, and protected forests. As Chinese investment increases, conflict involving land grabbing is likely to increase if legal procedures and protections remain unenforced or underenforced.

Social and Environmental Impacts

In the same vein, Chinese investment projects in natural resources extraction have produced harmful effects on the environment, in part because they often bypass proper impact assessment processes. Due to corruption and involvement of powerful individuals, projects are frequently endorsed without careful considerations about the impacts on the environment and local communities' livelihoods.

A prominent example is the Kamchay dam project, built by the Chinese Sinohydro Corporation in the late 2000s in Kampot province. The project was awarded in 2005 after a closed-door negotiation, and lacked a completed Environmental Impact Assessment (EIA) when its construction permit was issued. The 112-metre high dam cost US$330 million to build and was almost entirely financed by a US$600 million aid package that was pledged by Beijing to Phnom Penh in April 2006.[42] Bokor National Park, where Kamchay dam is constructed, is known to be rich in biodiversity. According to an estimate, the project area is home to 37 mammals, 68 bird species, 23 reptile species and 192 fish species. Another study in 2002 revealed that there are 10 endangered species, including Asian elephants, leopard cats and tigers living in the protected forest. More importantly, the forest is an important source for local residents, many of whom depend on forests products as their income source. Another environmental problem is water contamination. It is reported that the river's water quality deteriorated because of untreated sewage from construction activities, which "harm the local tourism industry, pollute irrigation water that feeds durian and rice fields, and contaminate Kampot Town's water supply, extracted just downstream of the dam site".[43]

A similar situation would apply to Cheay Areng dam project, mentioned above, if it proceeds. According to one estimation, more than 1,500 indigenous people would have to relocate to places far away from their farms and ancestors' forests. The proposed dam would choke the river water and ruin downstream habitats for wild fish upon which local communities depend, and would alter natural seasonal flow of the Stung Cheay Areng, damaging the river itself and harming hundreds of hectares of rice paddy. Likewise, the reservoir of the dam would inundate the habitats of 31 endangered species, such as Siamese Crocodile, tigers, Asian elephants, pileated gibbons, and one of the world's most prized freshwater fish, the Asian arowana.[44]

In parallel, there are significant impacts on the environment of the areas where Chinese companies and joint ventures are holding economic land concessions under the banner of agrobusiness and industrial development. Notwithstanding the Cambodian government's argument claiming that economic land concessions for agriculture will produce green cover, the replacement of natural forests by man-made landscape cannot sustain the ecological balance. The massive eucalyptus plantations and paper mill in Pursat, run by Chinese State Farms Corporation and Pheapimex (mentioned above), are a worry as they could damage the area's ecology and the nearby Tonle Sap. According to British forestry specialist Hardcastle, the project has not followed the legal requirements for environmental review. While Hardcastle noted that "planting eucalyptus would likely have little negative environmental impact on the area, because the forest had already been logged in recent years", he also warned that "the pulp mill could cause serious harm if chemical processes were used, rather than the semi-mechanical method that has become more commonly used".[45] The project exemplifies the risk that companies have come in simply to log the forests or extract natural resources, exiting the country with handsome financial gains without investing in meaningful forms of development.

Popular Protest and Human Rights Violations

A severe result of land grabbing and negative social and environmental impacts is popular protest, often through bringing petitions to local and central government officials to seek regulatory intervention and fair compensation. As rural communities are deprived of proper land title and regulatory systems are not fully enforced, protests are common when land concessions and investment projects encroach on people's lands and negatively affect their livelihoods. Since land concessions extend over large areas, local villagers' houses and farmlands frequently fall inside their boundaries. Worse than that, without proper demarcation or monitoring systems, some economic land concessions extend far beyond what they are granted by the law, as noted above. In some cases, economic land concessions are approved by provincial and municipal authorities, while central government and related ministries are not well-informed about the projects.

When conflicts break out and people stage protests or petition central government to demand intervention, the situation becomes even more volatile. In some cases, villagers and activists are threatened or brought to courts. Villagers frequently end up in detention. Despite Prime Minister Hun Sen's directive measure to suspend all new economic land concessions and review existing land concessions in May 2012, the move has done little to improve the situation of land grabbing and protest in general.[46] Ownership deprivation and land dispossession are violations of fundamental human rights.[47] What makes problems worse is that victims often face lawsuits filed against them by the companies. According to a new research on "Human Security and Land Rights in Cambodia":

> Land insecurity affected people's livelihoods and increased physical and psychological insecurity. The prominent cause of insecurity was poverty, followed by land grabbing, corruption, lack of food, lack of land for next generation, and inadequate access to healthcare. Forced and distress-based land sales were also a central cause of land insecurity.[48]

Though popular protests seem to be less prevalent in recent years, roadblocks and other street demonstrations were frequent headlines of both local and foreign newspapers, especially before the national election in 2013. According to the human rights organization Licadho, 2,246 families have been victimized by "a renewed wave of violent land grabbing" since early 2014.[49] Chinese companies have often been targeted by popular protest. In 2002, villagers affected by the Pheapimex joint venture with Chinese State Farms Corporation (mentioned above) blocked the road with logs and their bodies when officials began to mark boundaries of the concessional lands.[50] In 2011, 18 village representatives and villagers in Krokor district received court summonses filed by Pheapimex accusing the villagers of incitement and destruction of property and preventing development of the economic land concessions.[51] Again in 2012, villagers protested outside a makeshift Pheapimex company office in Krokor district, Pursat province, to demand solutions to land-related issues.[52]

In another case where Pheapimex-Fuchan Ltd., a joint venture with a Chinese company, holds 6,800 hectares of economic land concession in Pursat province, resistance involved road blockades and filing lawsuits challenging the government. In 2004 when the company sent hundreds of workers and heavy equipment to clear the forest, hundreds of villagers demonstrated along National Highway No. 5 to block the deployment. Villagers also organized into small groups to closely watch the movement of machinery. During the night, an incident happened when someone threw a grenade into a group of sleeping villagers, injuring eight people.[53]

Forests clearance activities of Wuzhisan LS in Pursat and Mondulkiri have also faced similar protestations. According to Jon Buckrell of Global Witness, "months of land-grabbing, bulldozing of spirit forests and destruction of crops [by Wuzhisan] have driven members of the Phnong minority to a series of public protests which are now being met with threats by the security forces".[54] Despite a series of protests and filing complaints, deforestation activities still continue, mostly in secret. Wuzhisan LS was also involved in a labour conflict. In May 2011, more than 300 workers working for the project demonstrated to claim unpaid wages.[55]

The Kamchay dam, introduced above, has also generated controversy and popular protest. In November 2008, more than 70 families from two villages blocked a local road to the construction site.[56] In May 2011, representatives of more than 200 families living around the affected areas held a protest in front of Kampot provincial hall, claiming that Sinohydro had blocked a road which people used to travel to collect bamboo and vines needed for their basket-weaving business.[57]

The Cheay Areng dam, also introduced above, is currently one of the most controversial hydropower projects involving a Chinese company. Areng valley is located in the 400,000-hectare remote Central Cardamoms Protected Forest, one of the country's last pristine natural forests in Koh Kong province. The valley is also home to an old ethnic Chong community that has settled in the area for centuries. Dam construction in the area was first proposed in 2006 when a

Chinese company called China Southern Power Grid (CSG) signed a Memorandum of Understanding with the Ministry of Industry, Mines and Energy (MIME) to conduct a feasibility study.

In late 2014 and early 2015, local villagers, mostly ethnic Chong, organized a series of protests and erected a makeshift roadblock on the only road into the valley. The roadblocks were raised to prevent groups of officials and Chinese engineers from entering the dam site.[58] The government, in contrast, condemned the blockade as a possible attempt to create an "autonomous zone". Areng Valley activist and the founder of the NGO Mother Nature, Alejandro Gonzalez-Davidson, was deported back in late February 2015 because of his activities to raise awareness about the project. The deportation prompted heated social debates and outcries across social media, and in March, another Areng Valley activist and a community representative, Ven Vorn, was summoned for questioning at Koh Kong Provincial Court.[59] However, the controversy and protests have been calmed down after Prime Minister's announcement that the project will not be moved ahead until the next mandate in 2018.

Conclusion

Drawing on its "Going Global" strategy, articulated in 2001, Chinese engagement across Southeast Asia is often aimed at ensuring a supply of natural resources for its own industrial development. As a close ally of Beijing, Cambodia exemplifies this pattern. Through its aggressive investment in agribusiness, mineral extraction, and hydro dam construction, China is now the largest investor in natural resource extraction in Cambodia. As the cases above suggest, Chinese investment in natural resources extraction tends to receive relatively favourable treatment from Cambodia's government. Close ties between Beijing and Phnom Penh, among other things, help explain the facilitative business environment Chinese investors find in Cambodia. Aid and economic investment are important tools that China uses to draw Cambodia's government in line with its interests.

Chinese investment in natural resource extraction has helped Cambodia maintain high GDP growth in recent decades. Such investment provides Cambodia with finance needed to build major infrastructure projects, opportunity to diversify economic activities away from agriculture, employment opportunities, and an improved business climate. While many developed countries hesitate to invest in the risk-prone sectors, China is a risk-taking financer that is one of Cambodia's current options. On the other hand, given poor regulatory implementation and local corruption, benefits have not always trickled down to local people as often claimed by the government. Furthermore, various Chinese investments on natural resource extraction have produced different impacts on the environment and society, as laws are circumvented and environmental impact assessment mechanisms are taken for granted. Negative environmental impacts have included the loss of biodiversity (including wildlife and agro-diversity), food insecurity (crop damage), deforestation (resulting from illegal logging and forest clearance), and loss of vegetation cover. The social impacts have also been important. Extensive economic land concessions and land grabbing have resulted in displacement, loss of traditional livelihoods, and violations of human rights.[60]

Exacerbated by the lack of transparency and corrupt practices, favourable treatment given to Chinese companies as part of the deference to Beijing causes significant loss in national budget and prompts public opposition. Favourable treatment given to Chinese investment and China's willingness to tolerate and work with uneven regulation at best, and corruption at worst, has made Chinese investment a target of popular opposition. As the largest holder of economic land concessions, mining licenses and hydro dam construction projects, Chinese companies have been closely linked to illegal land grabbing, destruction of natural resources, deforestation and human rights abuses. These social issues have become the sensitive topics in Cambodian domestic politics, posting challenges to the CPP's government.

While China appears to benefit significantly from this state of affairs, this chapter has shown how the results for Cambodia have been mixed. To ensure more inclusive development within Cambodia, monitoring mechanisms, regulatory frameworks and legal procedures must be improved and upheld in accordance with existing laws and guidelines. Chinese laws and regulations aimed at Chinese companies' activities abroad can help as well. These include Protection Guidelines for Overseas Investment and Cooperation, co-released by the Ministry of Commerce and the Ministry of Environmental Protection, the Guide on Sustainable Overseas Silviculture by Chinese Enterprises (a voluntary measure released by the State Forestry Administration and the Ministry of Commerce) and the sustainable financing benchmark study for the

Chinese banking sector initiated by World Wildlife Fund in cooperation with China Banking Regulatory Commission, which provides some preliminary assessment of Chinese banks' performance in environmental protection, sustainable use and management of natural resources, and green lending.[61] Cambodia, similarly, has proper laws and regulations that could ensure sustainable development and resources management. Challenges remain, however, because laws are not effectively enforced by both sides. In addition, corruption, nepotism and weak regulatory enforcement are features of the Cambodian business context, and have helped Chinese companies circumvent regulations as well as insulate them from legal consequences back home. In the future, both governments should collaborate further on information sharing and legal enforcement to reduce unsustainable and corrupt practices related to investment in natural resource extraction.

NOTES

1. Special thanks to Mike Dwyer for his editorial assistance with this chapter.
2. Nayan Chanda, *Brother Enemy: The War After the War* (New York: Collier Books, 1988).
3. Josh Kurlantzick, *Charm Offensive: How China's Soft Power is Transforming the World* (New Haven: Yale University Press, 2007); Darryl S.L. Jarvis and Anthony Welch, *ASEAN Industries and the Challenges from China* (Hamshire: Palgrave Macmillan, 2011).
4. Also see Yu Hongyuan and Tai Wei Lim in this volume.
5. Sigfrido Burgos and Sophal Ear, "China's Strategic Interests in Cambodia: Influence and Resources", *Asian Survey* 50, no. 3 (2010): 615–39, available at <http://www.jstor.org/stable/10.1525/as.2010.50.3.615>.
6. Narin Sun, "Cambodia's Hun Sen Slams U.S. Threats over Aid", *The Wall Street Journal*, 3 August 2013, available at <http://stream.wsj.com/story/latest-headlines/SS-2-63399/SS-2-293500/>.
7. ASEAN Secretariat, *ASEAN Community in Figures (ACIF)* (Jakarta: ASEAN Secretariat, 2014); Hongmei Hao, "China's Trade and Economic Relations with CLMV", in *Development Strategy for CLMV in the Age of Economic Integration*, edited by Chap Sotharith (Chiba: IDE-JETRO, 2008), pp. 171–208.
8. Xinhua, "Chinese Investment in Cambodia up in 2013", 18 January 2014, available at <http://www.globaltimes.cn/content/838148.shtml>.
9. Frik Els, "Slowdown. What Slowdown? China's Copper, Iron Ore Imports Set Records", mining.com, 13 October 2013, available at <http://www.mining.com/slowdown-what-slowdown-chinas-copper-iron-ore-imports-set-

records-99571>; Naomi Basik Treanor, "China's *Hongmu* Consumption Boom: Analysis of the Chinese Rosewood Trade and Links to Illegal Activity in Tropical Forested Countries", *Forest Trends*, December 2015, available at <http://www.forest-trends.org/documents/files/doc_5057.pdf>.

10. Jef Rutherford, Kate Lazarus, and Shawn Kelley, *Rethinking Investments in Natural Resources: China's Emerging Role in the Mekong Region* (Phnom Penh: Heinrich Böll Stiftung, WWF and International Institute for Sustainable Development, 2008).

11. John Agnew, "Sovereignty Regimes: Territoriality and State Authority in Contemporary World Politics", *Annals of the Association of American Geographers* 95, no. 2 (2005): 437–61.

12. Cambodia Investment Board, "Investment Trend", 2015, available at <http://www.cambodiainvestment.gov.kh/investment-enviroment/investment-trend.html>.

13. Thomas Lum, Hannah Fischer, Hulissa Gomez-Granger, and Anne Leland, "China's Foreign Aid Activities in Africa, Latin America, and Southeast Asia", *Congressional Research Service: Report for Congress* (25 February 2009).

14. Sophie Richardson, *China, Cambodia, and the Five Principles of Peaceful Coexistence* (New York: Columbia University Press, 2010).

15. Chanda, *Brother Enemy*; David P. Chandler, *The Tragedy of Cambodian History: Politics, War, and Revolution since 1945* (New Haven: Yale University Press, 1991).

16. Martin Vaughan, "Cambodia's Hun Sen Proves a Feisty Asean Chair", *World Street Journal*, 4 April 2012, available at <http://blogs.wsj.com/searealtime/2012/04/04/cambodias-hun-sen-proves-a-feisty-asean-chair/>.

17. Xinhua, "Cambodia Opens China-funded Bridge for Traffic", 24 January 2011, available at <http://www.chinadaily.com.cn/china/2011-01/24/content_11907394.htm>.

18. Caroline Hughes, "Transnational Networks, International Organizations and Political Participation in Cambodia: Human Rights, Labour Rights and Common Rights", *Democratization* 14, no. 5 (2007): 834–52; Caroline Hughes, "Cambodia in 2007: Development and Dispossession", *Asian Survey* 48, no. 1 (2008): 69–74; Andrew Robert Cock, "External Actors and the Relative Autonomy of the Ruling Elite in Post-UNTAC Cambodia", *Journal of Southeast Asian Studies* 41, no. 2 (2010): 241–65.

19. Kheang Un, "State, Society and Democratic Consolidations: The Case of Cambodia", *Pacific Affairs* (2006): 225–45.

20. See John D. Ciorciari, "China and Cambodia: Patron and Client?", IPC Working Paper no. 121, International Policy Center, Gerald R. Ford School of Public Policy, University of Michigan, 2013, available at <http://ipc.umich.edu/working-papers/pdfs/ipc-121-ciorciari-china-cambodia-patron-client.pdf>.

21. A prominent title given to influential business leaders who contribute significantly to the country and possess intimate connection with the government.
22. Ciorciari, "China and Cambodia: Patron and Client?".
23. Un. "State, Society and Democratic Consolidations", pp. 225–45.
24 Also see Ciorciari, "China and Cambodia: Patron and Client?".
25. Xinhua, "China Provides Military Trucks and Uniforms to Cambodia", *People Daily*, 7 February 2014, available at <http://english.peopledaily.com.cn/90786/8528898.html>.
26. Bopha Phorn, "Defense Minister Says 400 Personnel Will Soon Study in China", *Cambodia Daily*, 12 May 2014, available at <http://www.cambodiadaily.com/news/defense-minister-says-400-personnel-will-soon-study-in-china-58462/>.
27. Ciorciari, "China and Cambodia: Patron and Client?" .
28. Ibid.
29. Amy Kazmin, "China Boosts Cambodian Relations with $600m Pledge", *Financial Times*, 10 April 2006, available at <https://www.ft.com/content/127cb9fa-c7fa-11da-a377-0000779e2340>.
30. Sokha Cheang, "China Steps In with Lorries", *Phnom Penh Post*, 3 May 2010, available at <http://www.phnompenhpost.com/national/china-steps-lorries>.
31. May Titthara, "Kings of Concessions", *Phnom Penh Post*, 25 February 2014, available at <http://www.phnompenhpost.com/national/kings-concessions>.
32. Xinhua, "Chinese-built 338 MW Hydropower Dam in Cambodia Begins Operation", *Xinhuanet*, 12 January 2015, available at <http://news.xinhuanet.com/english/china/2015-01/12/c_133913369.htm>.
33. World Bank, *Seizing the Global Opportunity: Investment Climate Assessment and Reform Strategy for Cambodia* (World Bank Group, 2014).
34. Transparency International, "Corruption Perception Index 2014: Results", 2014, available at <https://www.transparency.org/cpi2014/results> (accessed 13 September 2017).
35. May Titthara, "China Reaps Concession Windfalls", *Phnom Penh Post*, 12 April 2012, available at <http://www.phnompenhpost.com/national/china-reaps-concession-windfalls>.
36. Rutherford, Lazarus, and Kelley, *Rethinking Investments in Natural Resources*.
37. Luke Hunt, "Cambodia's Well-Heeled Military Patrons", *The Diplomat*, 10 August 2010, available at <http://thediplomat.com/2015/08/cambodias-well-heeled-military-patrons/>.
38. Richard Sine and Phann Ana, "Cambodia's Largest Land Concession Poses Ominous Threat to Environment", *Cambodia Daily*, 6 April 2002, available at <https://www.cambodiadaily.com/archives/cambodias-largest-land-concession-poses-ominous-threat-to-environment-30950/>.

39. Environmental Justice Atlas, "Pheapimex–Fuchan Conflict, Cambodia", 20 April 2014, available at <https://ejatlas.org/conflict/pheapimex-fuchan-conflict-cambodia>.

40. May Titthara, "Fear Accompanies Summons Over Land Disputes", *Phnom Penh Post*, 20 June 2012, available at <http://www.phnompenhpost.com/national/fear-accompanies-summons-over-land-disputes>.

41. Cambodia Office of the High Commissioner for Human Rights, *Land Concessions for Economic Purposes in Cambodia: A Human Rights Perspective* (Phnom Penh: Cambodia Office of the High Commissioner for Human Rights, 2004).

42. Frauke Urban, Johan Nordensvard, Giuseppina Siciliano, and Bingqin Li, "Chinese Overseas Hydropower Dams and Social Sustainability: The Bui Dam in Ghana and the Kamchay Dam in Cambodia", *Asia & the Pacific Policy Studies* 2, no. 3 (2015): 573–89.

43. International Rivers, "Cheay Areng Dam", 2014, available at <http://www.internationalrivers.org/campaigns/cheay-areng-dam>. Also see Peter Zsombor and Narim Khoun, "Hydro Dam Does Little for Locals, Study Finds", *The Cambodian Daily*, 26 August 2015, available at <https://www.cambodiadaily.com/news/hydro-dam-does-little-for-locals-study-finds-92523>, and Urban, Nordensvard, Siciliano, and Li, "Chinese Overseas Hydropower Dams and Social Sustainability", pp. 573–89.

44. International Rivers, "Cheay Areng Dam".

45. Sine and Ana, "Cambodia's Largest Land Concession Poses Ominous Threat to Environment".

46. RFA's Khmer Service, "Half a Million Cambodians Affected by Land Grabs: Rights Group", 1 April 2014, available at <http://www.rfa.org/english/news/cambodia/land-04012014170055.html>.

47. Natalie Bugalski and Ratha Thuon, "A Human Rights Impact Assessment: Hoang Anh Gia Lai Economic Land Concessions in Ratanakiri Province, Cambodia", International Academic Conference, RCSD Chiang Mai University, Chiang Mai, 2015, pp. 1–24.

48. Alice Beban and Sovachana Pou, *Human Security and Land Rights in Cambodia* (Phnom Penh: Cambodia Institute for Cooperation and Peace, 2015).

49. RFA's Khmer Service. "Half a Million Cambodians Affected by Land Grabs: Rights Group".

50. Sine and Ana, "Cambodia's Largest Land Concession Poses Ominous Threat to Environment".

51. Titthara, "Fear Accompanies Summons Over Land Disputes".

52. May Titthara, "Pheapimex Under Fire Again", *Phnom Penh Post*, 17 January 2012, available at <http://www.phnompenhpost.com/national/pheapimex-under-fire-again>.

53. Environmental Justice Atlas, "Pheapimex–Fuchan Conflict, Cambodia".

54. Jon Buckrell, "Natural Resource Governance – A Test of Political Will for the Cambodian Government and the International Donor Community", *Global Witness*, 2004, available at <https://www.globalwitness.org/archive/natural-resource-governance--test-political-will-cambodian-government-and-international/>.
55. Environmental Justice Atlas, "Pheapimex–Fuchan Conflict, Cambodia".
56. International Rivers, "Cheay Areng Dam".
57. Sokheng Vong and Sebastian Strangio, "Villagers Blockade Kampot Dam Quarry Site Over Airborne Rocks", *Phnom Penh Post*, 11 March 2009, available at <http://www.phnompenhpost.com/national/villagers-blockade-kampot-dam-quarry-site-over-airborne-rocks>.
58. Pichamon Yeophantong, "Cambodia's Environment: Good News in Areng Valley?", *The Diplomat*, 3 November 2014, available at <http://thediplomat.com/2014/11/cambodias-environment-good-news-in-areng-valley/>.
59. May Titthara and Alice Cuddy, "Areng Valley Dam Activist Summonsed", *Phnom Penh Post*, 27 March 2015, available at <http://www.phnompenhpost.com/national/areng-valley-dam-activist-summonsed>.
60. Environmental Justice Atlas, "Pheapimex–Fuchan Conflict, Cambodia".
61. Global Environment Institute, *Environmental and Social Challenges of China's Going Global* (Beijing: China Environment Press, 2013).

9

COMPLEX CONTESTATION OF CHINESE ENERGY AND RESOURCE INVESTMENTS IN MYANMAR

Diane Tang-Lee

Many Chinese large-scale investments, particularly those in the energy and natural resource sectors, have encountered nationwide protest in Myanmar. These protests have radically called for project terminations, which partly resulted in President Thein Sein's decision to suspend construction on the Chinese-backed mutli-billion dollar investment in the Myitsone dam in 2011. This incident has been widely regarded as a turning point in what had hitherto been seen as close relations between China and Myanmar. Bringing the research gaze down to the local level, this chapter examines the context and conditions of interactions between key Chinese actors and Myanmar civil society. It argues that a restrictive legal and political context, organizational limitations, and dismissive attitudes towards civil society have engendered mutual suspicion and mistrust between local Myanmar activists and Chinese companies. This, in turn, has hindered the development of more formal or regular communication channels between actors and disadvantaged Chinese efforts to build reputation and trust within the wider Myanmar public.

Introduction

In recent years, under the democratizing environment in Myanmar, civil society[1] has mounted forceful campaigns against irresponsible foreign investment practices, which have included Chinese, Thai, Indian and South Korean companies. However, Chinese investments account for about a third of Myanmar's total cumulative foreign investment of US$62.6 billion, as of 2017.[2] Most of the protest has been in the energy and natural resource industries, whose propensities for extensive socio-environmental impacts have encountered the most resistance. Much of it has been labelled and, indeed, presented itself as "anti-Chinese". As this chapter demonstrates, however, the causes of protest in Myanmar are multiple and complex.

Protests have played a critical role in compelling Chinese interlocutors — including state-companies, the embassy in Yangon, the Chinese Ministry of Commerce, and Chinese scholars — to confront the demands of Myanmar civil society. This chapter will first provide a brief overview of Chinese investments in Myanmar, followed by an examination of anti-Chinese sentiments as they have emerged through complex historical relations, local politics and Chinese investment practices. It analyses how Chinese interlocutors, whose conventional diplomatic and business approach is to deal predominantly with host governments, perceive and interact with local civil society actors in Myanmar, especially as their protests make these interactions inevitable.

Examining the legal and political contexts in Myanmar that shape these interactions in particular ways, this chapter finds that although local activists were successful in pressuring the Myanmar government to suspend construction of the Chinese state-backed Myitsone dam in 2011 and the Letpadaung copper mining project in 2012 (resumed after a parliamentary investigation led by Aung San Suu Kyi), these activists were seen as illegitimate in the eyes of Chinese interlocutors. It argues that a restrictive legal and political context, organizational limitations, and dismissive attitudes towards civil society have engendered mutual suspicion and mistrust between local Myanmar activists and Chinese companies. This, in turn, has hindered the development of more formal or regular communication channels between the two sides and disadvantaged Chinese efforts to build reputation and trust within the wider Myanmar public.

Contesting Chinese Energy and Resource Investments

Under the rhetoric of mutually beneficial or "win-win" partnerships (*China Daily* 2011; Yu 2012), China has invested at least US$14 billion in Myanmar over the past few decades, putting China at the top of a list of 32 countries investing in Myanmar.[3] Notably, 54 per cent of these investments are in the energy and natural resource sectors.[4] Many of them are backed by Chinese state-owned enterprises (SOEs). Some of the largest have been the US$3.6 billion Myitsone dam, the Letpadaung copper mine, with an estimated investment of US$1.065 billion,[5] and the US$2.54 billion Sino–Myanmar oil and gas pipelines.[6]

While the Myanmar government will receive various forms of tax and fee revenues, most of the energy and natural resources to be extracted from these projects will go to China. Ninety per cent of the electricity generated by the Myitsone dam is expected to serve the neighbouring Yunnan Province in China. For the pipelines, the project agreement allows for 20 per cent of its natural gas to go to Myanmar for local consumption[7] (see Table 9.1 for a list of these projects). Together with these high-profile and controversial projects, which have elicited much public outcry, there were no less than 69 Chinese companies involved in 90 completed, current and planned projects in Myanmar's extractive and hydropower sectors in 2012, and about ten other Chinese companies involved in projects outside of resources, such as telecoms and infrastructure.[8] These projects have been able to thrive amid the

TABLE 9.1
Project Information and Groups Contesting the Three Major Chinese-backed Extractive Projects in Myanmar

Project	Chinese SOE's Ownership	Other Investors	Name of Joint Venture Entity	Estimated Project Cost	Other Project Information	Examples of Civil Society Groups[a] Contesting the Projects
Myitsone dam	China Power Investment Corporation (CPI): 80 per cent	Myanmar Ministry of Electric Power, Asia World Co.	Upstream Ayeyawady Confluence Basin Hydropower Co Ltd.	US$3.6 billion	Set to provide 21 gigawatts of energy to China's Yunnan Province (Mahtani 2014).	Burma Rivers Network[b]

TABLE 9.1 (*continued*)

Project	Chinese SOE's Ownership	Other Investors	Name of Joint Venture Entity	Estimated Project Cost	Other Project Information	Examples of Civil Society Groups[a] Contesting the Projects
Letpadaung copper mine	Wanbao Mining, a subsidiary of North Industries Corporation (Norinco), a weapons manufacturer	Union of Myanmar Economic Holdings Ltd. (UMEHL), a Myanmar military-owned conglomerate	Myanmar Wanbao Mining Copper Ltd.	US$1.065 billion (Sun 2013)	92,500 tonnes of ore will be mined daily, expected to produce 100,000 tonnes of copper per annum.	Save Letpadaung Mountain Committee[c]
Sino–Myanmar oil and gas pipelines	China National Petroleum Corp (CNPC): 50.9 per cent	Oil: Myanmar Oil and Gas Enterprise (MOGE) Gas: Daewoo, Indian Oil, MOGE, Korea Gas, GAIL	Oil: South-east Asia Crude Oil Pipeline Co. Ltd. Gas: South-east Asia Gas Pipeline Co. Ltd.	US$2.54–5 billion (Chen 2013)	About 800 km pipelines, which pass through 21 townships across Myanmar and terminate in China's Yunnan Province.[d]	Myanmar–China Pipeline Watch Committee,[e] Thazin Development Foundation,[f] Shwe Gas Movement,[g] Badeidha Moe Civil Society[h] Organization

Notes:
[a] Groups that advocate across these projects include: Economically Progressive Ecosystem Development (known as EcoDev), a well-recognized environmental NGO. Its co-founder and managing director, Win Myo Thu, is known for his relevant technical background and is Myanmar's leading environmental educator; Paung Ku, a civil society strengthening initiative; and JU Foundation, which tackles environmental issues in Myanmar.
[b] Comprised of organizations representing various dam-affected communities in Myanmar.
[c] Formed by people from nearby communities and is subsequently headed by a resident of the nearby town of Monywa.
[d] They are expected to deliver 22 million tonnes of oil and 12 billion cubic metres of natural gas to China per annum (Chen 2013) and expected to raise US$900 million (excluding the transport tariff) in foreign exchange earnings each year. A further US$900 million annually is expected to arise from the sale of natural gas to China over the next thirty years, bringing the total revenue generated by the project to US$1.8 billion each year (Zhao 2011, p. 100).
[e] Given the scale and spread of the Sino–Myanmar pipelines, which spanned the country from West to East, across different ethnic groups along the borders and the Burman heartland, around two dozen CSOs based across the country joined forces to form the Myanmar–China Pipeline Watch Committee (MCPWC).
[f] Based in Rhakine State, where the pipelines start. It conducted research on the project's impact on local residents and the environment.
[g] Founded by the Arakan Oil Watch (AOW), based out of Chiang Mai, Thailand, All Arakan Students and Youth Congress (AASYC), an exiled Arakanese group with offices in Bangladesh and Chiang Mai and Mae Sot in Thailand, and EarthRights International (ERI), founded by a Karen exile and two US lawyers to fight for human rights and environmental justice.
[h] Conducts research and organizes photo exhibitions demonstrate the negative impacts of foreign investments and special economic zones.

lack of competition from US and European transnational corporations because of government sanctions against Myanmar in previous decades.

Overall, Chinese economic interests in Myanmar include substantial investments in a range of hydroelectric power schemes and mineral extraction, as well as cross border trade that helps boost the economy of China's landlocked southwestern provinces. Chinese and international investors have been eying Myanmar's natural resource endowments, with proven natural gas reserves in excess of 7.8 trillion cubic feet[9] and proven oil reserves of more than 50 million barrels.[10] Given the quasi-civilian regime's opening up to the West, Myanmar is also a focal point of US–China competition for economic opportunities and strategic influence in the region.[11]

Beyond economic cooperation, Myanmar is of unique geopolitical importance to China. Its location between China and the Indian Ocean allows oil imports from the Bay of Bengal, Middle East and Africa to be transported through the Sino–Myanmar pipelines to China, as well as boosting the economy of China's impoverished southwest region.[12] The oil and gas pipelines, which began operations in 2015 and 2014 respectively, would help China resolve what Chinese officials and academics have referred to as their "Malacca Dilemma".[13] An estimated 74–80 per cent of China's total oil import is shipped through the narrow Strait of Malacca, near Singapore, where the US Navy has a strong presence, and through the South China Sea, where territorial disputes with Southeast Asian neighbours have been intensifying.[14]

Hence, some expect the Sino–Myanmar pipelines to ensure China's future energy security while expanding its geopolitical influence, which has been widely regarded as China's ambition for this project.[15] Alongside the pipelines, plans to build a railway would further open up a southwest corridor to the Indian Ocean, which is another route for crucial imports that could bypass the Strait of Malacca. This transport route is of strategic importance to China's overall westward strategy and energy import via the Indian Ocean. However, the plan has also been stalled since the suspension of the Myitsone dam, further reflecting the wider impacts of local opposition to that project.[16]

The reasons for opposition to the Myitsone dam, Letpadaung copper mine, and Sino–Myanmar pipelines are complex, yet they stem from similar root causes. The projects lack accountability and transparency in disclosing destination of revenues and benefits. Communities became victims to land grabbing, forced evictions, unacceptable living standards in relocation sites, inadequate compensation, and environmental destruction. The scale of impact of the Myitsone dam would be enormous as it would flood an area of the Irrawaddy River Valley equivalent to the size of Singapore,[17] leading to large-scale deforestation and forced displacement of 10,000 ethnic Kachin villagers.[18] A further aggravator was that only 10 per cent of the electricity generated was to remain in Myanmar, with the remainder to be exported to China.

Near the Letpadaung copper mine, confiscation of some 3,000 hectares of land resulted in forced relocation of around 200 households.[19] The dumping of large amounts of contaminated soil on disputed land[20] resulted in loss of livelihoods that has not been compensated by employment opportunities. Grievances are further aggravated by the disappearance of the Letpadaung Mountain, which contained sites of historical and religious significance to the local people.

Similarly, along the Sino–Myanmar oil and gas pipelines, there are numerous reports[21] of environmental destruction, including loss of farmland and crops with little or no compensation, leaving water contamination, and damaged forests.[22] These negative impacts came as a shock to many affected people who were not previously consulted about the projects.[23, 24]

Complex Contexts of Anti-Chinese Sentiment

The problems of protest and resistance to Chinese energy and resource projects cannot be solely attributed to Chinese investors. In addition to the specific environmental, social and governance problems of the projects mentioned above, anti-Chinese sentiment has also emerged as a reflection of Myanmar's weak governance and ethnic tensions.

Familiar charges of corruption, rent-seeking, and lack of transparency and accountability are by no means unique to Chinese-backed projects. Corrupt rent-seeking behaviour flourished under the sixty years of military regime in Myanmar. The Revenue Watch Institute ranked Myanmar last among 58 nations evaluated for resource governance, scoring lowest in institutional and legal setting, reporting practices, and enabling environment, and second lowest in safeguards and quality controls.[25] Environmental and Social Impact Assessments (ESIA), mandatory in many countries, are not required under the laws of Myanmar. Ambiguities in land tenure and usage rights are also deep-seated problems in Myanmar, which have contributed to land grabs through logging, special economic zones, agribusinesses, and China's opium-substitution programme in northern Myanmar, among other commercial activities.[26]

Furthermore, extractive activities disproportionately affect Myanmar's ethnic minority groups, which are concentrated in Myanmar's border areas where most of the country's remaining natural resources are found. Border communities, including those affected by the Myitsone dam, bear the brunt of the social and environmental costs of these projects, such as forced relocation to places far away from the original settlement, where land is often not arable, and sometimes without receiving any compensation.[27] The Myitsone dam is located in the Kachin state, where a 17-year cease-fire between the ethnic militia and the Myanmar army collapsed in 2011, shortly before the President announced the dam's suspension. The Kachin Independence Army (KIA), the largest non-state armed group in Myanmar, controls parts of the territory near the Myanmar border with China. Over the past two decades, the independence movement has evolved into a reasonably functional de facto government operating in Kachin. The Kachin Independence Organization (KIO), the political arm of KIA, has "established a modest social safety net in the form of a Kachin-language school system, free basic health care, and reliable utilities".[28]

Therefore, Chinese companies' preferences to deal with Myanmar's central government and their affiliated military networks from the previous junta — based on Beijing's principles of non-interference — tend to exacerbate ethno-political tensions. With the justification that "ethnic groups are part of separatist movements that lack formal political recognition",[29] project consultations and negotiations tend to exclude or marginalize the groups that are most affected by them. As ethnic minority groups make up 30 to 40 per cent of Myanmar's estimated population of 60 million, problems are bound to emerge.[30] In some cases, ethnic conflicts have been worsened by the insensitivity of investment projects. The Myitsone dam construction, for example, has indirectly exacerbated tensions between the Myanmar government and the KIO, which controls many areas near the dam, as the military and the KIO compete for control over areas with Chinese investment.[31]

The Myanmar companies involved in these controversial China–Myanmar joint ventures have eroded public trust and raised questions about their legality. For the oil and gas pipelines, the state-owned Myanmar Oil & Gas Enterprise (MOGE) is believed to be a primary vehicle through which resource revenues are channelled towards the Myanmar military.[32] For the Myitsone dam, the Myanmar counterpart, Asia World Group, is a company that emerged from a drug cartel led by the recently deceased Lo Hsing Han and which is currently under the leadership of his son Stephen Law, who is also involved in a wide range of industrial and construction-related investments.[33] For the Monywa copper mine, the Union of Myanmar Economic Holdings Ltd (UMEHL) is a Myanmar military-owned conglomerate, while the Chinese SOE Norinco is a weapons manufacturer. This arrangement has aroused suspicion around possible copper-for-arms agreements between the two companies.[34]

Apart from the problems resulting from these extractive projects themselves, Myanmar people's grievances towards China also stem from deeper discontent relating to China's economic cooperation with and perceived protection of the former military regime. China continued to support the military junta despite western sanctions and shielded the junta from international condemnations by exercising its veto power in the UN Security Council. Local people perceive China as providing diplomatic support for their corrupt and repressive government. A case in point is the Shwe gas project. Myanmar had received a higher bid from India for the price of gas than China offered. However, in 2007, China vetoed a Security Council resolution calling for the Myanmar military to stop the persecution of minority and opposition groups. Many observers believed that this resulted in Myanmar finally opting to sell the gas to China at a lower price.[35]

Anti-Chinese sentiments are also deep-seated as a result of other issues, such as Chinese immigrants' perceived domination of local businesses and being "at the top of the pyramid" in northern Myanmar and Mandalay, which is Myanmar's second largest city and built in what was the heartland of successive kingdoms.[36] There are also often complaints about cheap and low-quality imports from China.[37] These words of an interviewee illustrate perceptions of Chinese interests threatening Myanmar culture and livelihoods:

> In a place near Pangwa, Chinese people speak in Chinese, use Chinese currency... our cultural values are destroyed, it is our territory but Chinese people invaded it there... In Mandalay, there are lots of Chinese people investing with corrupt money.

> *– Interview with a local NGO staff member, Yangon, 7 March 2013*

Negative sentiments towards the Chinese are not simply a reflection of their investment in energy and resource projects. They result from a complex mixture of socio-political and historical factors. These factors help to explain the complex contexts out of which anti-Chinese sentiment has arisen in response to Chinese investments in the energy and resource sectors.

Mutual Mistrust and Suspicion

The scale, frequency, longevity and impact of protests in Myanmar have far exceeded what would generally be allowed within China, given the Chinese government's effective "stability maintenance" efforts for dissolving protests.[38] How then do Chinese interlocutors respond to this situation? When asked whether Myanmar activists were able to voice their concerns directly to Chinese interlocutors, some expressed that they have no channels through which to do so. They reported that only a few select Civil Society Organizations (CSOs) have had a chance to meet with the Chinese companies face-to-face, but such communications have been informal and irregular. This section analyses the context and constraints faced by Myanmar civil society to explain the mutual mistrust that has emerged and lack of effective communication between local Myanmar activists and Chinese interlocutors.

Lack of Legal Status due to State Repression

One of the primary constraints facing Myanmar civil society, which affects their interactions with the Chinese, is their lack of formal legal status. Indeed, many Myanmar CSOs, including the official (government recognized) participants in the multi-stakeholder group (MSG) of the Myanmar Extractives Industry Transparency Initiative (MEITI) have not registered with the government. Under the military-era association law, only recently repealed in 2014, many CSOs had no means by which to register officially and legally. A manager of a Chinese SOE the author spoke with in Yangon in 2014 expressed his hesitation and suspicion towards these "so-called" NGOs partly because they have no official registration with the government, making them technically illegal. His comments suggest that lack of legal status is a major impediment to activist groups. It discredits them in the eyes of Chinese SOEs and hinders a more productive dialogue between the two sides.

In the five decades of military rule that ended in 2011, civil society organizations, including trade unions, student unions, and advocacy groups, were declared illegal and suppressed by the authorities under various laws, most notably the colonial-era Unlawful Associations Law 1908. Under that law, any association can be declared unlawful by the head of state. Opposition groups and peaceful opposition activists alike have been imprisoned on the basis of this law.[39] In December 2011 the Parliament enacted the Law Relating to Peaceful Assembly and Peaceful Procession, which permits peaceful assembly for the first time in decades. A further amendment in 2014 obliges the authorities to grant permission for peaceful demonstrations unless there are "valid reasons" not to do so. However, the amended law still provides for the arrest and imprisonment of peaceful protesters for failing to seek prior permission. It has been used to stifle dissent and hamper the ability of affected communities to voice their concerns on the implications of resource extraction for their livelihoods. In 2013, ten people who protested in Rakhine State against the Sino–Myanmar pipelines were arrested and charged with demonstrating without a permit under Section 18 of this legislation. The number of arrests has only increased in 2014.[40]

Another repressive law is the Law Relating to Formation of Associations and Organizations (Law No. 6/88), which was issued by the military regime known as the State Law and Order Restoration Council (SLORC) after it seized power following a coup in 1988. These restrictions effectively banned any civil society organization from registering unless it maintained close ties with the government. Groups that failed to register risked being deemed illegal and members could be penalized under the Unlawful Associations Act. In 2014, along with the democratization reforms, parliament repealed this law and enacted the Law Relating to the Registration of Organizations, which sets out the requirements for NGO registration in its place. More importantly, it does not contain any restrictions or criminal punishments for unregistered NGOs. It has been praised as being "very progressive" by civil society.[41]

Despite some recent progress, arrests are still common and activists face unending uncertainty as to the limits of their newfound political freedom. One activist expressed that during their campaign against a potential Chinese-backed project, a government official asked him whether his organization was legally registered. He felt threatened by the question to silence himself (Interview, Yangon, 3 March 2014). Another activist, who works on raising awareness about a Chinese project among local villagers, said that in order to defend herself from arrest, she would carry copies of the section of the constitution that protects her freedom of expression (Interview, Yangon, 6 February 2013). Operating under considerable state repression during the successive military and quasi-civilian regimes, Myanmar civil society's activities have not been protected by law and, thus, they have also been be perceived to be "illegal" by Chinese interlocutors.

Vulnerability to Dismissive Attitudes towards Civil Society

In addition to questions on legal status, CSOs are also limited by their relative lack of capacity, which could be used as part of a strategic discourse aimed to discredit activists. Although local civil society in Myanmar is now vibrantly proliferating, it is a relatively new phenomenon. Many CSOs were formed as service delivery organizations after the Cyclone Nargis hit in 2008. It was only later that they started to form activist groups to campaign against social and environmental injustices. Examples include the Myanmar–China Pipeline Watch Committee and the Thazin Development Foundation. They have conducted research into project impacts on local residents and the environment. They have also used their research findings as evidence of environmental destructions, human rights violations, and lack of transparency in disclosing project plans to the public.

Given that these activities were in recent years, most activists have not had the opportunity to be formally trained, except perhaps for a few elite environmental NGOs.[42] Prior to 2011 and 2012, much activism in Myanmar was driven by its exiled or displaced communities based in the border regions and northern Thailand. At that time, exiled and international media played the role of transmitting the voices of society, and only recently has local media gained importance and become accessible to the public (Interview with a local NGO, Yangon, 2 March 2014). However, those inside the country have also only recently been exposed to these opportunities to express their concerns in the local media. Not long ago activist CSOs operated "without electricity or internet access, lacking of information and knowledge, [and] people were afraid to talk and could get arrested (for dissent)" (Interview with an activist on the Sino–Myanmar pipelines, Yangon, 9 March 2013). Furthermore, in a country that is rapidly changing, understandably, many CSOs find it challenging to keep up with newly emerging problems (Interview with a local environmental NGO staff, Yangon, 7 March 2014).

CSOs' relative newness and limited resources make them more vulnerable to dismissive attitude. They are more prone to being discredited on the basis of lacking a sound scientific evidence base for their advocacy, an accusation that many companies make to confrontational NGOs in general. A manager of a Chinese SOE expressed his view that the reports of Myanmar CSOs are biased and do not convey the real facts of the projects. He shunned their reports as less professional than the official ESIA report. His company accepted interviews from *The New York Times* and the BBC, which he believed to be more impartial than reports from Myanmar CSOs. He also said that he welcomed opinions from NGOs if their objective was for public benefit, but that Myanmar lacked such NGOs capable of making objective recommendations to them (Interview, Yangon, 7 March 2014). While there may be elements of truth, it reflects a deeper structural condition that the lack of open communication led to mistrust and suspicion.

Many SOE managers and policy advisors interviewed expressed frustration over civil society demands as unreasonable. While they were cognisant of the need to understand the demands of these CSOs, they also worry that the newfound freedom of expression under democratization is being misused. One academic at the Institute of Southeast Asian Studies, Yunnan Academy of Social Sciences suggested that a lot of NGOs might still be immature and rely on misinformation, or be taken advantage of by people with political motives (Interview, Yunnan, 8 August 2013). Similarly, a leader of a Myanmar registered NGO also expressed that "people may not be mature enough to enjoy such freedom. So when there are campaigns, people just join in. No one knows what the root causes are, they just jump into the hot issues" (Interview, Yangon, 5 March 2013). Chinese interlocutors' assumptions about their lack of capacity hinder more productive engagement and communication. It leads to further mutual misconceptions and suspicion, which will be further explored below.

Perceived Lack of Legitimacy

Some NGO activists are also perceived to be illegitimate because their funding source and motivations are unclear to Chinese interlocutors. Similar to the rhetoric of the Chinese authorities that activities undermining China's interest (especially one-party rule) are instigated by the West, Chinese SOE officials and academics involved in Myanmar also often expressed concerns that some of the NGOs rely on western funding. This made them question the backgrounds and political motivations of those groups. Dong Yunfei, Admin Manager of Wanbao, pointed out that opposition groups such as 88 Generation[43] influenced and mobilized project-affected villagers with their own political motivations (Interview, Yangon, 27 February 2014). Another Chinese SOE manager accused them of fabricating accounts in order to damage the reputation of Chinese companies (Interview, Yangon, 7 March 2014). His response is worth quoting in full:

> We are very suspicious towards the legitimacy of some so-called NGOs and also their background and financial sources. They do not evaluate the project objectively, and they only point out the negative points and problems. If they were more objective, we can accept it. They criticized the material and technology used for the project and published the criticisms in the newspapers. What are their intentions? There is only one intention – to stop the project. They could suggest what else we could do, but what they do is only to destroy our reputation.
>
> – *Interview, Yangon, 7 March 2014*

A report published by the Chinese Academy of Social Sciences, a government think-tank, claims that "western-funded" activist groups are wrongly accusing Chinese companies of causing environment destruction and creating social problems to confine China's economic influence in countries along the Mekong River.[44] They "bear Western ideology and are deeply influenced by Western politics [... and] they tend to over-emphasise the significance of environmental protection, while ignoring Mekong countries' demand for economic development, threatening the sovereign rights of these countries".[45] Similarly, Xiong Liying, a researcher at the Yunnan Academy of Social Sciences believes that the US and European media irresponsibly exaggerated and sensationalized the negative impacts of the Myitsone dam, and US and EU organizations have increased their presence in Myanmar since 2011. Myanmar activists draw their opinions and evidence from US and European media to oppose the Myitsone dam.[46]

Shortly after the suspension of the Myitsone dam construction, the vice-secretary general of the China Society for Hydropower Engineering, Zhang Boting, wrote articles, posted on the website of China Power Investment Corporation (CPI), the Myitsone dam investor, that reflect his assertion that Western powers are determined to contain China, and anti-Myitsone dam organizations have become their tools and leading protagonists. He points to Wikileaks for evidence that the US has supported organizations that obstructed the Myitsone dam.[47] Jiang Heng, a senior at the Chinese Academy of International Trade and Economic Cooperation (CAITEC, a research institute) of MOFCOM cautions that this kind of judgement shadowed by conspiracy theory, although supported by evidence, has inspired further anti-Chinese and pro-US sentiments among the Myanmar educated class, and only conceals the many inherent tensions in the project. It is more constructive to conduct self-assessment and improve on investment practices, such as by winning the respect of the Myanmar public.[48] Although progressive opinions such as this are present among policy influencers and some company officials, the legitimacy of Myanmar activists is discounted by the perception that they are supported by the West and hence driven by "impure" motivations. Needless to say, dismissing the very real concerns of Myanmar civil society as nothing more than infiltration of foreign ideas is not conducive to constructive engagement.

As Howell remarks, the suspicion of Chinese state/Party officials towards foreign civil society organizations is an extrapolation of civil society development within China, "where the registration processes are onerous and where individuals skirt around the edges, contesting the boundaries, to organise around shared concerns".[49] Furthermore, given that many registered social organizations are government-organised NGOs (GONGOs) in China functioning as an extension of the government, Chinese interlocutors will also view western NGOs receiving government funding as similarly working on behalf of their governments.[50]

Myanmar NGO leaders express that funding from western governments is to support general democratic and human rights principles, but never to target a specific country (Interviews, Yangon, 27–28 May 2015). For instance, the US Embassy small grants programme application form for Myanmar civil society organizations lists the following objectives: to advance democracy and civic engagement, to promote the rule of law and human rights, to strengthen civil society capacity for advocacy and networking, as well as to promote conflict resolution and peace building. Negative socio-environmental impacts of foreign investments can be framed as human rights issues, and CSOs receive funding on that basis. However, China is by no means the only advocacy target; other countries such as Japan and Thailand are also the subjects of criticisms from rights groups, such as Physicians for Human Rights and EarthRights International, for their Special Economic Zones, albeit to a lesser extent because of the smaller scale of their overall investments.

Radical Approach

A common frustration among Chinese interlocutors is that most of the campaigns called for project termination, which seems radical to companies and investors, but is perhaps the only way that activists see fundamental issues of socio-environmental justice can be resolved. Myanmar civil society actors have mounted forceful campaigns despite being subject to subordination, and without much help of pressure "from above". It has been predominantly local and exiled activists who have driven the advocacy activities in Myanmar. This is not one of the typical cases where transnational activists connect with international regimes and create a transnational structure in which International Organizations (IOs) and great powers are convinced to pressure target states from above.[51] Pressure from IOs and great powers were largely absent in the series of campaigns against the three resource extraction projects funded by China; that is, they were predominately movements "from below". There were certainly some transborder and transnational linkages, and it is recognized that the divide between the international and national realms is undoubtedly artificial.[52] Yet, the activism examined in this chapter has been largely "local" in the sense that the relevant campaigns have not involved any IOs. Neither were foreign government agencies nor state actors involved, through means such as public shaming or sanctions, in supporting Myanmar activists' demands of the termination of the project and/or environmental justice.

Myanmar civil society was outside the international epistemic community that consists of "states, interstate institutions, and certain privileged non-state actors who provide experts", which "do not embrace the concerns of the more radical NGOs and grassroots movements, which have often emerged as a result of direct experience of environmental problems".[53] They do not tap on the transnational power of International

Non-Governmental Organization (INGO) networks. Yet, the activism flourished outside of formal institutions established in the North, many of which target technocratic problem-solving. Perhaps precisely because they were free from an agenda imposed by external actors or the North, they were able to be true to their emancipatory aims to challenge existing power structures, as well as "the very premises of an incremental and technical problem-solving approach",[54] rather than follow the northern agenda and be forced to fit inside their pre-determined, formal institutionalized framework. That is one of the reasons why most of the campaigns were able to be as bold and radical as they were in calling for the termination of the project, in which scenarios with deeper issues of socio-environmental justice might be resolved, instead of merely aiming to work within existing structures to devise actual and workable ways to influence policy. Certainly, actions aimed at disrupting or stopping resource projects can also be a way to increase bargaining power in future negotiations.[55]

For many activists, it is perhaps not the case that they first set a goal to be "radical" and then decided to call for project termination until power violations and human rights abuses were addressed. They pursue this path because their immediate survival is threatened. Indeed, social movements against natural resource exploitations around the world often stem from resistance in defence of livelihood, to protect assets; hence security by "challenging the structures, discourses and institutions that drive and permit exploitation and dispossession" from land and water sources.[56] Particularly in the Global South, those who resist are more concerned with "immediate existential 'environmental security' priorities, such as food and water", as opposed to longer-term issues such as wildlife conservation and climate change (which characterizes many northern environmental movements).[57] These immediate survival threats are further exacerbated by poverty and in many cases, authoritarian governance. It is only through fundamental struggles for justice, rather than incremental policy changes, that they might be able to secure their very survival.[58] Hence, their campaigns often involve calling for a stop to infrastructure projects that further entrench power disparities. In so doing, they are not only addressing environmental threats, but are also challenging prevailing socio-political orders.

It is therefore by no means a unique or surprising phenomenon in Myanmar that Chinese investments are radically opposed by the public.

Research has found that many environmental movements in Southeast Asia oppose large-scale resource projects, "particularly those that involve appropriation of the local resource base by interests of state, capital, and dominant social groups in the name of national development".[59] In the South, "almost all environmental activists challenge existing power relations and are radical within their own political milieu".[60]

What this radical approach means for their interaction with Chinese interlocutors is that on the one hand, activists achieved success in securing at least a temporary suspension of the Myitsone dam construction, of which the longer-term impact is the heightened awareness among Chinese regulatory bodies and companies as to responsible investment practice. On the other hand, such a radical approach rendered it highly difficult to engage in constructive policy discussions with the Chinese interlocutors. Indeed, the more radical CSOs risk marginalizing themselves. Moreover, they do not benefit from the legitimacy that they could have borrowed from transnational activists for their cause,[61] as they worked largely outside of the networks of transnational activists and international epistemic community. Despite having impacts in terms of project suspensions, the pressuring, "naming and shaming" tactics many CSOs use have undermined their legitimacy in the eyes of the Chinese interlocutors, and may jeopardize any potential for direct communication with the companies. CSOs should also be important stakeholders of investment projects, apart from project communities. It is beneficial for all stakeholders to engage with and be consulted by the companies. However, the issue frame and radical tactics they employ and the companies' responses (or lack thereof), often set their relationship on a difficult course.

As observed in the authors' interviews with Chinese interlocutors and Myanmar civil society actors, there is very little trust between the parties. Several well-educated civil society elites commented that it is now very difficult to speak in a positive light about China or to cooperate with China at this stage because of the overwhelming anti-Chinese sentiments. People do not want to be seen to be pro-China and there is lack of trust of China (Interviews, Yangon, 1–6 March 2014). A Chinese SOE manager based in Yangon expressed to me that there is no Myanmar local CSO that could make a constructive recommendation to them (Interview, Yangon, 7 March 2014). At a different level, the Chinese NGOs that work closely with the Chinese state, namely the China Foundation for Poverty Alleviation and Global

Environmental Institute, has had meetings and seminars with the elite few in Myanmar civil society, namely Myanmar Development Resource Institute (MDRI), one of the five organizations that has the privilege of advising the Myanmar government,[62] and EcoDev, led by a prominent environmentalist, which comprises former employees of the Forestry Department.[63] With these exceptions, the majority of the civil society actors do not have any direct channels to voice their demands to the Chinese state and SOEs, nor to engage in policy dialogue.

Chinese state-backed projects have responded to civil society demands by setting up corporate social responsibility (CSR) projects. There are also efforts to increase project transparency through setting up websites and Facebook pages. However, the approach employed by Myanmar grassroots activists has not in itself led to fundamental changes to policy processes, or the enactment of laws that ensure responsible and participatory business practice, which are crucial for creating an environment in which stakeholders are incorporated into the decision-making at all stages of development projects. Without the resources and recognition (from Chinese interlocutors) that international epistemic communities may enjoy, their ability to cooperate with Chinese policymakers and companies to crystallize policies of responsible investment is undermined.[64] For Chinese companies, the lack of formal or regular communication channels between the two sides is disadvantageous for their efforts to build reputation and trust among the Myanmar public. It is unfavourable for companies to exclude environmental activists as legitimate stakeholders of their public engagement or consultation processes, given the wealth of local knowledge and connections that environmentalists possess.

Conclusion

Although trust and communication between the vast majority of Myanmar civil society and Chinese interlocutors are still limited, Chinese state-backed companies have shown increased awareness and efforts in public engagement and investing in CSR projects.[65] The campaigns have not only pushed for suspension of the Myitsone dam construction (although the suspension might be only temporary), they continue to reverberate and thus see lasting impact. In view of mounting pressure from Myanmar society, there is an array of rare and unprecedented initiatives being undertaken by the Chinese Ministry of Foreign Affairs,

Embassy in Myanmar, GONGOs, such as China Foundation for Poverty Alleviation (CFPA) to foster "people-to-people exchanges", including with the largest opposition party, Aung San Suu Kyi's National League of Democracy (NLD), journalists, and the public. Interestingly, the Embassy in Myanmar and the companies responsible for the Myitsone dam and Letpadaung copper mine maintain active Facebook pages (even though Facebook is banned in China). All of the companies involved in the three extractive projects, namely the Myitsone dam, the Letpadaung copper mine, and Sino–Myanmar oil and gas pipelines, commenced publishing CSR reports and Environmental and Social Impact Assessments (ESIAs). These new measures challenge the traditional foreign policy and foreign investment norms of China.

What is noteworthy is that the power that led to these changes is mustered by a group of people who operate under a military-backed, quasi-civilian state that has only recently been given more space to engage in civil society activism, particularly around environmental issues. On top of having to deal with the uncertain political environment, the activities organized by grassroots civil society actors are still perceived by their target, the Chinese interlocutors, to be ill-informed, uneducated, illegitimate, and illegal. Moreover, their advocacy is aimed at influencing China, a state that places significant political constraint on its own domestic NGOs, preventing them from building any confrontational or politically sensitive campaign.

Nonetheless, the civil society actors in Myanmar have formed strong networks across geography, ethnic groups, and the urban/rural divide. They utilize their capability and knowledge, although at times limited, to gather evidence of negative social and environment impacts of the investment projects. They leverage access to local, exile and international media outlets to get their messages out. Some have been able to form strong advocacy networks with exile activists in neighbouring countries and INGOs.[66] Cooperating with ethnic and religious leaders, artists and literary figures, these grassroots civil society actors continue to campaign against negative business practices of other projects.

These efforts are highly coordinated and are delivering strong messages, but they have not gained their legitimacy and credibility among their targets (such as the Chinese state officials and SOEs). They have managed to organize themselves and take advantage of the opening up of political space to voice their demands overtly (even forcefully), but they have yet to break the hierarchical barriers between them and the key actors they seek to influence. However, it is precisely owing to their position outside of formal global governance structures that has shaped the radical frame and tactics they employ. There are tangible behaviour adjustments on the part of Chinese companies in terms of improving public channels of communication and CSR efforts, but most Myanmar CSOs are still excluded from the negotiation of these projects as legitimate stakeholders. This needs to be corrected if Chinese companies are to secure societal acceptance, which is indispensable for the continuation of the existing projects and the commencement of new ones.

NOTES

1. Based on the Myanmar context, the author defines civil society as "actors, voluntary associations and networks operating in the space between the family/clan, the state in its various incarnations and the for-profit market" (Petrie and South 2013). Civil society organization (CSO) is a term commonly used in Myanmar. The author uses this term to refer to a wide range of organizations, including 1) local and national level NGOs, 2) community-based organizations (CBOs), 3) small and volunteer-based organizations (Petrie and South 2013).
2. HKTDC Research, "Myanmar: Market Profile", 2017, available at <hkmb. hktdc.com/en/1X09SI4E/hktdc-research/Myanmar-Market-Profile> (accessed 28 September 2017).
3. Ibid.
4. Chenyang Li, *Annual Report on Myanmar's National Situation (2011–2012)* (China: Social Sciences Academic Press, 2013), p. 116. [Text in Chinese]
5. Yun Sun, "Chinese Investments in Myanmar: What Lies Ahead? Great Powers and the Changing Myanmar", *Stimpson Issue Brief* No. 1 (2013): 5.
6. Ibid., p. 8.
7. Ibid., p. 9.
8. Robinson, "Myanmar Cleans House — China's Worst Nightmare?".
9. World Economic Forum, Asian Development Bank and Accenture, *New Energy Architecture: Myanmar*, 2013, available at <http://www.adb.org/sites/default/files/publication/30265/new-energy-architecture-mya.pdf> (accessed 5 September 2015).
10. UK Trade and Investment, *Opportunities for British Companies in Burma's Oil and Gas Sector*, 2015, p. 3, available at <https://www.gov.uk/government/publications/opportunities-for-british-companies-in-burmas-oil-and-gas-sector> (accessed 5 September 2015).

11. Jurgen Haacke, "Myanmar: Now a Site for Sino–US Geopolitical Competition? The New Geopolitics of Southeast Asia", *IDEAS Report SR015* (2012): 53–60. Yun Sun, "China's Tug of War", *The Irrawaddy*, 9 April 2013, available at <http://www.irrawaddy.org/burma/chinas-tug-of-war.html> (accessed 4 September 2015).

12. Bertil Lintner, "Same Game, Different Tactics: China's 'Myanmar Corridor'", *The Irrawaddy*, 13 July 2015, available at <http://www.irrawaddy.org/magazine/same-game-different-tactics-chinas-myanmar-corridor.html> (accessed 13 July 2015).

13. James Reilly, "China's Economic Statecraft: Turning Wealth into Power", Lowy Institute for International Policy, University of Sydney, 2013, available at <http://www.lowyinstitute.org/publications/chinas-economic-statecraft-0> (accessed 3 September 2015).

14. Jacob Gronholt-Pedersen, "Myanmar Pipelines to Benefit China New Oil, Gas Supply Routes are Set to Help Slake Nation's Growing Thirst for Energy; Local Tensions Rise", *The Wall Street Journal*, 12 May 2013, available at <http://www.wsj.com/articles/SB100014241278873243265045784669515586644848> (accessed 2 September 2015).

15. Boyuan Chen, "Myanmar Pipeline Project Gives China Pause for Thought", *China.org.cn*, 21 June 2013, available at <http://www.china.org.cn/business/2013-06/21/content_29188744.htm> (accessed 27 August 2015).

16. Sun, "China's Tug of War".

17. Greg Torode, "Myanmar Work Will Continue, Vows National Endowment for Democracy", *South China Morning Post*, 1 October 2012, available at <www.scmp.com/news/asia/article/1050977/myanmar-work-will-continue-vows-national-endowment-democracy> (accessed 9 September 2015).

18. Thiha Tun, "Chinese Investor Assures Transparency If Myanmar Restarts Dam Project", *Radio Free Asia*, 26 December 2013, available at <http://www.rfa.org/english/news/myanmar/dam-12262013140753.html> (accessed 9 September 2015).

19. John Buchanan, Tom Kramer, and Kevin Woods, *Developing Disparity: Regional Investment in Burma's Borderlands* (Amsterdam: Transnational Institute, 2013), p. 39. Nwet Kay Khine, "Foreign-Investment-Induced Conflicts in Myanmar's Mining Sector: The Case of the Monywa Copper Mine", *Perspectives Issue 1: Copper, Coal, and Conflicts: Resources and Resource Extraction in Asia* (June 2013): 48.

20. Ibid.

21. For a list of reports and statements compiled by Myanmar civil society organizations, see <http://www.earthrights.org/campaigns/civil-society-reports-statements-shwe-and-myanmar-china-oil-and-gas-projects> (accessed 9 September 2015).

22. Nyein Nyein, "NGOs Call For Suspension of Shwe Gas Project", *The Irrawaddy*, 3 October 2012, available at <http://www.irrawaddy.com/burma/ngos-call-for-suspension-of-shwe-gas-project.html> (accessed 2 May 2015).

23. Andrew D. Kaspar, "Burma's Extractive Industries Not Digging Deep Enough with Reforms: Report", *The Irrawaddy*, 17 July 2013, available at <http://www.irrawaddy.org/natural-resources/burmas-extractive-industries-not-digging-deep-enough-with-reforms-report.html> (accessed 3 September 2015).

24. See Buchanan et al. (2013) and Cox (2015) for details about these disruptions and grievances caused to project-affected populations.

25. Revenue Watch Institute, "The 2013 Resource Governance Index: A Measure of Transparency and Accountability in the Oil, Gas and Mining Sector", 2014, available at <http://www.resourcegovernance.org/sites/default/files/rgi_2013_Eng.pdf> (accessed 3 September 2015).

26. Buchanan, Kramer, and Woods, *Developing Disparity: Regional Investment in Burma's Borderlands*.

27. Ibid., p. 45.

28. Tylter Stiem, "Burma: The War Goes On", *World Policy Blog*, 31 December 2014, available at <http://www.worldpolicy.org/blog/2014/12/31/burma-war-goes> (accessed 15 September 2015).

29. Bernt Berger, "China Still Has Its Wrong in Myanmar", *Asia Times*, 10 September 2013, available at <http://www.atimes.com/atimes/Southeast_Asia/SEA-01-100913.html> (accessed 2 September 2015).

30. Buchanan, Kramer, and Woods, *Developing Disparity: Regional Investment in Burma's Borderlands*, p. 14.

31. Ibid., p. 21.

32. Kaspar, "Burma's Extractive Industries Not Digging Deep Enough with Reforms: Report".

33. Berger, "China Still Has Its Wrong in Myanmar".

34. Ibid.

35. Adam Simpson, *Energy, Governance and Security in Thailand and Myanmar (Burma): A Critical Approach to Environmental Politics in the South* (UK: Ashgate Publishing Ltd., 2014), p. 83.

36. Thant Myint-U, *Where China Meets India: Burma and the New Crossroads of Asia* (Basingstoke: Macmillan, 2011), pp. 36–44.

37. Fan Hongwei, "Enmity in Myanmar against China", *ISEAS Perspective 2014 #08* (Singapore: Institute of Southeast Asian Studies, 17 February 2014). Maung Aung Myoe, *In the Name of Pauk-Phaw: Myanmar's China Policy since 1948* (Singapore: Institute of Southeast Asian Studies, 2011). Min Zin, "Burmese Attitude Toward Chinese: Portrayal of the Chinese

in Contemporary Cultural and Media Works", *Journal of Current Southeast Asian Affairs* 31, no. 1 (2012): 115–31.

38. Ching Kwan Lee and Yonghong Zhang, "The Power of Instability: Unraveling the Microfoundations of Bargained Authoritarianism in China", *American Journal of Sociology* 118, no. 6 (2013): 1475–1508.

39. Myanmar Centre for Responsible Business (MCRB), *Civil Society Organisations and the Extractives Industries in Myanmar — A Brief Overview*, 2014.

40. Ibid.

41. Paul Vrieze, "Civil Society and MPs Draft 'Progressive' Association Registration Law", *The Irrawaddy*, 21 October 2013, available at <http://www.irrawaddy.org/burma/csos-mps-draft-progressive-association-registration-law.html> (accessed 5 September 2015).

42. Economically Progressive Ecosystem Development (known as EcoDev) is an example of a well-recognized environmental NGO. Its co-founder and managing director, Win Myo Thu, is known for his relevant technical background and is Myanmar's leading environmental educator.

43. The 88 Generation student group was formed after the 1988 pro-democracy uprising, when government troops opened fire on mass student demonstrations in Yangon, leading to the deaths of thousands of people. The group's members spent years in prison and many were only released in early 2012. Their main focus of work is on democratization, peace and reconciliation (Snay Lin 2013).

44. Zhi Liu and Guangsheng Lu, *Report on the Cooperation in the Greater Mekong Sub-region 2012–2013* (China: Social Sciences Academic Press, 2013). [Text in Chinese]

45. Li, *Annual Report on Myanmar's National Situation (2011–2012).*

46. Le Zhang, "*Zhuanjia cheng miandian jiaoting misong dianzhan yin dangdi jushi buwen*" ["Expert Says the Suspension of Myitsone Dam Construction is due to Local Instabilities"], *The Beijing News*, 11 October 2011, available at <http://dailynews.sina.com/bg/chn/chnpolitics/sinacn/20111011/14162832525.html> (accessed 6 September 2015).

47. Boting Zhang, "Exploring the Issues Related to the Suspension of Myitsone Dam Construction", Sciencenet.cn, 25 April 2012, available at <http://blog.sciencenet.cn/blog-295826-563583.html> (accessed 6 September 2015). [Text in Chinese]

48. Heng Jiang, *Out of the Mine Fields and Blind Areas of Overseas Investment Security — Conflict Risk Assessment and Management* (Beijing: China Economic Publishing House, 2013), pp. 101–6.

49. Jude Howell, "Shifting Global Influences on Civil Society: Times for Reflection", in *Global Civil Society: Shifting Powers in a Shifting World*, edited by Heidi Moksnes and Mia Melin (Uppsala: Uppsala University, 2012), p. 50.

50. Ibid., p. 55.

51. Thomas Risse and Kathryn Sikkink, "The Socialization of International Human Rights Norms into Domestic Practices: Introduction", in *The Power of Human Rights: International Norms and Domestic Change*, vol. 66, edited by Thomas Risse, Ropp, Stephen C. Ropp, and Kathryn Sikkink (Cambridge: Cambridge University Press, 1999), pp. 1–38.

52. Margaret E. Keck and Kathryn Sikkink, *Activists Beyond Borders: Advocacy Networks in International Politics*, vol. 6 (Ithaca, NY: Cornell University Press, 1998), p. 4.

53. Lucy Ford, "Challenging the Global Environmental Governance of Toxics: Social Movement Agency and Global Civil Society", in *The Business of Global Environmental Governance*, edited by David L. Levy and Peter J. Newell (Cambridge, MA: MIT Press, 2005), p. 308.

54. Ibid.

55. Ciaran O'Faircheallaigh, "Negotiating Cultural Heritage? Aboriginal–mining Company Agreements in Australia", *Development and Change* 39, no. 1 (2008): 46.

56. Anthony Bebbington, Denise Humphreys Bebbington, Jeffrey Bury, Jeannet Lingan, Juan Pablo Muñoz, and Martin Scurrah, "Mining and Social Movements: Struggles Over Livelihood and Rural Territorial Development in the Andes", *World Development* 36, no. 12 (2008): 2888–2905, 2890.

57. Simpson, *Energy, Governance and Security in Thailand and Myanmar (Burma)*, p. 5.

58. Ibid., p. 32.

59. Philip Hirsch, "Dams, Resources and the Politics of Environment in Mainland Southeast Asia", in *The Politics of Environment in Southeast Asia: Resources and Resistance*, edited by Philip Hirsch and Carol Warren (London: Psychology Press, 1998); Simpson, *Energy, Governance and Security in Thailand and Myanmar (Burma)*.

60. Simpson, *Energy, Governance and Security in Thailand and Myanmar (Burma)* p. 32.

61. Risse and Sikkink, "The Socialization of International Human Rights Norms into Domestic Practices: Introduction", p. 5.

62. Nirmal Ghosh, "From Dissident to Contributor: New Think Tanks Help Myanmar's Transition to Democracy", *The Straits Times*, 21 June 2013, available at <http://timesofindia.indiatimes.com/impact-journalism-day/more-stories/From-dissident-to-contributor-New-think-tanks-help-Myanmars-transition-to-democracy/articleshow/20694549.cms> (accessed 2 September 2015).

63. Kyaw Yin Hlaing, "Associational Life in Myanmar: Past and Present", in *Myanmar: State, Society and Ethnicity*, edited by Narayanan Ganesan and Kyaw Yin Hlaing (Singapore: Institute of Southeast Asian Studies, 2003), p. 164.

64. A related issue is the lack of direct links between civil society in China and Myanmar, making it more difficult for Myanmar civil society to call Chinese actors to account, and to influence and discussion and implementation of policies.

65. Diane Tang-Lee, "Corporate Social Responsibility (CSR) and Public Engagement for a Chinese State-backed Mining Project in Myanmar: Challenges and Prospects", *Resources Policy* 47 (2016): 28–37.

66. Simpson, *Energy, Governance and Security in Thailand and Myanmar (Burma)*. Pichamon Yeophantong, "China, Corporate Responsibility and the Contentious Politics of Hydropower Development: Transnational Activism in the Mekong Region", The Global Economic Governance Programme, University of Oxford, 2013.

10

ANTI-CHINESE PROTEST IN VIETNAM
Complex Conjunctures of Resource Governance, Geopolitics, and State–Society Deadlock

Jason Morris-Jung and Pham Van Min

Since the turn of the millennium, Chinese economic activity in Vietnam has been growing in scale, diversity and geographical coverage. An increasingly important area of this growing activity has been in natural resource and energy sector projects. More recently, they have also been a source of public discussion and controversy, generating some of the most spectacular incidents of domestic protest in the post-war era. However, as this chapter argues, to view these incidents simply as historical animosity towards China misses a lot. Rather, they reflect complex conjunctures of multi-level governance problems and contextual factors, among them Vietnam's on-going geopolitical tensions with China in the South China Sea. This chapter examines recent trends and public concerns emerging around China's growing economic activity in Vietnam, and then explores in more detail two case studies of popular resistance to large resource sector projects with Chinese involvement, namely a controversy over bauxite mining in the late 2000s and attacks on Chinese workers at a massive steel factory in Central Vietnam in 2014.

Introduction

Since market reforms in the 1980s, foreign investment has been and continues to be a driving force for Vietnam's economic development.[1] Since normalization of bilateral relations in 1991, Chinese economic investments in Vietnam have increased dramatically. More recently, they have also shown a growing interest in Vietnam's rich mineral and energy resources. Geographical proximity makes Vietnam particularly interesting to Chinese resource companies, especially for minerals that have high transportation costs. Furthermore, shared histories, cultures and socio-political organizations have facilitated trade and business relations between these two nations.

On the surface of it, Chinese demand for natural resources and Vietnam's need for foreign investment is a perfect match. Yet one does not have to dig very deep below the surface to find a much more problematic situation. Recently, the General Director of the Department of Geology and Minerals of Vietnam publicly complained that up to 60 per cent of mineral licenses in northern Vietnam showed "Chinese traces".[2] It was as if, he suggested, "the Chinese were standing behind our back and controlling our own mining industry". While the widespread opposition towards Chinese involvement in bauxite mining in the late 2000s was perhaps a first signal of unrest, the riots that targeted mainland Chinese workers at a Taiwanese-owned steel factory in Central Vietnam in 2014 demonstrated just how grave the matter can become.

However, to understand these incidences simply as to view them as a replay of historical animosity towards China misses a lot. Rather, they reflect complex conjunctures of on-going political struggles and historical grievances within the Vietnamese party-state. Chief among them are weaknesses in resource governance (or just governance, more generally speaking), anxieties about the regional implications of the rise of China (especially as it concerns Vietnam's territorial claims on the South China Sea), and an authoritarian political system that provides few channels for Vietnamese citizens to address their grievances and influence policies. Hence, to take these expressions of anti-Chinese sentiment at face value would be overly reductive. More importantly, it would fail to address the many underlying problems that have given rise to and, for reasons discussed below, have found expression in resistance towards large resource sector projects with Chinese involvement.

This chapter begins with a brief overview of China's growing economic activities in Vietnam in recent times and the public suspicions that have emerged around them. It then examines two of the most spectacular expressions of popular resistance to Chinese involvement in large resource sector projects. They are the controversy around bauxite mining in 2008 and 2009, which was the first major expression of popular resistance towards a Chinese contracting company and its expanding force of Chinese workers in Vietnam; and the riot at the Formosa steel complex in 2014, which was the most extreme expression of anti-Chinese sentiment in the market reform era. The case studies are based on field work in Vietnam from 2009 to 2011 conducted by the first author on the bauxite mining controversy and preliminary interviews with residents in Ky Anh District of Ha Tinh Province in 2014, as well as reviews of domestic media coverage and online discussions.

China's Growing Economic Presence in Vietnam

Chinese economic activities in Vietnam have been growing rapidly over the past two and a half decades, reflecting China's growing portfolio of overseas investments in Southeast Asia and around the world. While Chinese investment has given an important boost to the Vietnamese economy, it has also been met with apprehension and uncertainty, reflecting both historical misgivings towards Chinese domination and more recent tensions between the two nations around territorial claims in the South China Sea.

The contemporary era of Chinese investment in Vietnam started when the two countries normalized bilateral relations in 1991. The new era was marked by a joint-venture enterprise to open a restaurant in Hanoi, with an investment capital of US$200,000.[3] Since 1991, Chinese investment in Vietnam can be roughly divided into three main periods, based on its expansion of scale (*viz.*, number of projects and capital values), diversification of sectors, and extension of geographical coverage.

During the first period from 1991 to 2001, Chinese investment in Vietnam was relatively small when compared to that of other countries. China ranked only 20th among sixty foreign investors in Vietnam. By the end of 2001, the number of Chinese projects was 110 with a total registered capital of only US$221 million. The average capital value per project was around US$1.5 million, while projects worth more than US$10 million were few. Their lifespans were generally between ten and fifteen years.

The second period from 2001 to 2011 witnessed a rapid expansion in both the number of projects and their total registered capital value. By 2009, the cumulative number of Chinese projects reached 657 with a total registered value of US$2.67 billion, or more than ten times what it was in 2001.[4] The average capital per project increased from around US$2.5 million in 2001 to US$4.3 million in 2007. Key factors promoting Chinese investment were the ASEAN–China Framework Agreement on Comprehensive Economic Cooperation in 2002 and both China's and Vietnam's accession to the World Trade Organization (WTO) in 2002 and 2007 respectively. The level of Chinese investment decreased in 2009 and 2010, however, due to the 2008 global financial crisis. For instance, only forty-eight new projects were licensed in 2009, equivalent to only one half of the previous year, and the total registered capital was also only about one third.[5] Nonetheless, by the end of 2010, China had climbed to 11th among foreign investors in Vietnam.

The most significant developments in Chinese investment happened in the third period, from 2011 to the present. In 2013, 110 new Chinese foreign direct investment (FDI) projects were licensed by the Vietnamese government, with a total registered capital worth US\$2.34 billion.[6] The dramatic increase in capital value was mainly due to the coal-fired power plant Vinh Tan 1 in Binh Thuan province. As the fifth foreign-invested power plant in Vietnam since the country opened its doors to foreign investment, this build-operate-transfer (BOT) project was valued at US\$2.0 billion. By the end of 2014, the number of Chinese cumulative FDI projects reached 1,082 with a total registered capital value of US\$7.9 billion.[7] These figures further bumped China up to 9th among 101 foreign investors in Vietnam.[8] Even so, Chinese FDI currently accounts for only around 3.0 per cent of the total registered capital from all foreign investors in Vietnam.

The two sectors that currently attract most Chinese investment in terms of capital value are processing and manufacturing, on the one hand, and producing and distributing electricity, gas and air-conditioning, on the other, as illustrated in Table 10.1. These

TABLE 10.1
Top Five Sectors of Chinese Investment in Vietnam, 2014[9]

Order	Sector	Number of Projects	Total Capital (US$ billion)	Percentage in Total
1.	Processing & manufacturing industry	704	4.13	53
2.	Producing and distributing electricity, gas and air-conditioning	3	2.04	28
3.	Construction	98	0.559	7
4.	Real estate	14	0.46	6
5.	Hotel and restaurant	12	0.29	4

two sectors account for more than US$6.0 billion out of nearly US$8.0 billion, equivalent to 75 per cent of total Chinese investment in Vietnam.

Yet, FDI is not the only measure of China's growing economic activity in Vietnam. Chinese contracting companies have been an important, though more difficult to trace, area of involvement. These companies have also contributed to greater diversification into resource sectors and geographical coverage of Chinese economic activity in Vietnam.

Even as Chinese investment accounted for only 1.5 per cent of the total FDI in Vietnam between 1991 and 2010, the Vietnam Economic Forum reported that Chinese investors controlled 90 per cent of all Engineering, Procurement and Construction (EPC) contracts.[10] Most of these contracts were in the areas of electricity, mining and petroleum. Chinese investors served as EPC contractors in 23 out of 24 projects in cement production (96 per cent), 15 out of 20 coal-fired power plants (75 per cent), 2 out of 2 bauxite projects (100 per cent) and 3 out of 3 coal extraction and processing projects (100 per cent).[11] These figures suggest a new interest from China in Vietnam's resource and energy sectors. The new trend has also been responsible for a geographical expansion of Chinese investment projects across Vietnam.

In the first period, Chinese enterprises were present mainly in cities and provinces with good infrastructure, transportation and cheap labour. This was because Chinese enterprises were largely invested in the textile, garment, and hotel and restaurant sectors. With China's more recent interests in the resource and energy sectors, however, Chinese enterprises and contractor companies are now present in 55 out of 63 provinces in Vietnam. They have installed themselves in northern border provinces, such as Lao Cai, Lang Son, Quang Ninh, Cao Bang, and Ha Giang, and in the Central Highlands in the provinces of Lam Dong and Dak Nong. These provinces are among Vietnam's poorest regions with underdeveloped infrastructure and transportation. Yet they have attracted Chinese enterprises with their wealth of natural and especially mineral resources.

Since the turn of the millennium, China's economic presence in Vietnam has become increasingly important, diverse and geographically widespread. Even as Chinese investment accounts for only a minor percentage of FDI in Vietnam, China's dominance among foreign contracting companies inside Vietnam, especially in the area of resource and energy sector projects, has raised many public concerns.

Anti-Chinese Sentiment in Resource Sector Projects

Chinese involvement in resource and energy sector projects, whether as direct investors or contractor companies, has been especially controversial. Domestic reporting (and its echoes in the Vietnamese language foreign media, such as the BBC and Radio France Internationale), expert opinion, and a proliferation of online commentary have given expression to public distrust towards China's growing economic interests in Vietnam. This section provides an overview of key areas of concern, while the next section examines in more detail two specific cases of popular resistance towards Chinese involvement in resource sector projects.

One key area of concern has been a suspicion that China seeks to "buy up" Vietnam's natural resources. Despite increases in Chinese investment since the early 2000s, more than 70 per cent of China's total investment value is concentrated on natural resources.[12] Chinese investors have often managed to increase their stocks in mining projects to control output.[13] This has helped foment rumours that Chinese investors provide money for domestic corporations to invest in the mining sector so that they can then more easily export the mineral products of the domestic company back to China.[14] For instance, the General Department of Geology and Minerals of Vietnam recently reported that 5,000 licenses for exploiting minerals were given to 2,000 enterprises in 2010. However, 60 per cent of these enterprises were suspected of clandestinely cooperating with Chinese investors.[15]

Moreover, higher prices for minerals in China has allegedly led to problems of illegal export. Vietnamese officials have estimated that nearly half of the country's total annual output in coal (some ten million tons) is illegally exported to China, while 90 per cent of the coal from the country's most important coal deposit in Quang Ninh province is suspected of illegal exportation to China.[16] More recently, the Chairman of the Vietnam Foundry and Metallurgy Science and Technology Association, Pham Chi Cuong, noted disparities between Chinese and Vietnamese customs data that indicated 3.1 million tons of iron had unofficially made its way from Vietnam to China in 2013.[17] The Chairman further suggested that the difference was equivalent to three trillion VND (US$141.2 million) in lost tax revenues and five hundred billion VND in export tariffs. The Minister of Industry and Trade has acknowledged that illegal export activities occur mainly in the northern provinces along the Chinese border and through trading routes in the South China Sea.[18] Illegal exports are thus believed to contribute to a wider trend of China "buying up" Vietnamese resources, which hinders Vietnam from developing its own resources with higher value production.

A second area of concern has been that Chinese contractors often fail to ensure project quality or meet agreed timeframes for the projects. While this concern is more general to Chinese invested projects, it has particular ramifications for the resource and, especially, energy sectors. The Vietnam Energy Association reported that all energy projects are between three months and three years behind schedule due to the limited capacity of Chinese investors.[19] Chinese-invested projects, such as Hai Phong 1 and 2, Cam Pha 1 and 2, Quang Ninh 1 and 2, Mao Khe, Thai Nguyen, Vinh Tan and Duyen Hai 1 are critical to the national power development plan for 2010 and 2020. However, all of these projects have also failed to meet their initial deadlines. Failure to meet the deadlines not only incurs additional costs, but also hinders infrastructure development and energy security for Vietnam.

A third area of especially controversial concern has revolved around a common practice of Chinese contractor companies to import masses of Chinese labour rather than, say, hire local Vietnamese labour. The number of Chinese workers has reportedly increased from 21,217 in 2005 to 75,000 in 2010, while the actual number is expected to be much higher.[20] In many big projects contracted by Chinese investors, Chinese labourers outnumber local workers. For instance, the 4,200 Chinese migrants working at the Quang Ninh Thermal Power Plant account for 95 per cent of its total workforce. In the bauxite mining projects in Dak Nong province, Chinese workers accounted for 85 per cent of the workforce.[21] Chinese companies argue that they prefer to recruit Chinese labour because of their common language and more advanced skills. However, media reports have also suggested that many Chinese workers are unskilled or low-skilled labourers.[22]

Concerns have also been raised about the number of Chinese workers that are illegal. Of the 4,000 Chinese workers at the Hai Phong Thermal Power Plant, only 7 per cent were found to be legally registered to work in Vietnam.[23] In the bauxite mining projects in Lam Dong and Dak Nong, less than 20 per cent were reported as legally registered. Reports in the domestic media of Chinese workers violating local laws and tensions with local workers have further fuelled public concern.[24] These tensions have been in part blamed for — but by no means excuse or justify — the riots that broke out at the Formosa Steel Complex in 2014.

A fourth area of concern has revolved around Vietnam's growing trade deficit with China and the challenge this has created for domestic industries. Chinese contractors often import iron, steel, materials, spare parts and accessories from China, even though they could also source these goods from Vietnam. As many EPC projects are funded by Chinese preferential loans and export buyer's credits, their contract agreements specify that Chinese investors can or must import materials from China. The import of Chinese products has led to an extremely low rate of local contents in Chinese EPC projects. The rate of local contents in EPC projects contracted by investors from G-7 countries accounts for about 25 per cent, while this same number in Chinese-invested projects is only about 3 per cent. In numerous cases, that number is nil.[25]

In 2007, Vietnam's trade deficit with China was US$9.0 billion. It increased to US$16.4 billion in 2012 and US$23.7 billion in

2013.[26] In 2013, Chinese materials imported for FDI projects in Vietnam accounted for about 70 per cent of the total import value.[27] The import of Chinese products has also challenged domestic industries. Chinese contractors often give little mechanical work to local contractors. For instance, Chalco gave Vietnamese subcontractors a contract of only about US$8.0 million in the US$466 million Nhan Co project and a contract of US$2.5 million in the US$499 million Tan Rai project. According to Nguyen Van Thu, chairman of Vietnam Association of Mechanical Industry (VAMI), the Chinese EPC contractors have defeated the domestic mechanical industry.[28] He has also suggested that if Chinese investors continue to win EPC projects, Vietnam's mechanical industry will soon go bankrupt.

Last but not least, historical fears of China wanting to dominate or even reclaim Vietnam as one of its provinces have stoked the ancient spectre of China's 1,000 years of domination over Vietnam. This nationalist historical trope has fuelled popular concerns about China seeking to control or sabotage the Vietnamese economy as a means to weaken the nation politically. For many Vietnamese, the stand-off between Manila and Beijing over the Scarborough Shoal in 2012 served as a warning to how China might use economic clout to advance its territorial claims.[29] Two years later, during Vietnam's own stand-off with China in the South China Sea, many foreign and domestic newspapers reported that Beijing had ordered its state-owned enterprises to stop bidding for new investments in Vietnam.[30] In the aftermath of this incident, a series of discussions were held, some of them coordinated by the Vietnamese government, to consult with economists, scholars, and retired government officials on how Vietnam could escape from China's economic orbit.[31]

These events and discussions have had a real impact for a number of Chinese investment projects. For instance, in 2013, Thua Thien Hue province licensed a US$250 million project to the World Shine Hong Kong Co. Ltd., through its Vietnam-based unit, The Dieu Co. The project was designed for Cua Khem Cape, where the Hai Van Mountain meets the South China Sea and the border runs between Da Nang and Thua Thien Hue provinces.

According to the license, The Dieu Co. would develop a 200-hectare resort complex with a 450-room five-star hotel, 2,000-seat international convention centre and a five-storey block of 200 deluxe departments.[32] However, Colonel General Nguyen Van Rinh, former deputy national defense minister, protested that the Cape plays a crucial role in national defense and suggested that better locations are available for such a project.[33] The chairman of Da Nang People's Committee, Van Huu Chien, sent an official document to the governmental ministries and agencies calling for the re-consideration of the project in Hai Van Mountain due to its strategic position in national defense.[34] Numerous National Congress deputies also raised their concerns about the project. As a result, Thua Thien Hue province revoked the license issued to the The Dieu Co. Company and stopped the project.[35]

Similarly, Nam Dinh province announced in 2014 that Foshan Sanshui Jialiada (China), Luenthai (Hong Kong) and Vinatex (Vietnam) agreed to a US$400 million investment project to develop Rang Dong industrial park for textile and garment companies.[36] After the announcement of Nam Dinh province, the popular website *Bauxite Vietnam* uploaded a series of posts that speculated on China's real intentions behind its investments in Nam Dinh province.[37] According to *Bauxite Vietnam*, China's main interest was to connect Nghia Hung port in Nam Dinh with Hainan Island. Chinese investors were also expected to request investment in Nghia Hung port to reduce transportation costs, which *Bauxite Vietnam* warned posed a serious threat to Vietnam's national security because such an investment could enable the Chinese military capacity to control the Tonkin Gulf.[38]

Although resource and energy sector projects describe only a part of Chinese economic activity in Vietnam, they have been among the most controversial. Public scrutiny of, as well as speculation over, these projects has contributed to sentiments of distrust and suspicion towards Chinese companies and investors. In the next section, we examine the underlying factors and conditions in two particularly spectacular cases, as well as the complex set of problems they express at local, national and regional levels.

Chinese Resource Conflicts and Cooperation in Vietnam

The controversy over Vietnamese government plans for mining massive reserves of bauxite in the Central Highlands in the late 2000s was the first to draw widespread public attention to the growing presence of Chinese workers and contracting companies in Vietnam. The controversy over a Taiwanese steel factory in Central Vietnam, which involved extensive use of Chinese migrant workers, became the most unsettling in 2014 when these expressions turned violent. This section highlights complex levels of domestic struggle present in both cases. They reflect at least as much on-going problems in resource governance and domestic state–society relations as they do the new challenges emerging in the China–Vietnam bilateral relation.

The Central Highlands Bauxite Mining Controversy

Government plans to mine some 5.4 billion tons of bauxite in the Central Highlands and process it into alumina (a low-value feedstock for the production of aluminum) generated an unprecedented public opposition to a major government policy in late 2008 and early 2009. A key issue in the discussion was a Chinese state-owned company contracted by the government to design and build the mines and refineries, as well as reasonable suspicions that their end goal was to export to China for aluminium smelting.

The public outrage over bauxite mining began when in late 2008 a domestic NGO set about initiating a wider policy dialogue with the local government, the state-owned Vietnam Coal and Minerals Corporation (Vinacomin) (the project owner for all of the bauxite-alumina projects in the Central Highlands), and its informal network of Vietnamese scientists and experts. These discussions highlighted concerns familiar to other resource extraction projects in Vietnam such as deforestation, soil erosion and river siltation. Experts argued that bauxite mining would generate net economic losses because of its detrimental impacts on regional agriculture, forestry and hydropower development.[39] They would also increase risks of droughts and floods for the Central Highlands and the lowland coastal regions below it, as well as displace local populations and ethnic minority groups.[40]

Processing bauxite into alumina on the highland plateau would generate tens or even hundreds of millions of tons of a caustic bauxite residue, commonly known as "red sludge". Red sludge can contaminate the soil and water source and it is a known hazard to human health in its dry form ("red dust"). Vinacomin proposed to store the caustic sludge in its untreated form in cesspools on the highland plateau indefinitely. As one expert noted, however, while this method might be suited to the flat, arid and sparsely populated regions of northern Australia — the world's largest bauxite producer — it spelt disaster in the high elevation, monsoon climate, and densely populated Central Highlands of Vietnam.

Not only did these concerns reflect a wide range of socio-ecological risks, they also suggested serious inadequacies in the government's ability to plan for and manage risks. Such types of concerns were not, however, new. Rather, they reflected enduring problems in Vietnam with large-scale development projects and resource governance. They also highlighted widespread problems of transparency and lack of public accountability in state institutions.

These discussions were amplified considerably when the then ninety-eight-year-old military hero of the People's War, General Vo Nguyen Giap, wrote a letter to the Prime Minister to protest bauxite mining in January 2009. While reiterating social, environmental and economic concerns, General Giap also mentioned what had been probably already known but not yet raised publicly by activists. General Giap drew attention to the Chinese state-owned company contracted to design and build the projects, and the hundreds and possibly of Chinese workers being brought in to the Central Highlands to work on them.[41]

While it may have took somebody of the public stature of General Giap to let this cat out of the bag, his remarks were followed by a barrage of public criticism of the government and party regime on bauxite mining — most of which now occurred online because the government had banned the domestic press from further discussion on the topic. Online discussions questioned the quality and standards of Chinese technology, equipment, and social and environmental management practices. They speculated on how Chinese workers would take away jobs from local workers and create inter-cultural tensions. Conversely, other commentators worried that Chinese workers might marry into local communities and establish "Chinese villages" throughout the Central Highlands, or they might even be Chinese soldiers in disguise, as explicitly suggested by an open letter penned by another renown revolutionary era general.[42]

Increasingly, commentators connected these ideas with the Central Highlands' strategic military role during the Vietnam War and the current territorial conflicts with China in the South China Sea. One prominent Vietnamese writer noted how massive land concessions and development projects in Laos and Cambodia further enabled China to accumulate bodies and equipment along Vietnam's borders, which could create a national security threat if China ever found reason to attack Vietnam militarily, which might result from their competing claims on the South China Sea.[43] Soon these massive mining projects were not seen simply as a social or environmental problem, but rather — as one group of prominent Vietnamese intellectuals expressed it — as concerning nothing less than the "fate of the nation".[44]

Such statements not only highlighted rising tensions with China, but also opened up new possibilities for expressing disapproval of the party regime and its alleged complicity with the Chinese leadership. Even as many claims about China's role and intentions in bauxite mining were highly speculative, they rallied diverse groups together and enabled them to challenge state authorities in ways that had previously been rarely seen. In sum, a closer look into the history and development of the bauxite mining controversy shows that it reflected a complex conjecture of enduring frustrations with Vietnamese governance, political leadership and a rising China. The controversy, however, also opened up new possibilities for domestic contestation that have had surprising and widespread implications to this day (Morris-Jung 2015; London 2014).

The Formosa Ha-Tinh Integrated Steel Complex: When Anti-Chinese Sentiment Turns Violent

In May 2014, riots broke out at the Taiwanese-owned Formosa Ha Tinh steel factory in Central Vietnam. In addition to ransacking and burning down buildings, rioters attacked crews of Chinese workers at the factory and nearby dormitories, resulting in hundreds of injuries and at least one Chinese death.[45] These events followed two other riots in the environs of Ho Chi Minh City one day earlier. The riots broke out during nationwide demonstrations against a Chinese oil rig that had been stationed inside of Vietnam's Exclusive Economic Zone on the South China Sea. These attacks on Chinese workers were by far the most violent expression of anti-Chinese sentiment in Vietnam since the border war of 1979. However, closer examination of the events leading up to and following the riots also show that, like the bauxite mining controversy, they were a product of on-going problems in Vietnamese governance and politics.

The Formosa Ha Tinh Integrated Steel Complex is located in the Ky Anh District of Ha Tinh province. Ha Tinh is one of the poorest provinces in Vietnam, located along a narrow band of climatically harsh and, with climate change, increasingly vulnerable coastal provinces in northern Vietnam. Ky Anh is a coastal district with a population of about 200,000 people, most of whom depend on agriculture and fishing for a livelihood. The National Highway No. 1 also runs through Ky Anh, which has helped to generate additional commercial activity.

Prior to the construction of the Formosa factory, industrial development in Ky Anh was very limited. This situation changed dramatically when Taiwanese investors at Formosa Plastics promised to invest nearly US$30 billion in a massive integrated steel factory over two phases from 2011 to 2020. With a projected annual capacity of 22.5 million tons per year, the complex would be the largest steel factory in Southeast Asia and among the largest in the world.[46] Plans for the complex included six blast furnaces, a 2,100 MW coal-fired power plant, a railway docking station, and two 32-berth seaports, one for importing iron ore, coal and other materials and the other for exporting steel and steel-related products. The massive complex extends over 21 km² of coastal land and another 12 km² into the sea, which have been conceded to Formosa on a seventy-year lease. Only 60 km to the south of the Formosa factory is the 550 million ton Thach Khe iron mine, which has proposed to boost production to 5–7 million tons per year.

Since 2013, the Formosa Steel Complex has also been the subject of a vocal public discussion. Discussions in the domestic press and online have raised wide-ranging concerns over the social and environmental impacts of the massive development, its management of waste and pollution, its displacement of local farming and fishing communities, and questioned its contributions to the local economy and jobs. In 2016, some of these concerns were sadly validated when more than 100 tons of dead fish were washed up on the shores of four central Vietnamese provinces over the course of a month as a result of waste discharge from the Formosa factory (Morris-Jung 2016).

However, as with the bauxite mining controversy, these concerns were amplified by the attention given to the many Chinese contractor companies and migrant Chinese workers building the factory. Domestic sources reported thousands of Chinese workers living and working in Ky Anh, raising familiar concerns about taking away local jobs, creating tension with local communities, and establishing "Chinese villages".[47] While the Deputy Minister in the Ministry of Labour[48] suggested that only 1,913 Chinese workers were working in Ky Anh, the Head of the Labour Department in Ha Tinh suggested that Formosa had already asked permits for 10,000 foreign workers, of which 90 per cent were Chinese, and the Head of Immigration reported that he was preparing permits for 8,400 foreign workers to enter the Ky Anh area.[49] However, the inability of state authorities to provide consistent figures on how many Chinese workers were actually in Ky Anh and, of those, how many held legal working papers seemed to reveal only the government's ineptitude in managing the situation or, perhaps, its unwillingness.

A lack of information and transparency around the Formosa project also raised suspicions around "special" policies offered to the foreign investors. They allegedly included an extensive list of tax and import fee exemptions, the natural resource tax exemption, and privileges related to land leasing, borrowing money, use of foreign currency, and use of foreign vessels in Formosa's deep sea port. Former chairman for the Vietnam Steel Association, Pham Chi Cuong, suggested that the "incentives granted to Formosa Ha Tinh are unprecedented".[50] This aura of suspicion and apparent disregard for local concerns was not helped by the massive wall of reinforced concrete topped with glass shards that surrounded the perimeter of the entire Formosa construction site. As one blogger expressed, what actually happens behind this wall "nobody can know".[51]

As in the bauxite controversy, anxieties around the Chinese presence in Ky Anh were heightened by discussion of its geostrategic location. Located on the narrowest band of the national territory, less than a hundred kilometres from the coast to the inland border with Laos, commentators suggested that a military occupation of the Ky Anh region would split Vietnam in two. One blogger outlined the perilous triangle that China was developing with its presence in Ky Anh. This triangle would be completed by a 2.1 km strip of coastal land conceded to a Chinese food production company 190 km to the south in Quang Binh Province and its deep sea port; and the Chinese naval base on Hainan Island only about 130 km from either of the deep sea ports in Ha Tinh or Quang Binh (see Figure 10.1).[52] This blogger reflected popular fears that if ever China were to have reason to attack Vietnam, Chinese forces would be able to mobilize quickly the Chinese naval base and geographically split Vietnam in two. Such concerns were articulated by Vietnamese economist Bui Tien Than:

FIGURE 10.1
A Perilous Triangle on the South China Sea

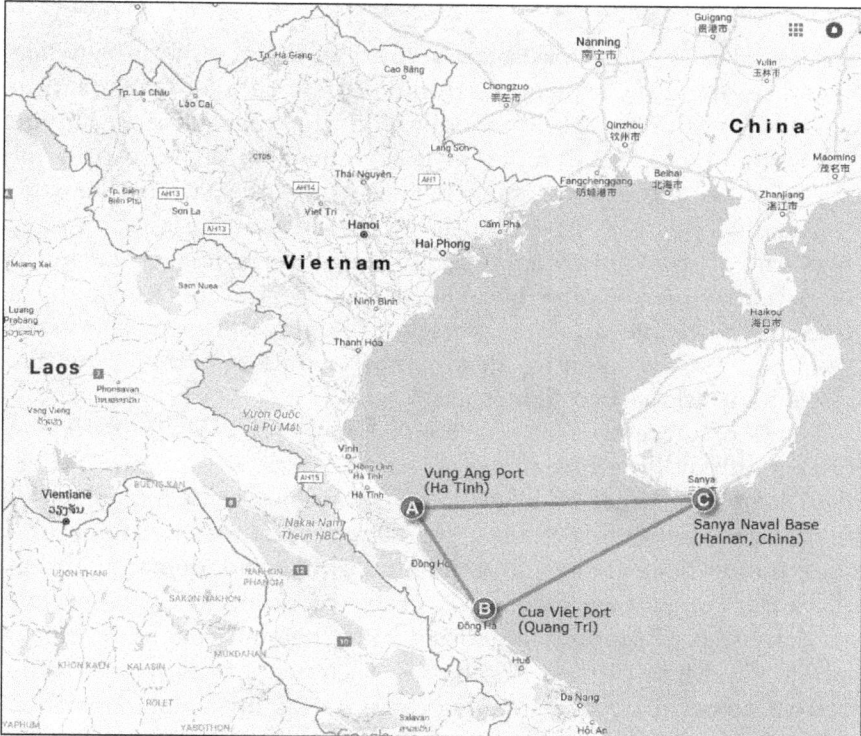

Note: For purposes of graphic clarity, this image has been re-drawn based on this blog posting:
Nguy n H u Quý, "Trung Qu c ang có âm m u gì Hà T nh và Qu ng Tr ?" [Does China Have a
Secret Plot in Ha Tinh and Quang Tri?], *Bauxite Vietnam*. Blogger.com, 1 March 2014, available
at <http://boxitvn.blogspot.sg/2014/02/trung-quoc-ang-co-am-muu-gi-o-ha-tinh.html>. Special
thanks to Justin Fong for preparing the image.

Source: Adapted from Map data © 2017 Goggle, ZENRIN.

Vung Ang – Ha Tinh is very close to Hainan Island. If China builds
a port in Vung Ang connecting it to Hianan Island, the Tonkin Gulf
will become China's lake. And what then would happen to maritime
transportation between the North and South of Vietnam? In addition,
other perils to national defense exist. It is only about fifty kilometers
from Vung Ang to the Lao border. As a result, what could we do

if China moved from Lao [where a Chinese company also operates many large land concessions] to Vung Ang? They could cut Vietnam in two halves (Nam Nguyên 2014).

The crisis in this discussion arrived in early May 2014, when the state-owned Chinese National Offshore Oil Corporation (CNOOC) stationed a billion dollar oil rig, the Haiyang Shiyou 981, within 120 nautical miles of the Vietnamese coast to drill for petroleum in the South China Sea. By the International Law of the Sea, the rig was located inside Vietnam's Exclusive Economic Zone and on its continental shelf. Beijing was evidently aware of the potential provocation because the rig was accompanied by a protective barricade of eighty vessels, including seven marine ones. Over the next two months, that number would rise to 122 marine vessels and include two fighter jets.

News of the events on the South China Sea quickly generated widespread indignation across Vietnam. Domestic media outlets filled their front pages and headlines with reports and commentaries on the crisis. Online activists posted open letters, petitions and other public declarations to condemn China and demand urgent action from their own leaders. A group of prominent intellectuals and activist organizations organized mass demonstrations in Hanoi and Ho Chi Minh City on May 11th.[53] Their sentiments were broadly reflected across the country and in certain areas involved cooperation with government organizations.

Two days later, however, peaceful demonstration turned to violence as riots broke out in Special Industrial Zones around Ho Chi Minh City, notably in the adjacent provinces of Binh Duong and Dong Nai. Some demonstrators also vandalized factories in Ba Ria-Vung Tau province, also adjacent to Ho Chi Minh City. One day later riots broke out at the Formosa factory in Ky Anh. Demonstrators began at Ky Anh capital and then marched towards the gates of Formosa some ten kilometres away. Brawling ensued between demonstrators and Chinese workers, while buildings at the Formosa factory and nearby dormitories where Chinese workers were housed were ransacked, looted and burned. These riots were the first time in the post-war era that Vietnamese demonstrators had rioted against foreign factories in multiple provinces.

However, nearly as surprising as the riots themselves has been the lack of official explanations on how or why they happened. Both official and unofficial sources reported groups of young men and women working the crowds and inciting violence. Yet even after more than 1,000 arrests and 400 indictments by domestic security, no coherent explanation has been put forward. Domestic reporting on the riots by both provincial and national newspapers showed strong signs of a centrally controlled script. Accounts in provincial and national newspapers,[54] from the government's mouthpiece *Thong Tan Xa Viet Nam* to one of the country's most outspoken newspapers *Tuoi Tre*, adhered to a common narrative of patriotic demonstrations that were taken advantage of by a few unruly elements in the crowd. However, these articles also failed to enlighten on who exactly these unruly elements were, how they were organized, who might have supported them, or what were their motivations. Even among the few dozen persons that were sentenced to jail terms because of their participation in the riots, domestic reporting is unclear on whether they were the same people that actually incited the riots.

Needless to say, no official public investigation was carried out on these extraordinary incidents. Neither has any NGO, human rights organization, or independent researcher that we are aware of been able to investigate them in any detail. Instead, the riots themselves remain shrouded in mystery. Questions about why police or military forces were unable to respond to these events within a twenty-four-hour period continue unexplained. Rumours emerged that the government itself may have had a hand in seeding the violence, either to emphasize the urgency of the oil rig crisis or perhaps to suppress the peaceful demonstrations that portrayed state leaders as weak and compromised.[55] The lack of information combined with police intimidation, however, only contributed to, rather than resolved, questions on Vietnam's political leadership and governance. It revealed the wide-ranging and multi-level governance problems at the root of incidents like the violent events that took place in Ky Anh. Without more effective channels for the public to express its concerns, they risk boiling over into such incidents as these.

Conclusion

A strong Chinese economy has been important to Vietnamese economic development, most notably in terms of two-trade, whose volume now exceeds US$60 billion per year, but also in Chinese direct investment and contracting. However, these trends have also increasingly entangled the fate of the Vietnamese economy with China. One key area of concern is that the Vietnamese economy becomes trapped in low value production in such sectors as resource production because of its dependence on the Chinese economy. These concerns are amplified further by governance issues, especially in the area of resource production and industrial development, which had led to such catastrophes as the massive die-off in Central Vietnam in 2016.[56] These dilemmas have no easy solutions, yet they are the responsibility of all parties involved.

The demonstrations and protests that have emerged around Chinese involvement in resource sector projects must be seen not simply as crude expressions of anti-Chinese sentiment but also for the multiple and complex problems to which they have given expression. A key theme in both of the cases discussed in this chapter was the lack of effective channels for Vietnamese citizens to convey their concerns to government and either an inability or unwillingness of state leaders to respond to them.

As the Vietnamese public has become increasingly assertive and resourceful in voicing its demands in recent years, the ruling regime seems to have become only more staunch in preserving its Leninist political system. The state-society deadlock results from a tradition of discouraging Vietnamese citizens from speaking out openly against their leaders, which state authorities have punished as behaviour seeking to undermine an independent Vietnamese nation. In this regard, rallying against China has been a way of creating political space to express more critical views of government and the party-state while emphasizing their own loyalty to the Vietnamese nation.

NOTES

1. Vo Thi Thanh and Nguyen Anh Duong, "Revisiting Exports and Foreign Direct Investment in Vietnam", *Asian Economic Policy Review* 6 (2011): 112–31.

2. Đăng Nam, "Doanh Nghiệp Việt Đứng Tên Cho Chủ Trung Quốc Khai Thác" [Vietnamese Mining Companies Fronting for Chinese Owners], *Tuổi Trẻ Online*, 18 January 2014, available at <http://vietstock.vn/2014/01/doanh-nghiep-viet-dung-ten-cho-chu-trung-quoc-khai-thac-1351-328672.htm>.

3. Nguyen Dinh Liem, "China's FDI in Vietnam: 20 Years in Retrospect", *Journal of Chinese Studies* 8, no. 156 (2014): 31–45.

4. Do Tien Sam and Han Thi Hong Van, "Vietnam–China Economic Relations: 2009–2010", in *China and East Asia: After the Wall Street Crisis*, edited by Peng Er Lam, Yaqing Qin and Mu Yang (Singapore: World Scientific Publishing Co. Pte. Ltd., 2013), pp. 225–39.

5. Ibid.

6. Vietnam General Statistics Office, "2013 Foreign Direct Investment", available at <https://gso.gov.vn/default.aspx?tabid=716>.

7. Foreign Investment Agency, "Vietnam–China Foreign Investment Development", Ministry of Planning and Investment, available at <http://fia.mpi.gov.vn/tinbai/2067/Tinh-hinh-hop-tac-dau-tu-Viet-Nam-Trung-Quoc>.

8. Top four foreign investors in Vietnam are regional neighbours Japan, Singapore, South Korea and Taiwan.

9. Compiled by the authors from the General Statistics Office (<http://www.gso.gov.vn/default.aspx?tabid=716>) and Foreign Investment Agency, Ministry of Planning and Investment (<http://fia.mpi.gov.vn/chuyenmuc/14/Tinh-hinh-dau-tu>).

10. Vietnam Economic Forum, "Chinese EPC Contractors Won 90% of the Top Project in Vietnam", VietnamNet, available at <http://community.vef.vn/2010-07-31-trung-quoc-trung-thau-90-cong-trinh-thuong-nguon-cua-viet-nam-?print=1>.

11. *Laodong Newspaper*, "Mối Nguy Từ Các Dự Án Tổng Thầu Epc Rơi Vào Tay Nhà Thầu Trung Quốc" [Dangers from Chinese Investors' Control Over EPC Projects], Vietnam General Trade Union, available at <http://laodong.com.vn/kinh-doanh/moi-nguy-tu-cac-du-an-tong-thau-epc-roi-vao-tay-nha-thau-trung-quoc-205587.bld>.

12. Pham Sy Thanh, "Ba Mối Lo Trong Quan Hệ Kinh Tế Việt Nam – Trung Quốc" [Three Mảo Concerns in Vietnam–Sino Economic Relations], *Finance Review*, 23 June 2014.

13. Dantri, "Khoáng Sản 'Đội Nón' Sang Trung Quốc" [Minerals "Illegally Exported" to China], *Dantri.com*, 16 November 2012.

14. Ibid.
15. Datviet, "Không Kiểm Soát Được Trung Quốc Mượn Danh Người Việt Khai Khoáng" [Unable to Control Chinese Enterprises under the Guise of Vietnamese in Mineral Investment], available at <http://baodatviet.vn/chinh-tri-xa-hoi/tin-tuc-thoi-su/tong-cuc-truong-tong-cuc-dia-chat-khoang-san-khong-kiem-soat-duoc-tq-muon-danh-nguoi-viet-khai-khoang-2365131/>.
16. K. Chi, "China Attempts to Control Vietnam's Mineral Industries", *VietnamNet Bridge*, 25 January 2014, available at <http://english.vietnamnet.vn/fms/business/94502/china-attempts-to-control-vietnam-s-mineral-industries.html>.
17. In 2013, Vietnam's customs records show that Vietnam exported 1.25 million tons of iron ore to China at US$48.72 per ton, while Chinese customs shows that China imported 4.5 million tons of iron ore at US$84.75 per ton. See *Tuoi Tre News*, "Vietnam Metallurgy Association Raises Alarm over Illegal Iron Ore Exports to China", 8 February 2014, available at <http://tuoitrenews.vn/business/21414/vietnam-metallurgy-association-raises-alarm-over-illegal-iron-ore-exports-to-china>.
18. Government Inspectorate of Vietnam, "Xuất Lậu Quặng, Khoáng Sản - Bộ Trưởng Vũ Huy Hoàng Thừa Nhận Trách Nhiệm" [Illegal Mineral Exports: Minister Vu Huy Hoang Takes Responsibility], available at <http://thanhtra.com.vn/chinh-tri/doi-noi/xuat-lau-quang-khoang-san-bo-truong-vu-huy-hoang-thua-nhan-trach-nhiem_t114c67n70780>.
19. VnExpress, "Hàng Loạt Dự Án Điện Của Nhà Thầu Trung Quốc Chậm Tiến Độ" [Series of Energy Projects by Chinese Contractors Behind the Schedule], available at <http://kinhdoanh.vnexpress.net/tin-tuc/vi-mo/hang-loat-du-an-dien-cua-nha-thau-trung-quoc-cham-tien-do-2716006.html>.
20. Nguyen Van Chinh, "Recent Chinese Migration to Vietnam", *Asian and Pacific Migration Journal* 22, no. 1 (2013): 7–30.
21. Ibid.
22. "Foreign Labourers in Vietnam", *Radio Free Asia*, 4 August 2009; Thanh Nien Online, "Thousands of Illegal Chinese Labourers in Camau", *Thanh Nien*, 9 August 2011.
23. Nguyen Van Chinh, "Recent Chinese Migration to Vietnam".
24. Baomoi.vn, "Công Nhân Tq Đánh Công Nhân Việt Nam: Lộ Nhiều Bất Cập Trong Quản Lý" [Chinese Workers Fought against Vietnamese: Management Inadequacy], available at <http://vietbao.vn/Xa-hoi/Cong-nhan-TQ-danh-cong-nhan-Viet-Nam-Lo-nhieu-bat-cap-trong-quan-ly/2131538768/157/>.
25. Vietnam Energy, "Nội Địa Hóa Nhà Máy Nhiệt Điện Thấp, Nguyên Nhân Từ Đâu" [Low Local Contents in Thermal Power Projects: What are the Causes?], available at <http://nangluongvietnam.vn/news/vn/dien-luc-viet-nam/noi-dia-hoa-nha-may-nhiet-dien-thap-nguyen-nhan-tu-dau.html>.

26. Xinhua, "China Remains Vietnam's Biggest Trade Partner in 2013", available at <http://www.chinadaily.com.cn/business/chinadata/2014-01/29/content_17264283.htm>.

27. Central Institute for Economic Management, "Thực Trạng Sự Phụ Thuộc Của Kinh Tế Việt Nam Vào Trung Quốc" [Facts on Vietnam's Economic Dependence on China], available at <http://www.vnep.org.vn/vi-vn/Hoi-nhap-kinh-te-quoc-te/Thuc-trang-su-phu-thuoc-cua-kinh-te-Viet-Nam-vao-Trung-Quoc.html>.

28. Vietnam Business Forum, "Diversifying Machinery Supply Sources", available at <http://vccinews.com/news_detail.asp?news_id=30784>.

29. Dau Tu Online, "Trung Quốc Bắt Đầu Chơi Con Bài Kinh Tế Với Việt Nam" [China Uses Economic Clout to Vietnam], available at <http://baodautu.vn/trung-quoc-bat-dau-choi-con-bai-kinh-te-voi-viet-nam-d1294.html>; Huang Keira Lu, "State Firms Barred from Vietnam Contract Bids", *South China Morning Post*, 9 June 2014; Dexter Robert, "China's State Enterprises Told to Stop Investing in Vietnam", *Bloomberg Business*, 9 June 2014.

30. Ibid.

31. Pham Minh Ngoc, "'Thoát Trung' Nhưng Cũng Cần Cẩn Trọng" ["Escaping China" But be Cautious], *The Saigon Times*, 6 June 2014; Nguyen Nha, "Việt Nam Và Cơ Hội Thoát Trung Lần 4" [Vietnam and the Fourth Opportunity to "Escape" China], *BBC Vietnamese*, 6 June 2014; *VTC News*, "Cuộc Chơi Sòng Phẳng, Sao Kinh Tế Việt Nam Phải 'Thoát' Trung" [Fair Competition, Why Vietnam Has to "Escape" China?], available at <http://vtc.vn/cuoc-choi-song-phang-sao-kinh-te-viet-nam-phai-thoat-trung.1.493780.htm>; Do T. Thuy, "Vietnam's Moderate Diplomacy Successfully Navigating Difficult Waters", *East Asia Forum*, 16 January 2015.

32. *Tuoi Tre News*, "Vietnam Province Pulls Plug on Defense-Sensitive Chinese-Invested Resort Project", available at <http://tuoitrenews.vn/business/24306/vietnam-province-pulls-plug-on-defensesensitive-chineseinvested-resort-project>.

33. Ibid.

34. Dantri, "Đà Nẵng Kiến Nghị Thu Hồi Giấy Phép Dự Án Trên Núi Hải Vân" [Da Nang Calls for Revoking the Project in Hai Van Mountain], available at <http://dantri.com.vn/xa-hoi/da-nang-kien-nghi-thu-hoi-giay-phep-du-an-tren-nui-hai-van-992484.htm>.

35. *Tuoi Tre News*, "Vietnam Province Pulls Plug on Defense-Sensitive Chinese-Invested Resort Project".

36. Talk Vietnam, "Vietnamese Told Not to be Too Eager for FDI from China", available at <http://www.talkvietnam.com/2014/05/vietnamese-told-not-to-be-too-eager-for-fdi-from-china/>.

37. Hoang Mai, "Tại Sao Trung Quốc Lại Đầu Tư Lớn Vào Tỉnh Nam Định, Và Đâu Là Mục Tiêu Sâu Xa" [Why China Heavily Invests in Nam Dinh,

What are the Real Intentions?], available at <http://boxitvn.blogspot.com. au/2014/03/tai-sao-trung-quoc-lai-au-tu-lon-vao.html>; Hoang Mai, "Băn Khoăn Khi Trung Quốc Tiếp Tục Đầu Tư 400 Triệu Đô La Vào Tỉnh Nam Định" [Worried When China Continues to Invest US$400 Million into Nam Dinh], available at <http://boxitvn.blogspot.com.au/2014/03/ban-khoan-khi-trung-quoc-tiep-tuc-au-tu.html>.

38. Hoang Mai, "Băn Khoăn Khi Trung Quốc Tiếp Tục Đầu Tư 400 Triệu Đô La Vào Tỉnh Nam Định" [Worried When China Continues to Invest US$400 Million into Nam Dinh].

39. Nguyễn Thanh Sơn, "Đại Kế Hoạch Bô – Xít Ở Tây Nguyên Bị Phản Đối Quyết Liệt" [Fierce Opposition to the Great Bauxite in the Central Highlands], *Tuan Vietnam*, 24 October 2008, available at <http://www.tuanvietnam. net/2008-10-24-dai-ke-hoach-bo-xit-o-tay-nguyen-bi-phan-doi-quyet-liet>; Nguyễn Thanh Sơn, "Đại Dự Án Bô – Xít Tây Nguyên: Người Trong Cuộc Đề Xuất Gì?" [The Great Central Highlands Bauxite Project: What Does an Insider Recommend?], *Tuần Việt Nam*, 26 October 2008, available at <http://www.tuanvietnam.net/dai-du-an-bo-xit-tay-nguyen-nguoi-trong-cuoc-de-xuat-gi>.

40. Nguyên Ngọc, "Chương Trình Bauxite Ở Tây Nguyên và Các Vấn Đề Văn Hoá – Xã Hội" [The Central Highlands Bauxite Projects and Their Social and Cultural Issues], in *Tài Liệu Hội Thảo Khoa Học* [Workshop Proceedings] (Đắk Nông: UBND Dak Nong, Vinacomin and CODE, 2008), pp. 94–97.

41. Võ Nguyên Giáp, "Letter to Prime Minister, Hon. Nguyễn Tấn Dũng", 5 January 2009, available at <http://www.viet-studies.info/kinhte/Thu_VNGiap_ NTDung.pdf>; Võ Nguyên Giáp, "Đại Tướng Võ Nguyên Giáp Góp Ý về Dự Án Bô Xít Tây Nguyên" [General Vo Nguyen Giap Gives His Recommendations for the Central Highlands Bauxite Projects], *Tuan Vietnam*, 14 January 2009, available at <http://community.tuanvietnam.net/2009-01-14-dai-tuong-vo-nguyen-giap-gop- y-ve-du-an-bo-xit-tay-nguyen?print=1>.

42. Nguyễn Trọng Vĩnh, "Thư của thiếu tướng đại sứ Nguyễn Trọng Vĩnh" [Letter from Major General Ambassador Nguyễn Trọng Vĩnh], *Diễn Đàn*, undated, available at <http://www.diendan.org/viet-nam/thu-cua-thieu-tuong-111ai-su-nguyen-trong-vinh/>.

43. Nguyên Ngọc, "Ý Nghĩa Văn Hoá Xã Hội Của Chương Trình Bôxit Tây Nguyên — Diễn Đàn Forum" [Socio-Cultural Perspectives on the Central Highlands Bauxite Program], *Diễn Đàn*, 12 April 2009, available at <http:// www.diendan.org/viet-nam/y-nghia-van-hoa-xa-hoi-cua-chuong-trinh-boxit-tay-nguyen>.

44. Nguyễn Huệ Chi, Phạm Toàn, and Nguyễn Thế Hùng, "Kiến Nghị về Quy Hoạch và Các Dự Án Khai Thác Bauxite Ở Việt Nam" [Petition on Bauxite

Master Plan and Projects in Vietnam], 12 April 2009, available at <http://www.boxitvn.net/kien-nghi>.

45. Linh Thư, Xuân Quý, and Minh Thăng, "1 Người Chết Ở Hà Tĩnh" [1 Person Dead in Ha Tinh], *VietNamNet*, 15 May 2014, available at <http://vietnamnet.vn/vn/chinh-tri/175609/1-nguoi-chet-o-ha-tinh.html>.

46. S. Tung, "Formosa Wants to Increase Capital to $28.5 Billion", *VietnamNet Bridge*, 29 May 2013, available at <http://english.vietnamnet.vn/fms/business/75420/formosa-wants-to-increase-capital-to--28-5-billion.html>.

47. *VietnamNet*, "'Phố Trung Quốc' Ở Hà Tĩnh" ["Chinese Cities" in Ha Tinh], 8 May 2013, available at <http://vietnamnet.vn/vn/kinh-te/120201/-pho-trung-quoc--o-ha-tinh.html>; *VietnamNet Bridge*, "Thousands of Chinese Workers in Ha Tinh Lack Permits", 10 October 2014, available at <http://english.vietnamnet.vn/fms/society/113797/thousandsof-chinese-workers-in-ha-tinh-lack-permits.html>; Văn Định, "Tràn Lan Lao Động Trung Quốc Trái Phép", *Tuổi Trẻ Online*, 20 October 2013, available at <http://tuoitre.vn/chinh-tri-xa-hoi/575548/tran-lan-lao-dong-trung-quoc-traiphep.html#ad-image-0>.

48. Full name is the Ministry of Labour, Invalids and Social Affairs (MOLISA).
49. Dat Viet, "No One Knows How Many Chinese Workers are in Vietnam", *VietnamNet Bridge*, 8 September 2014, available at <http://english.vietnamnet.vn/fms/business/111125/no-one-knows-how-many-chinese-workers-are-in-vietnam.html>.

50. *Tuoi Tre News*, "Taiwanese Firm Keeps Demanding More Despite Huge Incentives from Vietnam", 7 September 2014, available at <http://tuoitrenews.vn/business/20860/taiwan-firm-keeps-asking-for-more-despite-huge-incentives-from-vietnam>.

51. Nguyễn Hữu Quý, "Trung Quốc Đang Có Âm Mưu Gì Ở Hà Tĩnh và Quảng Trị?" [What is China Plotting in Ha Tinh and Quang Tri?], *Bauxite Việt Nam*, 3 January 2014, available at <http://boxitvn.blogspot.com.au/2014/02/trung-quoc-ang-co-am-muu-gi-o-ha-tinh.html>.

52. Ibid.
53. *Bauxite Việt Nam*, "Lời Kêu Gọi Biểu Tình Yêu Nước Của 20 Tổ Chức Dân Sự Việt Nam" [Call to Protest of the 20 Civic Organizations of Vietnam], 8 May 2014, available at <http://www.boxitvn.net/bai/25964>; *Bauxite Việt Nam*, "Thông Báo về Mit Tinh Chống Trung Quốc Xâm Lược" [Announcement of Meeting to Protest Chinese Invasion], 10 May 2014, available at <http://anonymouse.org/cgi-bin/anon-www.cgi/>.

54. Newspapers reviewed for this chapter included provincial newspapers for Ha Tinh, Binh Duong and Dong Nai and national newspapers *Thong Tan Xa Viet Nam* and *Tuoi Tre*.

55. Jason Morris-Jung, "Reflections on the Oil Rig Crisis: Vietnam's Domestic Opposition Grows", *ISEAS Perspective* 2014 #43 (Singapore: Institute of Southeast Asian Studies, 30 July 2014).

56. Jason Morris-Jung, "Vietnam's New Environmental Politics: A Fish Out of Water?", *The Diplomat*, 19 August 2016, available at <http://thediplomat.com/2016/05/vietnams-new-environmental-politics-a-fish-out-of-water/>; Thu Huong Le, "Amid Fish Deaths, Social Media Comes Alive in Vietnam", *The Diplomat*, 19 August 2016, available at <http://thediplomat.com/2016/05/amid-fish-deaths-social-media-comes-alive-in-vietnam/>; *BBC News*, "Vietnam Protest over Mystery Fish Deaths", 1 May 2016, available at <http://www.bbc.com/news/world-asia-36181575>.

11

COMPLEXITIES OF CHINESE INVOLVEMENT IN MINING IN THE PHILIPPINES

Menandro S. Abanes

Amid the controversies and historically conflictive and divisive character of mining in the Philippines, China's entry and involvement in the industry have made it even more controversial, conflictive and divisive because of the territorial disputes in the West Philippine Sea (South China Sea). Over the past five years, there have been a surge of Chinese mining investments and a growing opposition to them in areas where they operate. Opposition to Chinese mining is highlighted by over a hundred arrests of Chinese nationals involved in illegal mining operations and suspension orders of Chinese firms or their dummies. So it raises the big question — why is it that Chinese mining seems to be unwelcome and yet it continues to pour in and expand? In this article, the complexities of Chinese involvement in Philippine mining are discussed and examined. The article starts with the current situation of mining in the Philippines and the relevant mining-related laws, which, apparently, are being exploited by Chinese mining investors to serve their interests. To contextualize the widespread opposition to mining in general, contributions of mining to local development and poverty incidence are evaluated. Recommendations on how to reduce the controversies, tensions and conflict brought about by Chinese mining in local communities are outlined.

Introduction

When the Philippine Supreme Court issued a Temporary Environmental Protection Order (TEPO) against ninety-four "small-scale mines" in the Province of Zambales in 2013, the media touted the decision as anti-China, rather than pro-environment or anti-mining. Many of the big small-scale mines enjoined by the TEPO to "perform or desist from performing an act in order to protect, preserve, or rehabilitate the environment"[1] are reported to be dummies of Chinese firms. The case was filed by concerned members of local communities of the province who complained that the mining operations were allegedly outside the designated allowable mining area, were polluting the environment, were unregulated and untaxed.[2] Local and national government officials were also respondents to the case.

Mining in general in the Philippines has been controversial and divisive. On the one hand, mining operations have been described by civil society groups and the Catholic Church as destructive to traditional livelihoods. On the other hand, mining has also catalyzed vibrant economic activities through job creation, infrastructure works and business establishments in economically stagnating localities. Thus, government officials and businessmen tend to favour and endorse mining projects. As a result of these clashing views, conflicts and dilemmas over mining often occur because of fundamentally different development frameworks advocated by pro-mining and anti-mining groups.[3] If the mining operations are geared towards sustainability, mining companies must not overlook local communities in their operations because they are key to sustainability and social acceptability of mining in their localities.

Mainly emanating from the need to stimulate the economy and from the value of protecting the environment, the dilemmas have become more intense because of increased Chinese involvement in the mining industry. In 2012, the Mines Geosciences Bureau (MGB) of the Department of Environment and Natural Resources (DENR) officially listed fourteen Chinese firms with mining production sharing agreements, only five of which had processing permits and two with exploration permits. However, investigative media reports have suggested that at least twenty-five Chinese firms have been involved in mining extraction and other mining-related operations through Filipino counterparts, subsidiaries and dummies.[4]

The recent surge in Chinese mineral investments in the Philippines has coincided with a surge in anti-Chinese sentiment due to heightened territorial disputes over Scarborough Shoal, which China has claimed and controlled since 2012. The widely-documented reclamation activities by China in the disputed Spratly Islands have further exacerbated hostile popular sentiments towards China. This has provided a context for the type of anti-Chinese sentiment that has also arisen in mining projects involving Chinese entities. Over the past five years, hundreds of Chinese nationals were arrested in various parts of the Philippines conducting illegal mining operations by authorities aided by information from local community members.[5]

Amid the controversies and historically conflictive character of mining in the Philippines, China's entry and involvement in the industry have made it even more controversial and conflictive because of the territorial disputes in the West Philippine Sea (South China Sea). It raises the question — why does Chinese mining appear so unwelcome in the Philippines and yet it continues to expand? In this article, the author examines these complex tensions and dilemmas emerging around Chinese mining involvement in the Philippines. The following pages present the current situation of mining in the Philippines and the relevant mining-related laws, which, the author argues, are being exploited by Chinese mining investors to serve their interests. The contributions of mining to local communities are evaluated and the complex realities characterized by tensions and dilemmas on which Chinese mining operates are examined. In conclusion, the author suggests that, to reduce tensions with local communities, Chinese mining can be more cognizant of the Philippine laws, responsive to the conditions of the host local communities, and more inclusive and transparent in their operations.

Mining in the Philippines

The country's DENR ranks the Philippines as the 5th most mineralized country in the world (3rd in gold reserves, 4th in copper and 5th in nickel). That is why the Philippine mining industry has continued to attract billions of dollars of foreign investments. However, mining investments and productions remain low compared to the tremendous potential of untapped and available mineral resources.

TABLE 11.1
Mining Industry Statistics

	2010	2011	2012	2013	2014+
Total mining investments	US$1.05 billion	US$1.20 billion	US$0.92 billion	US$1.45 billion	US$0.69 billion
Large-scale mining gross production	PhP 63.0 billion	PhP 88.5 billion	PhP 97.1 billion	PhP 98.2 billion	PhP 137.6 billion
Small-scale mining gross production	PhP 48.9 billion	PhP 34.6 billion	PhP 2.3 billion	PhP 1.1 billion	PhP 1.0 billion
Non-metallic mining gross production	PhP 33.3 billion	PhP 41.1 billion	PhP 45.6 billion	PhP 57.8 billion	PhP 66.1 billion
Total exports of minerals and products	PhP 1.93 billion	PhP 2.84 billion	PhP 2.34 billion	PhP 3.41 billion	PhP 4.01 billion
Employment in mining	197,000	211,000	250,000	250,000	235,000
Taxes, fees and royalties	PhP 13.41 billion	PhP 22.39 billion	PhP 19.44 billion	PhP 22.82 billion	PhP 21.41 billion

Source: Mines and Geosciences Bureau (MGB), available at <http://www.mgb.gov.ph>.

Since 2010, large-scale mining production has been rising steadily. This spike continues in spite of the termination of several large-scale mines in 2014, such as Rapu-Rapu Minerals Inc. (RRMI) and Canatuan Mining Project of TVI Resource Development. In fact, mining production rose by 40 per cent from PhP 98.2 billion (US$2.28 billion)[6] in 2013 to PhP 137.6 billion (US$3.2 billion) in 2014 (see Table 11.1). According to MGB production of nickel ore during this period accounted for 58 per cent of the total production, followed by gold with 24 per cent and copper accounting for 16.5 per cent. The main destination of nickel and copper productions was China.

Between 2011 and 2012, small-scale mining production dipped enormously by 93 per cent, from PhP 34.6 billion (US$804 million) to PhP 2.3 billion (US$53.4 million). Under the People's Small-scale Mining Act of 1991, all gold production by small-scale miners should be sold to the Central Bank. However, media reports suggest that small-scale miners prefer to sell on the black market or smuggle their gold to Hong Kong and mainland China to avoid the new taxes imposed by the Central Bank on gold sales.[7] In 2013 and 2014, gross gold production from small-scale mining sold to the Central Bank went further down from PhP 1.1 billion (US$25.5 million) to PhP 1 billion (US$23.2 million) respectively.

On its involvement in Philippine mining, China launched its resource-led diplomacy in the 2000s, which aggressively and strategically developed diplomatic relations with many African and Southeast Asian countries, including the Philippines, "to secure its consumption of minerals in the long run".[8] China has spotted huge potential in mining in the Philippines, which has become one of the countries with major Chinese mining investments.[9] In fact, the official visit by former Philippine President Benigno Aquino III to China in 2011 sealed four mining agreements with Chinese companies, such as Jinchuan Group Ltd., Oriental Peninsula Resources Group Inc. and Eramen Minerals Inc., amounting to US$14 billion investments within the next five years, 2011–16.[10]

This development is one of the reasons why the large-scale mining productions have consistently increased over the years. But such developments could not be said for small-scale mining production as official records from the government agency, MGB, which is tasked to oversee and monitor mining productions, show declining production. Have productions in small-scale mining truly decreased since 2012?

How could we account for the glaring contrast between rising large-scale productions and falling small-scale mining productions? Mining laws, particularly on small-scale mining, in the country might provide some insights on the issue of this discrepancy.

Relevant Mining-related Laws

Mining conflict is not new to the Philippines. Mining laws over the past thirty years have attempted to reduce conflicts and regulate the industry more effectively, but these laws have struggled with on-the-ground enforcement and inadvertently created new forms of tension, which Chinese and other small-scale mining companies have been able to exploit.

Under the Regalian Doctrine adopted by the 1987 Philippine Constitution, the State has ownership rights over all public lands and mineral resources, including their exploration, development and utilization. More often than not, the State needs investments, technical expertise and state-of-the-art technology to be able to explore, develop and utilize vast mineralized areas. Thus, the State can enter into agreements with national and transnational mining firms for such undertakings. It is recognized that the country "has remarkably comprehensive national legislation and regulatory provisions that address indigenous rights, environmental concerns, and social benefits related to mining".[11] There are laws and regulations that mainly govern mining operations in the country, such as the Mining Act of 1995, People's Small-scale Mining Act of 1991, and Executive Order No. 79 (EO 79).

The Mining Act of 1995 has provisions mandating mining firms which apply for permits (e.g. Environmental Compliance Certificate) to prepare and implement comprehensive environmental plans to avoid or mitigate against negative effects of mining, during and after its operations.[12] It also requires mining firms to seek social acceptability of their large-scale projects with local communities through public consultations and hearings. Without ascertaining whether or not the large-scale mining projects are socially acceptable to the affected local communities, no mining permit can be issued. However, this mandatory provision seemed to be vulnerable to moneyed strategies and somehow could be skirted.

For example, touted as the first large-scale mining project to comply with the requirements of the Mining Act of 1995, the Rapu-Rapu Polymetallic Project in Bicol had two public hearings which became a show of force for both anti- and pro-mining groups. In each of these hearings, large crowds were observed, especially from the pro-mining side. There were reports that the mining firm had brought the crowd to the venue and hired barangay officials to facilitate the social acceptability of the project among the locals.[13] To bolster its social acceptability, the mining firm sought local government units (LGUs) that have jurisdiction over the planned mining area to endorse the project. This is the fear and concern of civil society groups because, with much capital and financial muscle, mining firms can simply buy the support and endorsement of local officials who hold sway over the decision-making of their constituents.

Small-scale mining operations, on the other hand, are covered by the People's Small-scale Mining Act of 1991. In this law, local government units can authorize and approve exploration, development and commercial operation of small-scale mining operations. However, many mining deals are able to skirt Philippine laws by using local rent-seekers who are surface land claimants and mining operators as dummies to access existing mining sites and operate in what could be considered as large-scale.[14] However by law, aside from being barred from using explosives and heavy equipment, small-scale mining operations only have a maximum allowable annual production of 50,000 dry metric tons of ore of metallic minerals and maximum capital investment of PhP 10 million (US$232,558). On the ground, however, many small-scale mining operations are able to extract more than the allowable productions. Having investments greater than what is allowed by law, these operations are financed by outsiders and do not generally comply with legal requirements and limitations.[15] The arrests of hundreds of Chinese nationals conducting illegal mining activities throughout the archipelago exposed the Chinese involvement in "small-scale" mining. They operated in cahoots with local rent-seekers — politicians and surface-land claimants. In the process, they excluded the majority of the locals from reaping their share of mineral wealth found in their own communities.

In 2012, a new mining regulation, EO 79, took effect. The new regulation limits the metallic minerals that can be the object of small-scale mining operations in designated people's small-scale mining areas (*Minahang Bayan*) to gold, silver and chromite. These three minerals can be mined by artisanal methods which do not necessitate the use of banned heavy equipment and explosives for small-scale mining operations. It also states that for new mining rights, there should be competitive public bidding to facilitate social acceptability of the new projects with local communities. However, instead of improving the small-scale mining industry, which is the objective of the new regulation, news reports pointed out that the new regulation has simply driven small-scale mining production towards the black market to avoid taxes. One of the most vibrant black markets is in Binondo (Chinatown in Manila) where the bulk of production goes to Hong Kong, the hub towards China.[16]

In spite of the comprehensive legal framework put in place, challenges to enforce and implement these legalities abound. Many mining deals and operations have remained outside of government regulation and supervision. While it is true that many investors and operators are deliberately attempting to avoid taxes and regulation by the government, a major obstacle to bring these deals and operations under government recognition is the fact that the government institutions tasked to oversee and supervise these are lacking resources and capacities to perform their tasks.[17] Moreover, these legal frameworks are efforts to formalize the small-scale mining industry, which is in reality structurally informal in arrangements and operations.[18] Here lies the difficulty of enforcing regulation when the operation itself is informal in nature.

Hence in the report by Pacific Strategies and Assessments (PSA) in 2011, Chinese mining investments and deals are hounded by issues of "unaccountability, misconduct, and corruption" and able to circumvent the mining laws with the help of corrupt politicians and local officials.[19] What happens on the operational level of small-scale mining involving Chinese is that they serve as financiers who make arrangements with rent-seekers (land claimants) and mining operators for initial investments and revenue-sharing schemes. These rent-seekers and operators who have their own miners and workers are backed by public officials who invoke the local government code to permit small-scale mining operations in their jurisdiction. The Chinese financiers at times bring their own miners to account for actual mining haul and productions. This explains why there have been hundreds of arrests of Chinese involved in illegal mining activities.

Setting the Context: What Has Mining Done to Local Communities?

Given the tens of billions of pesos in taxes, fees and royalties and hundreds of thousands of employed personnel in mining operations (see Table 11.1), one would think that areas that hosted mining operations would be economically better-off than non-mining areas. However, this is not supported by poverty statistics. In fact, as Table 11.2 shows, the economic sectors and their contributions to poverty incidence, the highest contributor in 2009 was the mining sector at 48.7 per cent, when all other sectors had contributed much lower to poverty incidence.[20] From 1988 to 2009, mining was the only sector of the economy that had an increasing contribution to poverty incidence from 27.8 to 48.7 per cent. This finding was also true for absolute numbers of people.[21] It means that almost half (48.7 per cent) of those who work in mining industry were in poverty in 2009. To illustrate this at a regional level, regions with large mining areas had high poverty incidences, such as the Caraga region (47.5 per cent), Bicol region (44.9 per cent) and Zamboanga Peninsula region (42.7 per cent), higher than the national average of 26 per cent.[22] In provincial levels, poverty incidences in mining areas, such as Masbate (54.2 per cent) in the Bicol region, Agusan del Sur (58.1 per cent) in the Caraga region, Zamboanga Sibugay (49.8 per cent) on the Zamboanga Peninsula, were higher than their regional averages.[23]

TABLE 11.2
Poverty Incidence by Sector
(population in percentage)

Sector	1988	1991	1994	1997	2000	2003	2006	2009
Agriculture	56.33	54.61	51.15	47.10	48.28	46.10	47.84	47.92
Mining	27.84	28.63	30.22	29.50	34.80	41.27	34.64	48.71
Manufacturing	24.29	22.13	15.71	13.72	14.96	14.51	16.19	17.79
Utilities	8.73	11.41	8.23	7.58	4.43	4.12	7.44	3.23
Construction	37.21	34.70	29.40	22.27	25.83	21.49	25.19	24.52
Trade	21.42	21.31	15.77	13.34	12.89	10.72	13.87	13.12
Transport and communication	27.28	20.89	18.45	14.33	15.16	12.79	15.62	18.25
Finance	10.21	9.27	4.85	3.60	7.37	4.83	4.13	2.54
Services	17.42	5.09	12.35	9.76	9.56	9.06	12.41	11.94

Source: Balisacan (2011).

Poverty statistics in these mining areas may not show a causal relationship between poverty and mining operations, but it does point to a correlation that challenges claims that mining reduces poverty and improves local standards of living.[24] An international report by Oxfam-America has found a strong correlation between extractive sectors (including mining) and poverty.[25] In a local study in Rapu-Rapu Island, mining was seen as the cause of poverty in Rapu-Rapu municipality which remained to be the poorest municipality in Albay Province in spite of hosting a large polymetallic project and the biggest private enterprise in Albay.[26]

While "the infusion of capital and technology in mining sector and the resulting creation of new values may appear to be poverty-alleviating influences with positive effects on the national economy", issues of equity and distribution along with associated social and environmental costs may offset these positive effects.[27] The pro-mining side highlights the advantages and opportunities, such as revenue and income generation and local economic development projects, whereas the anti-mining side points out environmental damages and a wide-range of issues, such as governance, corruption, health and safety.

> In both the Philippines and many other developing countries, the public debate over mining rests upon a weak knowledge base, and the statements of government and mining companies have little credibility with affected communities. Communities often know little about the actual mining process and are poorly prepared to judge the nature and seriousness of accidents, real or alleged. Both government and mining companies do a poor job of communicating and sharing information with the public-at-large. Indeed, the lack of transparency on the part of many mining companies is counter-productive and a threat to the viability of the industry. However, it is also true that anti-mining advocates often make exaggerated claims or inaccurate statements that detract from rather than enhance the quality of public debate.[28]

In Toledo City in Cebu Province, the positive impacts of mining gave "rise to enclaves of development within predominantly backward and stagnant areas", as if two separate and independent economies had co-existed in the city.[29] In the author's fieldwork in Rapu-Rapu Island, Albay Province in 2004–5, those barangays[30] covered by or close to mining operations exhibited considerable improvements in infrastructures and quality of life compared to other barangays in the island. However, even within those "improved" barangays, benefits and opportunities brought about by mining were unevenly distributed to locals. This inequality sparked the division within communities. On one side, there were those who were employed and earned their living directly or indirectly from mining, and on the other side, those who were displaced from their livelihood and suffered from mining, especially from mines' spillage incidents.

Lack of job opportunities, poverty, and unstable income from traditional livelihoods characterize the prevailing conditions of local communities that host many mining investments including those of the Chinese. Because of dire living situations, the locals have high hopes which are often detached from the intention of rent-seeking investments from outsiders.

Resistance of Local Communities against Chinese Mining[31]

Due to the questionable contribution of mining to poverty-alleviation and its destructive impacts on the environment and traditional livelihoods, it is perhaps not surprising that local resistance against mining is widespread. Resistance is not solely anchored to the lack of socio-economic benefits, but also on the unclear role and participation of local communities in the development agenda and on adverse medium- and long-term effects of mining operations on their communities after the end of a mine's life.

For example in Mindoro Island, the resistance against Intex Resources Corporation (IRC), which inked a financial deal with China-based MCC8 Group, to mine nickel in the island is led by a coalition of locals, civil society groups and Catholic Church known as *Alyansa Laban sa Mina* (ALAMIN) or alliance against mining. Since 2009, the coalition has been able to temporarily stop the mining exploration of IRC through the revocation of its environmental compliance certificate or mining permit by the Secretary of the DENR. The arguments of the coalition for resisting the planned mining project in Mindoro Island include the lack of social acceptability of the project, incursion in the ancestral domain area of indigenous peoples known as Mangyan, and overlap of proposed mining site with watershed area, thus endangering the ecological integrity of the island.[32]

While the national government welcomes foreign mining investments for economic development and increased revenues, many local communities and civil society groups harbour anti-mining sentiments. For the anti-mining groups, mining is not sustainable as it depletes natural resources. They also argue that it destroys the environment that supports the livelihood of people, thus displacing people from their roots and aggravating their poverty.

It is clear that there are local tensions in mining areas in the Philippines. Coupled with anti-Chinese sentiments and China's involvement in mining, these tensions become more complicated. Additionally, government and local officials are inclined to favour these mining investments, especially Chinese ones that are more financially generous and aggressive. However, those locally-elected officials who are up for re-election may find themselves in a dilemma because the locals who oppose mining are less likely to put them back to power.

In Brooke's Point, Palawan Province, Atty. Mary Jean Feliciano won the mayoralty election in 2013[33] with her staunch anti-mining stance against MacroAsia, majority-owned by Filipino-Chinese tycoon Lucio Tan, that earned the right to mine nickel in 1,114 hectares in the province and entered into financing agreement in 2007 with Jinchuan Group, the largest nickel mining company in China. She was supported by anti-mining groups and the indigenous peoples who rose up to defend their ancestral lands, a UNESCO-designated Biosphere Reserve.[34] They organized an advocacy campaign network, Ancestral Land/Domain Watch (ALDAW) defending and saving their ancestral lands and resources from mining and logging operations.[35] MacroAsia planned to conduct a full-scale operation in Mt. Gantong in Brooke's Point Municipality, Palawan. It received an endorsement from municipal council, but without the process of consultations from affected communities. Describing the protest of the planned operation by MacroAsia and the disregard of their public officials on the role of constituents in decision-making processes, the Center for World Indigenous Studies filed this report:

> On 7 June 2010, a 'Karaban' anti-mining rally has been organized. In the Palawan indigenous language, 'Karaban' is the bamboo quiver containing the blowpipes' darts. It is a symbol of ethnic identity; but, in this specific case, it signifies that people are willing to take whatever action is necessary to stop the penetration of mining companies on their traditional territories.[36]

In May 2011, according to ALDAW, a group of thirty tribal leaders who claimed to represent the indigenous peoples of the affected communities of MacroAsia's planned operations went to Manila to express and show support in the effort to get a document, Certificate of Precondition, indicating that all requirements have been met by the proponents and that it can legitimately commence mining operations in the areas traditionally occupied by indigenous peoples.[37] ALDAW questioned the right of these people to represent them, further exposing that these "fake" leaders were appointed by MacroAsia and the provincial National Commission on Indigenous Peoples (NCIP) to deaden ALDAW's advocacy against the proposed mining operations. To counter this adverse move and nullify the testimonies of the "fake" leaders supporting the proposed MacroAsia's mining operations, ALDAW decided to send its own delegation of leaders to the NCIP in Manila during the Commission's en-banc meeting. Among other goals in their visit to NCIP, they called for investigation of ongoing mining activities in the ancestral domain of indigenous peoples in Palawan without legitimate consent and consultation from affected tribal communities.[38] To raise greater public awareness and gain popular support of their resistance against extractive activities in their ancestral lands, ALDAW utilized social media tools, such as Facebook, Twitter, Youtube and Vimeo. On 23 February 2012, NCIP denied MacroAsia's request for Certificate of Precondition.[39] This means that MacroAsia could not proceed with its planned full-scale mining operations in Brooke's Point, Palawan and that ALDAW could declare their advocacy a success.

Another example is what happened in MacArtur Town of Leyte Province in 2012. Villagers composed mostly of farmers and fisherfolks, along with the Catholic Church through the Catholic Bishops Conference of the Philippines (CBCP), urged the national government to halt the mining operations and the continuous conversion of farm lands to mining sites by Chinese mining company, Nicua Mining Corporation.[40] Media reports indicated that rent-seeking landowners leased or sold their farmlands to Nicua Mining Corporation, thus displacing tenant-farmers from their livelihood and rendering them jobless and forced to work in the mine.[41] Fisherfolks who depend their livelihood on Lake Bito were also heavily affected by the mine due to reported incidences of fish kill on 12 May 2012, and gradual drying up of the lake. Their organization, Unahin Lagi Natin Ang Diyos-Bito Lake Fisherfolks Association (UNLAD-BLFA), took part in Solidarity Mission on 11–12 June 2012, together with the Catholic Church and civil society groups, and barricaded certain areas to prevent further damages to farms and Lake Bito from which they earn their living. One of the leaders of the Solidarity Mission, Fr. Edu Gariguez, the Executive Secretary of CBCP's National Secretariat for Social Action, was quoted challenging government officials: "to see for themselves how Chinese suck our magnetite sand and transport it with ease to China. We are fighting for sovereignty over Scarborough Shoal, but in front of us – face to face – they (Chinese) are claiming our lands."[42] Findings from the Bureau of Fisheries of Aquatic Resources (BFAR) showed that oil and grease from mining operations contributed to the massive fish kill. To consolidate their support from villagers, UNLAD-BLFA was renamed to UNLAD-BLFFA to include farmers who were affected by the mining operations.[43] In June 2012, the DENR ordered Nicua Mining Corporation to pay a fine of P50,000 (US$1,163) for contaminating the lake that led to the fish kill, but the president of UNLAD-BLFFA, Jesus Cabias, said that the villagers wanted the mining operations to stop.[44] Hence, with the support of Environmental Legal Assistance Center (ELAC), UNLAD-BLFFA through its president, Mr Cabias, filed an application to MGB-DENR for TEPO to halt the mining operations of Nicua Mining Corporation in MacArthtur, Leyte.[45] On 17 August 2012, the MGB-DENR ordered the suspension of the operations of Nicua Mining Corporation in MacArthur, Leyte citing violations of several provisions of the agreement with DENR, such as the employment of Chinese nationals from another Chinese company, Heng Sheng Mining Corporation, to work with Nicua and the non-authorization of DENR in the operations of mining companies in areas under Mineral Production Sharing Agreement (MPSA).[46] UNLAD-BLFFA welcomed the suspension order and continued its advocacy against mining. It has gained legal personality and has become a reputable people's organization in Leyte Province.

More importantly, the bulk of Chinese mining operations is with hundreds of thousands of small-scale miners operating in more than thirty provinces across the Philippines.[47] The regime of illegality in small-scale mining operations that evade taxation and regulation is pervasive. It is reported that more than 80 per cent of small-scale operations are without permit.[48] This kind of regime is ripe for Chinese mining firms to exploit. Indeed they do. Examples are the Jiangxi Rare

Earth & Metals Tungsten Group, Wei-Wei Group, and Nihao Mineral Resources Incorporated which own five mines disguised as "small-scale" in Zambales.[49] By having arrangements with local government officials and small-scale miners, many Chinese mining firms are able to operate. But they, more often than not, disregard the limitations set by law of their small-scale operations or conduct illegal mining operations in cahoots with local officials and small-scale miners.[50] With the aid of local communities, hundreds of Chinese nationals have been arrested for illegal mining activities in the country. In a small province, Camarines Norte alone, there were fifteen Chinese nationals who were nabbed for illegal mining.[51]

Opposition to mining by local communities is not unique to Chinese mining investments. The motivating reasons of local communities to protest against Chinese mining are generally similar to other mining investments by foreign and national entities. They can be summarized into three key reasons: (1) mining companies try to circumvent and, worse, violate laws that protect the environment from which many locals source their living. Examples are the cases in Mindoro with IRC and in Palawan with MacroAsia; (2) mining companies invest heavily on infrastructure projects, such as roads, school buildings, chapels, barangay halls, etc., but these projects are not responsive to the locals' basic needs. An example of this case is Leyte with Nicua Mining Corporation; (3) mining companies are not inclusive and transparent in their decision-making processes and operations. Examples are the cases in Mindoro with IRC and in Palawan with MacroAsia. Locals feel excluded from the fate of their own mineral-rich place which is now being directed and controlled by outsiders. Given the backdrop of anti-mining mobilizations by local communities, what sets apart Chinese investments from other foreign and national investments on mining is the sentiment-evoking territorial disputes and massive reclamation activities by China in the West Philippine Sea (South China Sea). Chinese mining investments then face double whammy from local communities in the Philippines. One, they are being seen as environmentally destructive and extractive operations, and the other is that they are being antagonistically treated as representatives of a rival country that has illegally claimed, occupied and taken away group of islands that is not rightfully its own.

Conclusion

Mining in the mineral-rich Philippines, as government records have shown, continues to attract foreign investments, while large-scale and small-scale mining gross productions have posted contrasting trends. Large-scale mining gross productions have steadily been growing whereas small-scale productions have steeply plunged in the last three years. This sharp fall in small-scale productions is mainly attributed to unregulated and unaccounted productions that were reportedly sold to black market and shipped mostly to China. The author has presented the mining laws instituted and aimed to regulate and account for mining operations and productions, but small-scale mining operators have learned how to circumvent and skirt relevant laws in cahoots with local officials with outside financing from mostly Chinese source. Local communities in defense of the environment from which they source their living and of their ancestral domain, along with civil society groups and the local Catholic Church, generally resist mining operations and activities in their localities. It is even more so when mining investors, financiers and workers are Chinese. In these cases, anti-mining mobilizations are conflated with anti-China sentiments because of the growing tension in the disputed islands. Thus in light of the case examples and of motivating reasons of local opposition to Chinese mining, to soften opposition to it, several recommendations could be put forward: (1) ensure Chinese firms comply with national and local laws. This is easier said than done, but there is no better way to proceed than with legal operations. Arrest and suspension orders from authorities will be avoided if laws are followed; (2) make Chinese investments responsive to locals' basic needs. As the saying goes, "A hungry stomach creates a lot of noise"; (3) make Chinese operations more inclusive and transparent to the locals. The anti-mining mobilizations, as previously shown in the examples in mining areas, were participated by locals who felt excluded and alienated from the major decisions affecting their homeland.

Apart from the West Philippine Sea (South China Sea) dispute-induced anti-Chinese sentiments, there is a prevailing notion from the locals that something associated with China is "substandard", meaning prone to accidents and disasters. This notion also contributes to the resistance of local communities and civil society groups against Chinese mining. They have become resigned to the fact that their communities may be accustomed to disasters, such as typhoons and earthquakes, but not another man-made Marcopper-like disaster which in 1996 rendered some villages in Marinduque uninhabitable.[52] They fear a similar scenario happening to their own communities with Chinese companies in their midst. This notion by the locals has contributed to the opposition to Chinese mining involvement in the Philippines.

The myriad of contexts presented here reflects the complex realities on which Chinese mining companies, either legally or illegally, operate. First is the existence of tension and division between anti-mining groups and public officials in mineral-rich communities. Second is the favourable stance of national and local government officials on investments, including mining. Third is the regime of illegal small-scale mining which is widespread in the country. Chinese mining firms have been adept at taking advantage of these circumstances for their own best interests.

NOTES

1. Rules of Procedure for Environmental Cases, Rule 1, Section 4 (d).
2. Jarius Bondoc, "SC Stops Zambales Mines; Chinese 'Invaders' Socked", *The Philippine Star*, 24 July 2013.
3. The clash of development frameworks is the basis of the conflict between pro-mining groups (mining firms and government units) and anti-mining groups (church, civil society and local communities). The pro-mining groups pursue economic development while the anti-mining advocates sustainable development. See Menandro Abanes, "People's Organizations (POs), Non-Government Organizations (NGOs) and the Catholic Church vs. the State and a Transnational Corporation (TNC): A Critical Discourse Analysis of Mining Issues in Rapu-Rapu Island, Albay", *Gibon* 5 (2005): 1–28.
4. Tom Stern, "Chinese Investments in the Philippines", *The Journal of Political Risk* 4 (2016), available at <http://www.jpolrisk.com/chinese-investments-in-the-philippines>; Asia Sentinel, "Indigenous Miners Help China Loot the Philippines", 13 November 2012, available at <http://www.asiasentinel.com/econ-business/indigenous-miners-help-china-loot-the-philippines>.
5. Roberto Romulo, "Small Scale Mining: Immeasurable Damage", *The Philippine Star*, 18 October 2013; also see Anthony Vargas, "14 Chinese National Arrested for Illegal Black Sand Mining Activity in Camarines Norte", 7 February 2014.
6. Exchange rate PhP 43 = US$1.

7. Rosemarie Francisco, "Special Report: Philippines' Black Market is China's Golden Connection", *Reuters*, 22 August 2012, available at <http://www.reuters.com/article/2012/08/23/us-philippines-gold-idUSBRE87M02120120 823#tYxtHGkAmWeUEidT.97>.

8. Pak Nung Wong, Kathleen Aquino, Kristinne Lara-De Leon and Sylvia Yuen Fun So, "As Wind, Thunder and Lightning: Local Resistance to China's Resource-led Diplomacy in the Christian Philippines", *South East Asia Research* 21, no. 2 (2013): 281–302.

9. Jill Shankleman, "Going Global: Chinese Oil and Mining Companies and the Governance of Resource Wealth", Wilson Center, July 2011, available at <https://www.wilsoncenter.org/sites/default/files/Shankleman_Going%20 Global.pdf>.

10. Jeanette Andrade, "$14-B Investments in Mining Eyed from China Within the Next 5 Years", *Philippine Daily Inquirer*, 7 September 2011, available at <http://business.inquirer.net/17327/14-b-investments-in-mining-eyed-from-china-within-the-next-5-years>.

11. Jeffrey Stark, Jennifer Li, and Katsuaki Terasawa, "Environmental Safeguards and Community Benefits in Mining: Recent Lessons from the Philippines", Working Paper No. 1, Foundation for Environmental Security and Sustainability, 2006.

12. This includes the Environmental Impact Statement (EIS) which "refers to the document(s) of studies on the environmental impacts of a project including the discussions on direct and indirect consequences upon human welfare and ecological and environmental integrity. The EIS may vary from project to project but shall contain in every case all relevant information and details about the proposed project or undertaking, including the environmental impacts on the project and the appropriate mitigating and enhancement measures" (DENR Administrative Order 96-40, series of 1996). The EIS is reviewed and validated by Environmental Impact Assessment Review Committee (EIARC) which is "a body of independent technical experts and professionals of known probity from various fields organized by the EMB/RED whose main tasks are to evaluate the EIS and other documents related thereto, and make appropriate recommendations to the EMB/RED regarding the issuance or non-issuance of ECCs" (DAO 96-37).

13. Menandro Abanes, "Under-Mining the Power of Communities: The Politics of Mining and Local Community in the Philippines", Grin Verlag, Munich, 2013.

14. There are varied and complex informal arrangements among stakeholders in small-scale mining in the Philippines. Often one finds a three-tier structure: on top are the financiers (outsiders) and surface land claimants, in the middle are the miners with their own specialists and diggers, and

at the bottom are the casual labourers. For detailed descriptions, see Boris Verbrugge and Beverly Besmanos, "Formalizing Artisanal and Small-scale Mining: Whiter the Workforce?", *Resources Policy* 47, no. 3 (2016): 134–41.

15. See ibid.
16. Francisco, "Special Report: Philippines' Black Market is China's Golden Connection", op. cit.
17. Boris Verbrugge and Beverly Besmanos, "Undermining the Myths about Small-scale Mining", Bantay Kita Philippines, 2015.
18. Verbrugge and Besmanos, "Formalizing Artisanal and Small-scale Mining", op. cit., pp. 134–41.
19. Tony Bergonia, "Chinese Mining Firms Skirt PH Laws", *Philippine Daily Inquirer*, 21 May 2011.
20. Arsenio Balisacan, "What Has Really Happened to Poverty in the Philippines: New Measures, Evidence, and Policy Implications", Discussion Paper No. 2011-04, UP School of Economics, 2011.
21. Caroline Howard, "Philippine Mining Industry: Boon or Bane?", ABS-CBN News, 2011, available at <http://www.abs-cbnnews.com/-depth/12/08/11/philippine-mining-industry-boon-or-bane>.
22. Christian Monsod, "Mining a Social Justice Issue", *ABS-CBN News*, 2012, available at <http://www.abs-cbnnews.com/-depth/03/03/12/christian-monsod-mining-social-justice-issue>.
23. National Statistical Coordination Board, *Poverty Statistics*, 2009.
24. Monsod, "Mining a Social Justice Issue", op. cit.
25. Michael Ross, "Extractive Sectors and the Poor: An Oxfam-America Report", 2001, available at <http://www.oxfamamerica.org/static/oa3/files/extractive-sectors-and-the-poor.pdf>.
26. Emerlina Regis, "Gold Mining Activities as Cause of Poverty of Local Communities in Gold Mining Areas", *Gibón* 4, no. 1 (2004): 81–146.
27. Risk Asia Consulting, "Fool's Good: The False Economic Promises of the Lafayette Mining Project in Rapu-Rapu", Greenpeace Southeast Asia, 2006.
28. Stark, Li, and Terasawa, "Environmental Safeguards and Community Benefits in Mining", op. cit.
29. John McAndrew, "Mining Industry Report: The Impact of Corporate Mining on Local Philippine Communities", ARC Publications, 1983.
30. Smallest political and administrative unit in the Philippines.
31. Chinese mining here is broadened to include not only actual engagement in mining operations, but also mining-related deals, be they financial or shipment, with Chinese firms.
32. E. Gariguez, "Reply to Ambassador's Investigation Report on OECD Complaint vs. Intex", 2010.
33. See <http://election-results.rappler.com/2013/region-4b/palawan/brooke's-point>.

34. UNESCO refers to biosphere reserves as "ideal sites for research, long-term monitoring, training, education and the promotion of public awareness while enabling local communities to become fully involved in the conservation and sustainable use of resources". See Biosphere Reserves, "The Seville Strategy and the Statutory Framework of the World Network", (UNESCO, Paris: Biosphere Reserves, 1996), available at <http://unesdoc.unesco.org/images/0010/001038/103849e.pdf>.

35. Melody Kemp, "Palawan Tribes Go Cyber to Keep Out Nickel Miner", *Asia Times*, 9 December 2011.

36. John Ahni Schertow, "Indigenous Peoples Unite Against Mining in Palawan", *IC Magazine*, 5 June 2010, available at <https://intercontinentalcry.org/indigenous-peoples-unite-against-mining-in-palawan/>.

37. Rainforest Rescue, "Philippines: Indigenous Peoples of Palwan Against Mining Operations", 2 June 2011, available at <https://www.rainforest-rescue.org/press-releases/3565/philippines-indigenous-peoples-of-palawan-against-mining-corporations-1>.

38. Ibid.

39. Forest Peoples Programme, "Philippines: ALDAW Update on the NCIP Resolution Denying MacroAsia (MAC) Corporation's Request for a Certificate of Precondition", 15 June 2012, available at <http://www.forestpeoples.org/topics/extractive-industries/news/2012/06/philippines-aldaw-update-ncip-resolution-denying-macroasia>.

40. GMA News Online, "CBCP Official Seeks Stop to Magnetite Mining in Leyte Rice Fields", 13 June 2012, available at <http://www.gmanetwork.com/news/story/261687/news/regions/cbcp-official-seeks-stop-to-magnetite-mining-in-leyte-rice-fields>.

41. Samar News, "Farmers and Fishers of Leyte Cry Freedom From Mining", 12 June 2012, available at <http://www.samarnews.com/news_clips20/news415.htm>.

42. Ibid.

43. Marissa Miguel Cano, "Unahin Natin Lagi ang Diyos Bito Lake Fisherolk and Farmers Association", in *MacArthur Leyte: Rising Above the Challenges of Mining and Super Typhoon Yolanda*, edited by Farah Sevilla, Check Zabala and Tess Tabada (Quezon City, Philippines: ATM National Secretariat, June 2014), available at <https://file.ejatlas.org/docs/MacArthur__Leyte_Post_Yolanda_publication.pdf>.

44. Joey Gabieta, "Villagers Still Hurting From Leyte Fishkills", *Philippine Daily Inquirer*, 4 August 2012, available at <http://newsinfo.inquirer.net/242303/villagers-still-hurting-from-leyte-fishkills>.

45. Philippine Indigenous Peoples Links, "Farmers, Fisherfolks File Petition for Environmental Protection Order Against Mining", 28 June 2012, available at

<http://www.piplinks.org/farmers,-fisherfolks-file-petition-environmental-protection-order-against-mining.html>.

46. Philippine Indigenous Peoples Links, "PH Suspends Blacksand Mining Ops in Region VIII", 29 September 2012, available at <http://www.piplinks. org/ph-suspends-blacksand-mining-ops-region-viii.html>.

47. Pacific Strategies and Assessments (PSA), "Exploitive Chinese Mining in the Philippines", Manila, 2011.

48. Verbrugge and Besmanos, "Undermining the Myths about Small-scale Mining", op. cit.

49. Bondoc, "SC Stops Zambales Mines; Chinese 'Invaders' Socked", op. cit.

50. Bergonia, "Chinese Mining Firms Skirt PH Laws", op. cit.

51. *Bikol Today*, "Police Crackdown on Illegal Mining in Camarines Norte Nets Chinese National", 26 June 2014.

52. Catherine Coumans, "Whose Development? Mining, Local Resistance, and Development Agendas", in *Governance Ecosystem: CSR in Latin American Mining Sector*, edited by Julia Sagebien and Nicole Lindsay (New York: Palgrave Macmillan, 2011), pp. 114–32.

12

CONCLUSION

Tai Wei Lim

Morris-Jung's introduction to this volume began with a macro analytical framework of China's resource encounters in Southeast Asia. It pulled together a wide diversity of issues covered in the individual chapters to introduce the central framework of this volume, which is a careful and detailed analysis of diverse issues centred on local and national development, environment, justice, power relations, territoriality, sovereignty and many others. Morris-Jung's chapter painted the complexities that all authors in the volume grappled with. How do scholarly works capture the highly-nuanced, contextual and complicated domestic and external factors behind the "new wave" of Chinese investments into the Southeast Asian region, which represents an interplay of both opportunities, challenges and risks? Morris-Jung correctly identified the multi-layered implications of these investments and subsequently each chapter contributor highlighted and surveyed these complex layers through case studies, local-area examples, contextual analysis and empirical data.

What Morris-Jung and the other chapter contributors have highlighted is the complicated ecology of state-owned enterprises (SOEs), private sector firms, local communities, regional elites, national politicians, Chinese diaspora and bureaucracies interacting with each other. In analysing them, we should not, as Morris-Jung reminds us, fall into a territorial trap of resource extraction based on unquestioned assumptions of (1) territory as geometrically bounded and inwardly focused and (2) the sovereign autonomy of the nation-state. This would unnecessarily delimit the imaginative as well as analytical spaces available, which the rest of the chapter contributors have explored with vigour and incision. Therein lies the most important contribution of this edited volume — a multi-perspective and multi-spatial analysis of the ecology of Chinese mining, resource extraction and investments in the diverse region of Southeast Asia.

Macro-regional Analyses

Philip Andrews-Speed, Mingda Qiu and Christopher Len's chapter "Mixed Motivations, Mixed Blessings: Strategies and Motivations for Chinese Energy and Mineral Investments in Southeast Asia" identified the specific motivations of Chinese enterprises and government in their engagements with Southeast Asia's energy and mineral resources, and examine some of their implications from their mix of corporate and state drivers. In this chapter, the authors argue that the increasing participation and investment of Chinese hydrocarbons, petrochemicals and commodities/resource firms may be showing differential outcomes in Southeast Asian countries due to varying incentives used, state priorities, global business experience, and sometimes less than optimal management. Although these firms attract Chinese funding to the region, it may also bring about negative impacts if project promises are not delivered. Both investing nation and project recipients may be disappointed. The divergence between ground reality and investment plans in this chapter support Morris-Jung's introductory argument that there is a need to deconstruct the uncritical assessment of investment perpetuated in mainstream narratives.

An important limitation of Chinese funding and investments is the fact that the close relationships between the state and private sector, state-owned or government-linked firms facilitated the conditions for monetary corruption and rent seeking. The Dunning's framework that the three authors used indicate the trend that Chinese resource, mineral and energy companies' investments and projects in Southeast Asia are strategically primed to open up doors to extract natural resources for

Chinese consumption or augment the private sector's output of the commodity, raw material or natural resource. Operating and investing outside China positions Chinese companies closer to the site of raw materials, foreign consumer markets, and bypasses domestic legal restraints on Chinese companies. In this case, the three authors argue that it is possible to de-privilege the notion of the sovereign nation-state as Chinese and other resource and mining firms have found a way to overcome the regulatory framework of the state.

Specific Case Studies

In terms of specific case studies, the chapter contributors can be categorized into two major categories — Chinese energy, resource and mining investments in mainland Southeast Asia and maritime ASEAN countries. China often conceptualizes its foreign relations in terms of concentric circles, addressing neighbouring regions as peripheral states (*zhoubian guojia*) and major powers as big nations (*daguo*). The chapters in the volume reflect this complexity of Chinese positioning of Southeast Asian states and Southeast Asian conceptualizations of China. The peripheral states or *zhoubian guojia* here really refers to the Indo-Chinese states and states of mainland Southeast Asia surrounding the boundaries of China. Chapter contributors to the volume also study the maritime states in Southeast Asia, located further away spatially but knitted up in a noodle bowl network of trade, investments and production networking. In other words, every Southeast Asian state has some degree of economic interdependence with China. The chapter contributors who have selected specific case studies to analyse in the volume perform the task of decoding some of these intertwined complexities in bite-sized scale by focusing on specific features of bilateral political and economic dynamics between China and Southeast Asian states.

Mainland Southeast Asia

Juliet Lu's chapter examines how narratives of a "Chinese model" that has shaped the rationalization and practices of Chinese rubber companies in Laos through the state-sanctioned Opium Replacement Program (ORP) that provides alternative agricultural livelihoods to opium cultivation. Lu's most important contribution is analysing the challenges and contradictions that arise when Chinese concepts and China's unique historical experience of development are transplanted to other contexts. Lu argues that there is divergence between implementation results and developmental narratives in the activities of Chinese rubber cultivation firms in Laos. Lu's study shows how agricultural cooperation in cash crop resources can take off when there is an alignment of interests between Laotian and Chinese economic interests exuding economic complementarity.

But, as Morris-Jung reminds readers, the story does not end with only first-level supply and demand economic complementarity. It needs to go further and probe deeper both narrative and practice of Chinese mining and resource development. The discourse of Chinese development is embedded in the investment practices of individual Chinese rubber companies in Laos and guided by the logic of the managers of these companies who justified and promoted rubber expansion. But, because of the power and influence of Chinese capital, Lu argues that Laos is now under the considerable influence of Yunnan Province as Chinese investments determine the shape and well-being of the Laotian economy, land use and its local and national politics. Lu's chapter is almost the quintessential account of economic development in the periphery of China, with the visible influence of Chinese economic power directly shaping the agenda of Laotian developmentalism. And as the chapter points out, a commonality emerges amongst the *zhoubian guojia* or peripheral states at China's borders which is the inevitable direct arrival and influence of Chinese state power and subnational state actors (in the case of Lu's chapter, the presence of Yunnan provincial government in Laos) riding on Chinese economic investments. As Laos is a close Chinese partner and sometimes ally, strong Chinese economic influence is managed carefully and diplomatically but this may not be the case in other Indo-Chinese states, as Morris-Jung and Pham Van Min's chapter suggests.

Jason Morris-Jung and Pham Van Min's chapter titled "Anti-Chinese Protest in Vietnam: Complex Conjunctures of Resource Governance, Geopolitics and State–Society Deadlock" is another chapter based on mainland Southeast Asia's Indo-China setting. The chapter examined an anti-Chinese public discourse that created distrust and suspicions towards Chinese investments in resource and energy sector projects. But Morris-Jung and Pham Van Min warn that outwardly directed nationalism and fears about the rise of Chinese dominance in Southeast

Asia is only part of the milieu that also includes growing public discontent with Vietnamese political control and party leadership at both local and national levels. The idea of an inseparable dual-tracked integrated domestic and external lens through which stakeholders view Chinese mining and resource investments in Southeast Asia permeates throughout all the chapters in this edited volume. Morris-Jung and Pham Van Min's case studies of the Chinese Chalco and Taiwanese Formosa are critiqued within this context of the complicated domestic and external factors. The domestic factors of inadequate regulation, corruption, and lack of transparency or accountability coupled with the growing and larger scale of Chinese and other foreign mining and resource operations in Vietnam has accentuated problems and issues of resource management, increasing their public visibility and impact on members of the Vietnamese public. Infringement into the local communities and public sphere is a common feature throughout all the country studies covered in the edited volume. It is a good example of critically examining the complex multi-layering of stakeholders' interests in mining and resource projects, involving not only the powerful state, but also incentivized subnational actors as well as the de-privileged stakeholders in the form of members of the public, local communities and non-government organizations. The deepest structural and systemic problems that Morris-Jung and Pham Van Min have identified are that at the core of the Vietnamese political system is an absence of an institutional outlet for members of the public to air their grievances and provide feedback for improving the existing situation.

Completing the CLV (Cambodia, Laos and Vietnam) analysis, Siem Pichnorak's chapter titled "The High Cost of Effective Sovereignty: Chinese Resource Access in Cambodia" examines the socio-economic impact of Chinese investments in Cambodia. He detects the complicity of Chinese firms holding rights and licenses in land, dam and mining commercial activities collaborating with local authorities to partake in non-legal activities like land grabbing, resources destruction, deforestation and human rights contraventions. These activities have in turned provoked unrest from local communities, civil society and international observers. Accusations based on the negative impact of these protests and illegal activities are integrated into Cambodian domestic political battles as opposition parties cite them to discredit the ruling party, according to the chapter contributor. This complex story and narrative of Siem's case study resonates with Morris-Jung and Pham Van Min's chapter where domestic political factors and external investments are drawn into a socio-economic milieu in almost every Southeast Asian country touched by Chinese and other foreign investments. Societal costs are very much at the core of this chapter, like almost every single-country study in this edited volume.

This brings readers back to Morris-Jung's introductory chapter which cautions against over-privileging the autonomy of the state. Due to its overwhelming sovereign presence and power, it masks the true costs of the investments as non-state actors like local communities and the citizenry which do not enjoy privileged or institutional access to authoritarian government agencies and feedback channels turn to the streets to make their voices heard. Coupled with the spectre of nationalism in Vietnam, it even became an anti-Chinese riot. The fact that disquiet is heard in the political sphere of Cambodian parliamentary politics shows that disenfranchised elites have also taken up the call for critique and change. Even in the case of China's current closest ally and partner within ASEAN, Chinese mining and resource investments are not unproblematic or without critics and opposition.

Siem's chapter points out that one of the reasons behind the greater presence of Chinese investments in Southeast Asia is due to China's "going out" strategy. Siem reiterates that, in Cambodia's case, corruption, nepotism and regulatory enforcement flaws have been exploited by Chinese firms heeding the call of "going out" with regulations and laws circumvented or avoided because of loopholes. To mitigate and prevent cases of such exploitation of national laws in Cambodia, Siem argues that both Cambodian and Chinese governments have to cooperate in sharing information and joint enforcement to reduce corrupt practices in resource investments, lest it infringes on the rights of the local communities. As Morris-Jung and Pham Van Min point out, in authoritarian political structures, feedback is either effectively suppressed or has the potential to become organically developed into sometimes uncontrollable xenophobic riots.

The last Indo-Chinese case study on Myanmar covered in the edited volume completes the CLMV (Cambodia, Laos, Myanmar and Vietnam) analysis. Tang-Lee examines why many Chinese large-scale investments, including those in the energy and natural resource sectors, ran into local resistance in Myanmar. The presence and emergence of local resistance is another common theme that weaves together the single-country case studies in the edited volume. It is a unifying feature found in case studies drawn from both maritime as well as mainland Southeast Asia. Local pressures on former President Thein Sein influenced the outcome of stopping work on the Chinese-funded Myitsone dam in 2011 which had geopolitical implications for Sino–Myanmar bilateral relations. This was a similar impact found in other chapters' case studies as well. They include the Chinese Yunnan province's strong influence on Laotian economic development (Lu's chapter), Chinese mining companies' economic influence on Filipino local governments of resource-rich areas (Abanes' and Camba's chapters), the outbreak of anti-Chinese riots in Vietnam (Morris-Jung and Pham), parliamentary opposition voices in Cambodia (Siem's chapter), tendency of the state to ignore civil society and mass media feedback (Springer's chapter). Tang-Lee focuses on the interactions between major Chinese players and Myanmar's civil society which is characterized by distrust and lack of communication, affecting the formal channels of communications between state agencies that can help augment the trust factor between the two countries.

Tang-Lee's important contribution to the edited volume is her observation that a new age of business practices implemented by strong governments and state-owned companies may be emerging in Myanmar. Due to increasing scrutiny from the Myanmar public, civil society and mass media, steps are implemented by the Chinese Ministry of Foreign Affairs, Embassy in Myanmar and semi-governmental organizations to increase exchanges between the peoples of both countries and also to become more encompassing by including parties located at various ends of the political spectrum. The firms engaged the three major Sino–Myanmar projects, including the Myitsone dam, the Letpadaung copper mine, and the Sino–Myanmar oil and gas pipelines, printed corporate social responsibility (CSR) reports and Environmental and Social Impact Assessments (ESIAs), all of which point towards a possible new model of Chinese firms conducting resource business in Southeast Asia. Tang-Lee argues that these new measures challenge the traditional foreign policy and foreign investment norms of China.

Maritime Southeast Asia

Maritime Southeast Asia historically enjoyed physical distance away from Chinese economic and geopolitical power and influence but, with globalization, distance is now overcome by not just physical infrastructure and transportation technologies but the presence of agents of trade and commerce, including the ethnic Chinese diaspora in the case of China. Alvin Camba's Filipino case study titled "The Direction, Patterns, and Practices of Chinese Investments in Philippine Mining" is a useful reference for an in-depth analysis of Chinese mining operations in the Philippines. Camba profiles the typical Chinese mining and resource extraction firms in the Philippines. Some features include observations in the following areas: (1) method of production; (2) method of accumulation; and (3) host country linkages. Chinese investments tend to take on artisanal small-scale mining (ASM) operations to evade attention of the Filipino state and unfriendly parties who are stakeholders in the maritime disagreements between China and the Philippines.

Camba also observes that small scale ASM mining draws on an eclectic array of techniques focused on manpower-intensive extraction, local area support and community mobilization, contextually-driven infrastructure and host country connections. International mining firms typically cultivate the partnership of state departments to carry out extraction activities. Chinese companies on the other hand depend on personal connections and networks of ethnic Chinese Filipino locals to work with local elites, regional officials and other local-area bureaucrats to put in place their projects, thereby avoiding national politics. Camba provides a useful historical narrative in detailing the rise of new Chinese elites, known as the *Taipans*, who started to control a significant chunk of the Philippine economy. The most important contribution made by this chapter is contextualizing Chinese investments in the economic sector of the Philippines at the subnational level against the backdrop of racial, ethnic and class differences.

Menandro Abanes' chapter titled "Complexities of Chinese Involvement in Mining in the Philippines" is another Filipino case study in this volume that is complementary with Camba's study because, like almost all the chapters in this volume, the author faces the contradictions of Chinese economic power and influence (like Camba who spotted the ability of Chinese mining companies to bypass national governments to work directly with local authorities). Abanes asks the question: why is it that Chinese mining appears unwelcome yet continues to expand in the Philippines? Abanes explains that resistance in the Philippines arose from absence of benefits from development for the local communities from a mining project, negative longer-term impact of mines on the community especially after they are closed.

Abanes argues that resistance is part of the complexity of the mining ecosystem in the local communities of the Philippines.

But at the same time, the Chinese mining companies have become highly skilful in exploiting the tensions and divisions between anti-Chinese mining firms and pro-mining local authorities eager to develop their mineral assets. Local authorities are typically supported by the national government which is also keen on mining investors developing their lands' resources. These form powerful motivations for non-legal mining activities to proliferate and take place in the Philippines. These motivations that drive local and national authorities to develop mining businesses became the same routes for Chinese mining companies to exploit to their advantage when investing in the Filipino mining industries. Abanes' state-centred perspective complements Camba's narrative which included the intermediaries of local ethnic Chinese as a bridge to aggregate state investments from China. Taking both studies together, it is possible to discern a multi-sectoral multi-stakeholder format through which Chinese resource and mining investments are made in Southeast Asia. Camba and Abanes' narratives of resource exploitation can be compared with Cecilia Han Springer's chapter on upgrading the value chain of bilateral trade. Again the story becomes familiar, it is a narrative of how states have mutually beneficial economic arrangements in extracting and exploiting resources but sometimes contravene the socio-economic needs of the local community or ignores feedback from mass media and civil society in the host country or international community.

Another archipelago, the largest one in Southeast Asia, Indonesia is the setting of case study by Cecilia Han Springer in the chapter titled "Energy Entanglement: New Directions for the China–Indonesia Coal Relationship" and is a major chapter focusing on bilateral economic relations based on an energy resource. Springer argues that this bilateral relationship is transformative because it has a global economic impact as Sino–Indonesian coal trade can determine prices of the commodity on the world market, given the enormous populations of China and Indonesia, and the sheer volumes of coal consumed and burnt in both countries causing environmental issues. Therefore, coal trade volumes, formats and decisions made by both countries have an immense impact on the global weather system. Beyond examining the scale of the trade, Springer's story also centres around the idea of value addedness, how

China and Indonesia have upgraded their relationship from downstream coal supply to upstream coal infrastructure investments.

Springer argues that 2015 was a milestone in bilateral Sino–Indonesian trade when political and economic changes caused both countries to draw down their volume of coal trade. For the Chinese, pollution and climate and environmental policies and coping mechanisms led to reduced demand for Indonesian coal and Indonesia devised new policies and curbs on coal exports to develop and nationalize its energy industry. But, in view of these changes, Springer argues that economic relations have become stronger rather than weaker because the bilateral coal trade has gone beyond the format of commodity supply and climbed up the value addedness towards coal infrastructure investments coming from China to augment Indonesian electricity generation. Engaging in higher value-added investments also bring about its own set of challenges. Springer argues that Chinese SOEs and the state may not be as amenable to the criticisms of the civil society, mass media and other organizations that act as informal watchdogs to curtail the excesses of such investment activities. Besides SOEs, Springer noted that private sector players from China and their activities are also less scrutinized by Beijing and other Chinese authorities. Consequently, there is a potential for Chinese electricity coal plants and/or low-efficiency projects to exacerbate the pollution situation in Indonesia rather than resolving it.

Springer's chapter is complemented by another chapter written by Zhao Hong and Maxensius Tri Sambodo in the chapter titled "Indonesia–China Energy and Mineral Ties: The Rise and Fall of Resource Nationalism?". Zhao and Sambodo's chapter continues the emphasis on Indonesian energy commodities in the volume. The chapter analyses various Indonesian worries and reactions to China's energy investments. The authors argue that the effectiveness of Chinese energy investment is dependent not only on China but also on reception in the investment destination and their national policies, laws and capabilities for implementing incoming investments, technology and management know-how. In other words, Zhao and Sambodo highlights Morris-Jung's argument that investment outcomes are not only dependent on the state but all stakeholders (experts, local labour, engineers, local community hosts) involved in receiving the investments and implementing them in the actual mine and resource sites. Local communities' willingness to accept Chinese resource investments is highlighted by Zhao and Sambodo and their priorities are often focused on the ability of investments to provide affordable energy, local employment and a fair competitive environment for local Indonesian companies. The upside to this relationship which provides a conducive bilateral state-level platform for augmenting economic cooperation is the fact that both countries are at an optimistic stage of official relations (elevated to comprehensive strategic partnership from 2005), which was also confirmed by Springer. Zhao and Sambodo argue that a motivating factor for managing resource investments is the alignment of aspirations in becoming major powers on the world stage (the *"daguo"* diplomacy mentioned earlier in this concluding chapter) riding on the back of better infrastructure for trade, more efficient and effective power facilities and better extraction of resources for domestic industrialization.

Conclusion

This concluding chapter began with Morris-Jung's overarching framework that defined the major debates running through all the chapters in this edited volume. It is followed by the tri-authored Andrew-Speed, Qiu and Len's chapter which gave a macro picture of China's resource footprints in Southeast Asia before moving on to examine the outcomes and arguments made by the single-country studies. Perhaps it is useful to end with a Chinese perspective on the issue. Yu Hongyuan and Lim Tai Wei's chapter probably stands out from the rest of the chapters as it examines the lens of Chinese resource, mineral and energy commodities extraction from a Chinese-centred perspective. This chapter analyses the key challenges that China is currently facing in securing supplies of strategic mineral resources from its neighbouring or "periphery" regions (translated as *"zhoubianguojia"* in Chinese hanyu pinyin romanization), which refer to those countries and regions surrounding the Chinese overland borders and coastal areas. The "periphery" is a necessarily loose categorization, but it has particular significance because of the particular geopolitical and close economic relations that China shares with these countries and regions historically. Southeast Asia is a necessarily important part of this periphery, where national and local-level protests against Chinese investment and escalating geopolitical tensions in the South China Sea make for complex political and economic relations.

From a Chinese perspective, in its economic development strategies to expand trade routes across Central Asia through the "New Silk Road Economic Zone" and the waters of Southeast Asia with "the 21st Century Maritime Silk Road", China wants to emphasize trade in strategic mineral resources in Central Asia, Southeast Asia and Australia. These world regions are the main importing regions for energy resources (e.g., coal, oil and gas) and minerals for Chinese state enterprises. Chinese state companies believe that collaborative development of strategic mineral resources with these countries can be used to promote mutual economic development and regional stability. What the subsequent chapters, especially the single-country studies, indicate is that Chinese intentions and plans may not always translate into their intended consequences and results. These chapters and the lessons learnt help to provide the contours to smoothen out the sharper edges of Chinese economic diplomacy conducted by the state, its SOEs as well as the private sector. The subsequent chapters also point towards some common features such as the need for more inclusivity in listening to the opinions of the international, regional and local mass media, non-governmental organizations, local communities and the civil society as they receive the direct impact of economic deals made by the Chinese and Southeast Asian national and local governments. To ensure sustainable long-term harmonious relations with the local communities in a win-win situation as the Chinese government professes, it may be useful to apply some of the lessons learnt in other world regions that global Chinese economic power is outreaching towards as well.

BIBLIOGRAPHY

Abanes, Menandro. "People's Organizations (POs), Non-Government Organizations (NGOs) and the Catholic Church vs. the State and a Transnational Corporation (TNC): A Critical Discourse Analysis of Mining Issues in Rapu-Rapu Island, Albay". *Gibon* 5 (2005): 1–28.

———. "Under-Mining the Power of Communities: The Politics of Mining and Local Community in the Philippines". Grin Verlag, Munich, 2013.

Adaro – Positive Energy. "Adaro Energy and Shenhua Signed MOU to Develop Initially a 2x300 MW Mine Mouth Coal Fired Power Plant in East Kalimantan", 24 November 2014. Available at <http://www.adaro.com/news/read/55/Adaro_Energy_dan_Shenhua_Menandatangani_Nota_Kesepahaman_Untuk_Mulai_Mengembangkan_Pembangkit_Listrik_Mulut_Tambang_Bertenaga_Batubara_dengan_Kapasitas_2_300_MW_di_Kalimantan_Timur> (accessed 23 November 2015).

Agnew, John. "The Territorial Trap: The Geographical Assumptions of International Relations Theory". *Review of International Political Economy* 1, no. 1 (1 April 1994): 53–80.

———. "Sovereignty Regimes: Territoriality and State Authority in Contemporary World Politics". *Annals of the Association of American Geographers* 95, no. 2 (2005): 437–61.

AidData. "Exim Bank Loans $240m to Indonesia for Takalar Steam Coal Power Plant". Available at <http://china.aiddata.org/projects/39373?iframe=y>.

Alden, Chris. *China in Africa*. London: Zed Books, 2007.

Alden, Chris, Daniel Large, and Ricardo Soares de Oliveira. *China Returns to Africa: A Rising Power and a Continent Embrace*. New York: Columbia University Press, 2008.

Allford, Jason and Morkti P. Soejachmoen. "Survey of Recent Developments". *Bulletin of Indonesian Economic Studies* 49, no. 3 (2013).

Alunan III, Rafael. "What's Yours is Mine". *Business World*, 30 July 2013.

Andrade, Jeanette. "$14-B Investments in Mining Eyed from China Within the Next 5 Years". *Philippine Daily Inquirer*, 7 September 2011. Available at <http://business.inquirer.net/17327/14-b-investments-in-mining-eyed-fromchina-within-the-next-5-years>.

Andrews-Speed, Philip. "China and Russia's Competition for East and Southeast Asia Energy Resources". In *The EU–China Relationship: European Perspectives: A Manual for Policy Makers*, edited by Kerry Brown. London: Imperial College Press, 2015, pp. 294–305.

Andrews-Speed, Philip and Roland Dannreuther. *China, Oil and Global Politics*. London: Routledge, 2011.

Andrews-Speed, Philip, Xuanli Liao and Roland Dannreuther. "The Strategic Implications of China's Energy Needs". *Adelphi Paper* 346 (2002).

Armony, Ariel C. and Julia C. Strauss. "From Going Out (*Zou Chuqu*) to Arriving In (*Desembarco*): Constructing a New Field of Inquiry in China–Latin America Interactions". *The China Quarterly* 209, no. 1 (March 2012): 4.

ASEAN Secretariat. *ASEAN Community in Figures (ACIF)*. Jakarta: ASEAN Secretariat, 2014.

Asia Pacific Economic Cooperation (APEC). "Annex C – APEC Accord on Innovative Development, Economic Reform and Growth". Available at <http://www.apec.org/Meeting-Papers/Leaders-Declarations/2014/2014_aelm/2014_aelm_annexc.aspx> (accessed 1 June 2016).

Asia Regional Integration Center, Asian Development Bank (ADB). "Regional Public Goods Asia-Pacific Partnership on Clean Development and Climate", 2015. Available at <https://aric.adb.org/initiative/asiapacific-partnership-on-clean-development-and-climate> (accessed 1 June 2016).

Asia Sentinel. "Indigenous Miners Help China Loot the Philippines", 13 November 2012. Available at <http://www.asiasentinel.com/econ-business/indigenous-miners-help-china-loot-the-philippines>.

Asthana, S.B. "The People's Liberation Army of China: A Critical Analysis". *Combat Journal* 30, no. 2 (1992): 1–8.

Baird, Ian G. "Turning Land into Capital, Turning People into Labor: Primitive Accumulation and the Arrival of Large-Scale Economic Land Concessions in the Lao People's Democratic Republic". *New Proposals: Journal of Marxism and Interdisciplinary Inquiry* 5, no. 1 (2011): 10–26.

Bali Update News. "Bali News: Old King Coal to Soon Power Bali", 1 November 2010. Available at <http://www.balidiscovery.com/messages/message.asp?Id=6457> (accessed 23 November 2015).

Balisacan, Arsenio. "What Has Really Happened to Poverty in the Philippines: New Measures, Evidence, and Policy Implications". Discussion Paper No. 2011-04, UP School of Economics, 2011.

Baomoi.vn. "Công Nhân Tq Đánh Công Nhân Việt Nam: Lộ Nhiều Bất Cập Trong Quản Lý" [Chinese Workers Fought against Vietnamese: Management Inadequacy]. Available at <http://vietbao.vn/Xa-hoi/Cong-nhan-TQ-danh-congnhan-Viet-Nam-Lo-nhieu-bat-cap-trong-quan-ly/2131538768/157/>.

Bauxite Việt Nam. "Lời Kêu Gọi Biểu Tình Yêu Nước Của 20 Tổ Chức Dân Sự Việt Nam" [Call to Protest of the 20 Civic Organizations of Vietnam], 8 May 2014. Available at <http://www.boxitvn.net/bai/25964>.

———. "Thông Báo về Mit Tinh Chống Trung Quốc Xâm Lược" [Announcement of Meeting to Protest Chinese Invasion], 10 May 2014. Available at <http://anonymouse.org/cgi-bin/anon-www.cgi/>.

BBC News. "China Decries Shenyang Pollution Called 'Worst Ever' by Activists", 10 November 2015. Available at <http://www.bbc.com/news/worldasia-china-34773556> (accessed 11 November 2015).

———. "Vietnam Protest over Mystery Fish Deaths", 1 May 2016. Available at <http://www.bbc.com/news/world-asia-36181575>.

Beban, Alice and Sovachana Pou. *Human Security and Land Rights in Cambodia.* Phnom Penh: Cambodia Institute for Cooperation and Peace, 2015.

Bebbington, Anthony. "Underground Political Ecologies: The Second Annual Lecture of the Cultural and Political Ecology Specialty Group of the Association of American Geographers". *Geoforum, Themed Issue: Spatialities of Ageing* 43, no. 6 (November 2012): 1152–62.

Bebbington, Anthony, Denise Humphreys Bebbington, Jeffrey Bury, Jeannet Lingan, Juan Pablo Muñoz, and Martin Scurrah. "Mining and Social Movements: Struggles Over Livelihood and Rural Territorial Development in the Andes". *World Development* 36, no. 12 (2008): 2888–2905, 2890.

Bebbington, Anthony, Leonith Hinojosa, Denise Humphreys Bebbington, Maria Luisa Burneo, and Ximena Warnaars. "Contention and Ambiguity: Mining and the Possibilities of Development". *Development and Change* 39, no. 6 (2008): 887–914.

Bello, Walden, Kenneth Cardenas, Jerome P. Cruz, Alinaya Fabros, Mary A. Manahan, Clarissa Militante, Joseph Purugganan and Jenina J. Chavez. "State of Fragmentation: The Philippines in Transition". *Focus on the Global South and Friedrich Ebert Siftung* (2014).

Berger, Bernt. "China Still Has Its Wrong in Myanmar". *Asia Times*, 10 September 2013. Available at <http://www.atimes.com/atimes/Southeast_Asia/SEA-01-100913.html> (accessed 2 September 2015).

Bergonia, Tony. "Chinese Mining Firms Skirt PH Laws". *Philippine Daily Inquirer*, 21 May 2011.

Bikol Today. "Police Crackdown on Illegal Mining in Camarines Norte Nets Chinese National", 26 June 2014.

Biosphere Reserves. "The Seville Strategy and the Statutory Framework of the World Network". UNESCO, Paris: Biosphere Reserves, 1996. Available at <http://unesdoc.unesco.org/images/0010/001038/103849e.pdf>.

Bondoc, Jarius. "SC Stops Zambales Mines; Chinese 'Invaders' Socked". *The Philippine Star*, 24 July 2013.

BP. *Statistical Review of World Energy 2015*. London: BP, 2015.

Bräutigam, Deborah. *The Dragon's Gift: The Real Story of China in Africa*. Oxford: Oxford University Press, 2009.

Bräutigam, Deborah and Haisen Zhang. "Green Dreams: Myth and Reality in China's Agricultural Investment in Africa". *Third World Quarterly* 34, no. 9 (2013): 1676–96.

Bräutigam, Deborah and Kevin P. Gallagher. "Bartering Globalization: China's Commodity-Backed Finance in Africa and Latin America". *Global Policy* 5, no. 3 (1 September 2014): 346–52.

Bräutigam, Deborah and Sigrid-Marianella Stensrud Ekman. "Briefing Rumours and Realities of Chinese Agricultural Engagement in Mozambique". *African Affairs* 111, no. 444 (1 July 2012): 10.

Bräutigam, Deborah and Tang Xiaoyang. "African Shenzhen: China's Special Economic Zones in Africa". *The Journal of Modern African Studies* 49, no. 1 (2011): 27–54.

———. "Economic Statecraft in China's New Overseas Special Economic Zones: Soft Power, Business or Resource Security?". *International Affairs* 88, no. 4 (1 July 2012): 799–816.

Bremmer, Ian and Robert Johnston. "The Rise and Fall of Resource Nationalism". *Survival* 51, no. 2 (1 May 2009): 149–58.

Bridge, Gavin. "Global Production Networks and the Extractive Sector: Governing Resource-Based Development". *Journal of Economic Geography* 8, no. 3 (2008): 389–419.

British Geological Survey. *World Mineral Production 2009–2013*. Nottingham: British Geological Survey, 2014.

Buchanan, John, Tom Kramer, and Kevin Woods. *Developing Disparity: Regional Investment in Burma's Borderlands*. Amsterdam: Transnational Institute, 2013.

Buckley, Chris. "China Burns Much More Coal Than Reported, Complicating Climate Talks". *New York Times*, 3 November 2015. Available at <http://www.nytimes.com/2015/11/04/world/asia/china-burns-much-more-coalthan-reported-complicating-climate-talks.html>.

Buckley, Lila. "Chinese Agriculture Development Cooperation in Africa: Narratives and Politics". *IDS Bulletin* 44, no. 4 (2013): 42–52.

Buckley, Peter J., L. Jeremy Clegg, Adam R. Cross, Xin Liu, Hinrich Voss and Ping Zheng. "The Determinants of Chinese Outward Investment". *Journal of International Business Studies* 38 (2007): 499–518.

Buckrell, Jon. "Natural Resource Governance – A Test of Political Will for the Cambodian Government and the International Donor Community". *Global Witness*, 2004. Available at <https://www.globalwitness.org/en/archive/natural-resource-governance--test-political-will-cambodian-government-and-international/>.

Bugalski, Natalie and Ratha Thuon. "A Human Rights Impact Assessment: Hoang Anh Gia Lai Economic Land Concessions in Ratanakiri Province, Cambodia". International Academic Conference, RCSD Chiang Mai University, Chiang Mai, 2015.

Bunker, Stephen G. and Paul S. Ciccantell. *Globalization and the Race for Resources*. Baltimore, Maryland: John Hopkins University Press, 2005.

Burawoy, Michael. *The Politics of Production: Factory Regimes Under Capitalism and Socialism*. London: Verso Books, 1985.

Burgos, Sigfrido and Sophal Ear. "China's Strategic Interests in Cambodia: Influence and Resources". *Asian Survey* 50, no. 3 (2010): 615–39. Available at <http://www.jstor.org/stable/10.1525/as.2010.50.3.615>.

Burke, Paul J. and Budy P. Resosudarmo. "Survey of Recent Developments". *Bulletin of Indonesian Economic Studies* 48, no. 3 (2012).

Cabello, Mateo. "Indonesia: Mining White Paper". *Oxford Policy Management*, November 2013.

Camba, Alvin A. "From Colonialism to Neoliberalism: Critical Reflections on Philippine Mining in the 'Long Twentieth Century'". *The Extractive Industries and Society* 2, no. 2 (2015): 287–301.

———. "Philippine Mining Capitalism: The Changing Terrains of Struggle in the Neoliberal Mining Regime". *Austrian Journal of South-East Asian Studies* 9, no. 1 (2016): 69–81.

Cambodia Investment Board. "Investment Trend", 2015. Available at <http://www.cambodiainvestment.gov.kh/investment-enviroment/investment-trend.html>.

Cambodia Office of the High Commissioner for Human Rights. *Land Concessions for Economic Purposes in Cambodia: A Human Rights Perspective*. Phnom Penh: Cambodia Office of the High Commissioner for Human Rights, 2004.

Cano, Marissa Miguel. "Unahin Natin Lagi ang Diyos Bito Lake Fisherolk and Farmers Association". In *MacArthur Leyte: Rising Above the Challenges of Mining and Super Typhoon Yolanda*, edited by Farah Sevilla, Check Zabala and Tess Tabada. Quezon City, Philippines: ATM National Secretariat, June 2014, pp. 2–9. Available at <https://file.ejatlas.org/docs/MacArthur__Leyte_Post_Yolanda_publication.pdf>.

Cao Yin and Guo Anfei. "Alternate Crops Replacing Opium Poppies". *China Daily*, 26 June 2012.

Central Institute for Economic Management. "Thực Trạng Sự Phụ Thuộc Của Kinh Tế Việt Nam Vào Trung Quốc" [Facts on Vietnam's Economic Dependence on China]. Available at <http://www.vnep.org.vn/vi-vn/Hoi-nhapkinh-te-quoc-te/Thuc-trang-su-phu-thuoc-cua-kinh-te-Viet-Nam-vao-Trung-Quoc.html>.

Chanda, Nayan. *Brother Enemy: The War After the War*. New York: Collier Books, 1988.

Chandler, David P. *The Tragedy of Cambodian History: Politics, War, and Revolution since 1945*. New Haven: Yale University Press, 1991.

Cheang, Sokha. "China Steps In with Lorries". *Phnom Penh Post*, 3 May 2010. Available at <http://www.phnompenhpost.com/national/china-stepslorries>.

Chen, Boyuan. "Myanmar Pipeline Project Gives China Pause for Thought". China.org.cn, 21 June 2013. Available at <http://www.china.org.cn/business/2013-06/21/content_29188744.htm> (accessed 27 August 2015).

Chen, Yuyu, Avraham Ebenstein, Michael Greenstone and Hongbin Li. "Evidence on the Impact of Sustained Exposure to Air Pollution on Life Expectancy from China's Huai River Policy". *Proceedings of the National Academy of Sciences* 110, no. 32 (6 August 2013): 12936–41.

Chiang, Bien and Jean Chih-yin Cheng. "Changing Landscape and Changing Ethnoscape in Lao PDR: On PRC's Participation in the Greater Mekong Subregion Development Project". In *Impact of China's Rise on the Mekong Region*. New York: Palgrave Macmillan, 2015, pp. 85–115. Available at <http://www.palgrave.com/br/book/9781137476210>.

China Daily. "CPI: Mutually Beneficial and Double Winning China–Myanmar Myitsone Hydropower Project". *China Daily*, 3 October 2011. Available at <http://www.chinadaily.com.cn/china/2011-10/03/content_13835493.htm> (accessed 2 September 2015).

China National Offshore Oil Corporation. "Key Operating Areas — Indonesia". Available at <http://www.cnoocltd.com/encnoocltd/AboutUs/zygzq/Overseas/1639.shtml> (accessed 21 May 2015).

China National Petroleum Corporation. "CNPC in Indonesia". Available at <http://www.cnpc.com.cn/en/cnpcworldwide/indonesia/PageAssets/Images/CNPC%20in%20Indonesia.pdf> (accessed 21 May 2015).

Chong, Terence. "Chinese Capital and Immigration into CLMV: Trends and Impact". *ISEAS Perspective* 2013 #50. Singapore: Institute of Southeast Asian Studies, 29 August 2013.

Ciorciari, John D. "China and Cambodia: Patron and Client?". IPC Working Paper no. 121, International Policy Center, Gerald R. Ford School of Public Policy, University of Michigan, 2013. Available at <http://ipc.umich.edu/working-papers/pdfs/ipc-121-ciorciari-china-cambodia-patronclient.pdf>.

Climate and Finance Policy Centre. "China's Outward Foreign Direct Investment in 2013". Available at <http://www.ghub.org/cfc_en/?p=591> (accessed 4 May 2015).

Cock, Andrew Robert. "External Actors and the Relative Autonomy of the Ruling Elite in Post-UNTAC Cambodia". *Journal of Southeast Asian Studies* 41, no. 2 (2010): 241–65.

Cohen, Paul. "Resettlement, Opium and Labour Dependence: Akha–Tai Relations in Northern Laos". *Development and Change* 31, no. 1 (2000): 179–200.

———. "The Post-Opium Scenario and Rubber in Northern Laos: Alternative Western and Chinese Models of Development". *International Journal of Drug Policy* 20, no. 5 (2009): 424–30.

Colberg, Claire Mai. "Catching Fish with Two Hands: Vietnam's Hedging Strategy Towards China". A thesis submitted to the Interschool Honors Program in International Security Studies, Stanford University, June 2014.

Coumans, Catherine. "Whose Development? Mining, Local Resistance, and Development Agendas". In *Governance Ecosystem: CSR in Latin American Mining Sector*, edited by Julia Sagebien and Nicole Lindsay. New York: Palgrave Macmillan, 2011, pp. 114–32.

Council on Foreign Relations. "Boston Review: Living with Coal: Climate Policy's Most Inconvenient Truth", 18 September 2009. Available at <http://www.cfr.org/coal/boston-review-living-coal-climate-policys-mostinconvenient-truth/p20230> (accessed 28 October 2015).

Cox, Anna. *Large-Scale Developments in Burma: Uncovering Trends in Human Rights Abuse*. Humanitarian and Relief Trust (HART-UK), 2015. Available at <https://www.hart-uk.org/wp-content/uploads/2012/10/HART-Report-Large-Scale-Developments-in-Burma-Uncovering-Trends-in-Human-Rights-Abuse.pdf> (accessed 9 August 2016).

Cullather, Nick. *Illusions of Influence: The Political Economy of United States–Philippines Relations, 1942–1960*. Stanford: Stanford University Press, 1994.

Đăng Nam. "Doanh Nghiệp Việt Đứng Tên Cho Chủ Trung Quốc Khai Thác". *Tuổi Trẻ Online*, 18 January 2014. Available at <http://vietstock.vn/2014/01/doanh-nghiep-viet-dung-ten-cho-chu-trung-quoc-khai-thac-1351-328672.htm>.

Dantri. "Khoáng Sản 'Đội Nón' Sang Trung Quốc" [Minerals "Illegally Exported" to China]. Dantri.com, 16 November 2012.

———. "Đà Nẵng Kiến Nghị Thu Hồi Giấy Phép Dự Án Trên Núi Hải Vân" [Da Nang Calls for Revoking the Project in Hai Van Mountain], 8 November 2014. Available at <http://dantri.com.vn/xa-hoi/da-nang-kien-nghi-thu-hoi-giay-phep-du-an-trennui-hai-van-992484.htm>.

Dat Viet. "Không Kiểm Soát Được Trung Quốc Mượn Danh Người Việt Khai Khoáng" [Unable to Control Chinese Enterprises under the Guise of Vietnamese in Mineral Investment]. Available at <http://baodatviet.vn/chinhtri-xa-hoi/tin-tuc-thoi-su/tong-cuc-truong-tong-cuc-dia-chat-khoang-san-khongkiem-soat-duoc-tq-muon-danh-nguoi-viet-khai-khoang-2365131/>.

———. "No One Knows How Many Chinese Workers are in Vietnam". *VietnamNet Bridge*, 8 September 2014. Available at <http://english.vietnamnet.vn/fms/business/111125/no-one-knows-how-many-chinese-workers-are-in-vietnam.html>.

Dau Tu Online. "Trung Quốc Bắt Đầu Chơi Con Bài Kinh Tế Với Việt Nam" [China Uses Economic Clout to Vietnam]. Available at <http://baodautu.vn/trung-quoc-bat-dau-choi-con-bai-kinh-te-voi-viet-nam-d1294.html>.

De Beule, Filip and Daniel van den Bulcke. "Locational Determinants of Outward Foreign Direct Investment: An Analysis of Chinese and Indian Greenfield Investments". *Transnational Corporations* 21, no. 1 (2012): 1–34.

De Castro, Renato Cruz. "The Obama Administration's Strategic Pivot to Asia: From a Diplomatic to a Strategic Constrainment of an Emergent China?". *The Korean Journal of Defense Analysis* 25, no. 3 (2013): 331–49.

Deng, Ping. "Why Do Chinese Firms Tend to Acquire Strategic Assets in International Expansion?". *Journal of World Business* 44 (2009): 74–84.

Devi, Bernadetta and Dody Prayogo. "Mining and Development in Indonesia: An Overview of the Regulatory Framework and Policies". International Mining for Development Centre, March 2013. Available at <http://im4dc.org/wp-content/uploads/2013/09/Mining-and-Development-in-Indonesia.pdf>.

Dewi Fortuna Anwar. "An Indonesian Perspective on the U.S. Rebalancing Effort Toward Asia". *NBR Commentary*, 26 February 2013. Available at <http://nbr.org/downloads/pdfs/outreach/Anwar_commentary_02262013.pdf> (accessed 26 March 2015).

Do Tien Sam and Han Thi Hong Van. "Vietnam–China Economic Relations: 2009–2010". In *China and East Asia: After the Wall Street Crisis*, edited by Peng Er Lam, Yaqing Qin and Mu Yang. Singapore: World Scientific Publishing Co. Pte. Ltd., 2013, pp. 225–39.

Do T. Thuy. "Vietnam's Moderate Diplomacy Successfully Navigating Difficult Waters". *East Asia Forum*, 16 January 2015.

Deutsch, Anthony. "Asia Giants' Scramble for Coal Reaches Indonesia". *Financial Times*, 9 September 2010.

Du, Julan, Kai Wang and Yongqin Wang. "Political Determinants of the Location Choice and Entry Mode of Chinese Outward FDI". Paper presented at the 6th Annual Joint Workshop on Socio-Economics co-sponsored by Fudan University, University of Paris 1 and FERDI, Paris, June 2014.

Ducourtieux, Olivier, Jean-Richard Laffort, and Silinthone Sacklokham. "Land Policy and Farming Practices in Laos". *Development and Change* 36, no. 3 (2005): 499–526.

Dunning, John H. "The Eclectic Paradigm as an Envelope for Economic and Business Theories of MNE Activity". *International Business Review* 9 (2000): 163–90.

————. "Towards a New Paradigm of Development: Implications for the Determinants of International Business Activity". *Transnational Corporations* 15, no. 1 (2006): 173–227.

Dunning, John H. and Sarianna M. Lundan. "Institutions and the OLI Paradigm of the Multinational Enterprise". *Asia Pacific Journal of Management* 25 (2008): 573–93.

Dupont, Alan and Christopher G. Baker. "East Asia's Maritime Disputes: Fishing in Troubled Waters". *The Washington Quarterly* 37, no. 1 (1994): 79.

Dwyer, Mike. "Turning Land into Capital: A Review of Recent Research on Land Concessions for Investment in Lao PDR". Land Issues Working Group, Vientiane, Laos, 2007.

Ebner, Julia. "The Sino–European Race for Africa's Minerals: When Two Quarrel a Third Rejoices". *Resources Policy* 43 (March 2015): 113.

The Economist. "Bauxite Bashers: The Government Chooses Economic Growth over Xenophobia and Greenery", 23 April 2009. Available at <http://www.economist.com/node/13527969> (accessed 28 March 2015).

————. "A Bleak Landscape: A Secretive Ruling Clique and Murky Land-Grabs Spell Trouble for a Poor Country", 26 October 2013. Available at <https://www.economist.com/news/asia/21588421-secretive-ruling-clique-and-murky-land-grabs-spell-trouble-poor-country-bleak-landscape>.

————. "Chinese Miner Tries to be Nice", 24 May 2014. Available at <http://www.economist.com/news/business/21602719-chinese-miner-triesbe-nice-kidnapped> (accessed 23 April 2015).

Economy, Elizabeth and Michael Levi. *By All Means Necessary: How China's Resource Quest is Changing the World*. Oxford: Oxford University Press, 2014.

The Editorial Committee of the Ministry of Land and Resources. *People's Republic of China, 2015 China Mineral Resources*. Beijing: Geological Publishing House, 2015.

Els, Frik. "Slowdown. What Slowdown? China's Copper, Iron Ore Imports Set Records". mining.com, 13 October 2013. Available at <http://www.mining.com/slowdown-what-slowdown-chinas-copper-iron-ore-imports-set-records-99571>.

Emel, Jody, Matthew T. Huber, and Madoshi H. Makene. "Extracting Sovereignty: Capital, Territory, and Gold Mining in Tanzania". *Political Geography* 30, no. 2 (2011): 70–79.

Energy Information Administration. "Indonesia", 5 March 2014. Available at <http://www.eia.gov/countries/analysisbriefs/Indonesia/indonesia.pdf>.

————. "Analysis of the Impacts of the Clean Power Plan", 2015.

Environmental Justice Atlas. "Pheapimex–Fuchan Conflict, Cambodia", 20 April 2014. Available at <https://ejatlas.org/conflict/pheapimex-fuchanconflict-cambodia>.

Ericsson, Magnus. "Mineral Supply from Africa: China's Investment Inroads into the African Mineral Resource Sector". *The Journal of the Southern African Institute of Mining and Metallurgy* 111 (July 2011): 497–500.

Esmaquel II, Paterno. "Binay: 'China Has Money, We Need Capital'". Rappler.com, 14 April 2015. Available at <http://www.rappler.com/nation/89880-binay-china-philippines-south-china-sea>.

Fan Hongwei. "Enmity in Myanmar against China". *ISEAS Perspective* 2014 #08. Singapore: Institute of Southeast Asian Studies, 17 February 2014.

Fatah, Luthfi. "The Impacts of Coal Mining on the Economy and Environment of South Kalimantan Province, Indonesia". *ASEAN Economic Bulletin* 25, no. 1 (1 April 2008): 85–98.

Fong-Sam, Yolanda. "The Mineral Industry of Laos". In *2012 Minerals Yearbook*. US: US Department of the Interior, US Geological Survey, November 2014.

Ford, Lucy. "Challenging the Global Environmental Governance of Toxics: Social Movement Agency and Global Civil Society". In *The Business of Global Environmental Governance*, edited by David L. Levy and Peter J. Newell. Cambridge, MA: MIT Press, 2005, pp. 305–28.

Foreign Investment Agency. "Vietnam–China Foreign Investment Development". Ministry of Planning and Investment. Available at <http://fia.mpi.gov.vn/tinbai/2067/Tinh-hinh-hop-tac-dau-tu-Viet-Nam-Trung-Quoc>.

Foreign Relations Council. "National Security Consequences of U.S. Oil Dependency". Available at <http://www.cfr.org>.

Forest Peoples Programme. "Philippines: ALDAW Update on the NCIP Resolution Denying MacroAsia (MAC) Corporation's Request for a Certificate of Precondition", 15 June 2012. Available at <http://www.forestpeoples.org/topics/extractive-industries/news/2012/06/philippinesaldaw-update-ncip-resolution-denying-macroasia>.

Foster, Vivien, William Butterfield, Chuan Chen, and Nataliya Pushak. *Building Bridges: China's Growing Role as Infrastructure Financier for Sub-Saharan Africa*. Trends and Policy Options no. 5. Washington, D.C.: World Bank, 2008.

Fox, Jefferson and Jean-Christophe Castella. "Expansion of Rubber (Hevea Brasiliensis) in Mainland Southeast Asia: What are the Prospects for Smallholders?". *The Journal of Peasant Studies* 40, no. 1 (2013): 155–70.

Francisco, Rosemarie. "Special Report: Philippines' Black Market is China's Golden Connection". *Reuters*, 22 August 2012. Available at <http://www.reuters.com/article/2012/08/23/us-philippines-gold-idUSBRE87M02120120823#tYxtHGkAmWeUEidT.97>.

Freeman, Duncan. "China's Outward Investment: Institutions, Constraints, and Challenges". Brussels Institute of Contemporary China Studies. *Asia Paper* 7, no. 4 (May 2013).

Friis, Cecilie and Jonas Østergaard Nielsen. "Small-Scale Land Acquisitions, Large-Scale Implications: Exploring the Case of Chinese Banana Investments in Northern Laos". *Land Use Policy* 57 (2016): 117–29.

Fujita, Yayoi and Kaisone Phengsopha. "The Gap between Policy and Practice in Lao PDR". In *Lessons from Forest Decentralization: Money, Justice and the Quest for Good Governance in Asia-Pacific*, edited by Carol J. Pierce Colfer, Ganga Ram Dahal and Doris Capistrano. London: Earthscan/CIFOR, 2008, pp. 117–32.

Gabieta, Joey. "Villagers Still Hurting From Leyte Fishkills". *Philippine Daily Inquirer*, 4 August 2012. Available at <http://newsinfo.inquirer.net/242303/villagers-still-hurting-from-leyte-fishkills>.

Gallagher, Kevin and Roberto Porzecanski. *The Dragon in the Room: China and the Future of Latin American Industrialization*. Stanford: Stanford University Press, 2010.

Gariguez, E. "Reply to Ambassador's Investigation Report on OECD Complaint vs. Intex", 2010.

Ghosh, Nirmal. "From Dissident to Contributor: New Think Tanks Help Myanmar's Transition to Democracy". *The Straits Times*, 21 June 2013. Available at <http://timesofindia.indiatimes.com/impact-journalism-day/more-stories/From-dissident-to-contributor-New-think-tanks-help-Myanmarstransition-to-democracy/articleshow/20694549.cms> (accessed 2 September 2015).

Gill, Bates, Michael Green, Kiyoto Tsuji, and William Watts. *Strategic Views on Asian Regionalism: Survey Results and Analysis*. Washington, D.C.: Center for Strategic and International Studies, February 2009.

Global Environment Institute. *Environmental and Social Challenges of China's Going Global*. Beijing: China Environment Press, 2013.

Global Indicators Database. Available at <http://www.pewglobal.org/database/indicator/24/country/101/> (accessed 15 May 2015).

GMA News Online. "CBCP Official Seeks Stop to Magnetite Mining in Leyte Rice Fields", 13 June 2012. Available at <http://www.gmanetwork.com/news/story/261687/news/regions/cbcp-official-seeks-stop-tomagnetite-mining-in-leyte-rice-fields>.

GoL. "State Land Leases and Concessions Inventory", edited by Deutsche Gesellschaft fur Internationiale Zusammenarbeit (GIZ), Ministry of Natural Resources and Environment (GoL). Vientiane: Centre for Development and Environment, 2011.

Goled, Geeta C. "Urban Mining: A Solution to China's Resource Crisis?". *Yale Environment Review*, 29 October 2015. Available at <https://environment.yale.edu/yer/article/urban-mining-a-solution-to-chinas-resource-crisis#gsc.tab=0> (accessed 1 June 2016).

Gonzales-Vicente, Ruben. "Mapping Chinese Mining Investment, with a Focus on Latin America". Paper presented at the China–Latin America meeting at UCLA Asia Institute, 15–16 April 2011.

———. "Mapping Chinese Mining Investment in Latin America: Politics or Market?". *The China Quarterly* 209 (March 2012): 37.

———. "Development Dynamics of Chinese Resource-Based Investment in Peru and Ecuador". *Latin American Politics and Society* 55, no. 1 (2013): 47.

Government Inspectorate of Vietnam. "Xuất Lậu Quặng, Khoáng Sản - Bộ Trưởng Vũ Huy Hoàng Thừa Nhận Trách Nhiệm" [Illegal Mineral Exports: Minister Vu Huy Hoang Takes Responsibility]. Available at <http://thanhtra.com.vn/chinh-tri/doi-noi/xuat-lau-quang-khoang-san-bo-truong-vu-huy-hoangthua-nhan-trach-nhiem_t114c67n70780>.

Greenpeace International. "Point of No Return", January 2013. Available at <http://www.greenpeace.org/international/en/publications/Campaignreports/Climate-Reports/Point-of-No-Return/> (accessed 3 May 2015).

Gronholt-Pedersen, Jacob. "Myanmar Pipelines to Benefit China New Oil, Gas Supply Routes are Set to Help Slake Nation's Growing Thirst for Energy; Local Tensions Rise". *The Wall Street Journal*, 12 May 2013. Available at <http://www.wsj.com/articles/SB10001424127887324326504578466951558644848> (accessed 2 September 2015).

Gunningham, Neil. "Managing the Energy Trilemma: The Case of Indonesia". *Energy Policy,* Decades of Diesel, 54 (March 2013): 184–93.

Guo, Yingjie, Shumei Hou, Graeme Smith and Selene Martinez-Pacheco. "Chinese Outward Directed Investment: Case Studies of SOEs Going Global". In *Law and Policy for China's Market Socialism*, edited by John Garrick. London: Routledge, 2012, pp. 131–43.

Gurtov, Mel. "China's Third World Odyssey: Changing Priorities, Continuities, and Many Contradictions". In *Handbook on China and Developing Countries*, edited by Carla P. Freeman. Cheltenham, UK: Edward Elgar Publishing, 2015, pp. 71–86.

Haacke, Jurgen. "Myanmar: Now a Site for Sino–US Geopolitical Competition? The New Geopolitics of Southeast Asia". *IDEAS Report SR015* (2012): 53–60.

Hanusch, Marek. "African Perspectives on China–Africa: Modelling Popular Perceptions and Their Economic and Political Determinants". *Oxford Development Studies* 40, no. 4 (2012): 492–516.

Hao, Hongmei. "China's Trade and Economic Relations with CLMV". In *Development Strategy for CLMV in the Age of Economic Integration*, edited by Chap Sotharith. Chiba: IDE-JETRO, 2008, pp. 171–208.

Hart, Gillian. "Denaturalizing Dispossession: Critical Ethnography in the Age of Resurgent Imperialism". *Antipode* 38, no. 5 (2006): 977–1004.

He, Gang and Richard Morse. "China's Coal Import Behavior and Its Impacts to Global Energy Market". In *Globalization, Development and Security in Asia*. World Scientific, 2013, pp. 69–85. Available at <http://www.worldscientific.com/doi/abs/10.1142/9789814566582_0032>.

Hickey, Dennis and Baogang Guo. *Dancing with the Dragon: China's Emergence in the Developing World*. Lexington Books, 2010.

Hicks, Charlotte, Saykham Voladeth, Weiyi Shi, Zhong Guifeng, Sun Lei, Pham Quang Tu, and Marc Kalina. "Rubber Investments and Market Linkages in Lao PDR: Approaches for Sustainability?". Sustainable Mekong Research Network, Bangkok, 2009.

Highlands Pacific. "Ramu Nickel". Available at <http://www.highlandspacific.com/current-projects/ramu-nickel> (accessed 27 May 2015).

Hirsch, Philip. "Dams, Resources and the Politics of Environment in Mainland Southeast Asia". In *The Politics of Environment in Southeast Asia: Resources and Resistance*, edited by Philip Hirsch and Carol Warren. London: Psychology Press, 1998, pp. 55–70.

HKTDC Research. "Myanmar: Market Profile", 2017. Available at <hkmb.hktdc.com/
    en/1X09SI4E/hktdc-research/Myanmar-Market-Profile> (accessed 28 September
    2017).
Hoang Mai. "Băn Khoăn Khi Trung Quốc Tiếp Tục Đầu Tư 400 Triệu Đô La
    Vào Tỉnh Nam Định" [Worried When China Continues to Invest US$400
    Million into Nam Dinh], 24 March 2014. Available at <https://boxitvn.blogspot.
    sg/2014/03/ban-khoan-khi-trung-quoc-tiep-tuc-au-tu.html>.
———. "Tại Sao Trung Quốc Lại Đầu Tư Lớn Vào Tỉnh Nam Định, Và Đâu
    Là Mục Tiêu Sâu Xa" [Why China Heavily Invests in Nam Dinh, What are
    the Real Intentions?], 26 March 2014. Available at <http://boxitvn.blogspot.
    com.au/2014/03/tai-sao-trung-quoc-lai-au-tu-lon-vao.html>.
Hongwei, Fan. "Enmity in Myanmar against China". *ISEAS Perspective* 2014 #08.
    Singapore: Institute of Southeast Asian Studies, 17 February 2014.
Hook, Leslie. "China Starts Importing Natural Gas from Myanmar". *Financial Times*,
    29 July 2013. Available at <http://www.ft.com/intl/cms/s/0/870f632cf83e-11e2-
    92f0-00144feabdc0.html#axzz43VwWZ15P> (accessed 17 November 2013).
Howard, Caroline. "Philippine Mining Industry: Boon or Bane?". *ABS-CBN News*,
    2011. Available at <http://www.abs-cbnnews.com/-depth/12/08/11/philippine-
    mining-industry-boon-or-bane>.
Howell, Jude. "Shifting Global Influences on Civil Society: Times for Reflection".
    In *Global Civil Society: Shifting Powers in a Shifting World*, edited by Heidi
    Moksnes and Mia Melin. Uppsala: Uppsala University, 2012, pp. 43–62.
Hsu, Sara. "China's Energy Insecurity Glaring in South China Sea Dispute". *Forbes*,
    2 September 2016. Available at <http://www.forbes.com/sites/sarahsu/2016/09/
    02/china-energy-insecurity-south-china-sea-dispute/>.
Huang Keira Lu. "State Firms Barred from Vietnam Contract Bids". *South China
    Morning Post*, 9 June 2014.
Hughes, Caroline. "Transnational Networks, International Organizations and Political
    Participation in Cambodia: Human Rights, Labour Rights and Common Rights".
    Democratization 14, no. 5 (2007): 834–52.
———. "Cambodia in 2007: Development and Dispossession". *Asian Survey* 48,
    no. 1 (2008): 69–74.
Humphreys, David. "New Mercantilism: A Perspective on How Politics is Shaping
    World Metal Supply". *Resources Policy* 39 (2013): 341–49.
Hung Ho-fung. "America's Head Servant?". *New Left Review* 60 (2009): 23.
———. *The China Boom: Why China Will Not Rule the World*. Columbia
    University Press, 2015.
Hunt, Luke. "Cambodia's Well-Heeled Military Patrons". *The Diplomat*, 10 August
    2010. Available at <http://thediplomat.com/2015/08/cambodiaswell-heeled-
    military-patrons/>.
Institute for Essential Services Reform. *The Framework for Extractive Industries
    Governance in ASEAN*. Jakarta: Institute for Essential Services Reform, 2014.

International Energy Agency (IEA). *World Energy Outlook 2012*. Paris: IEA, 2012.

———. *Key World Energy Statistics 2012*. Paris: Organisation for Economic Co-operation and Development, 2013. Available at <http://www.oecd-ilibrary.org/content/book/key_energ_stat-2012-en>.

———. *Southeast Asia Energy Outlook 2013*. Paris: IEA, September 2013.

———. "Energy Supply Security: Indonesia", 2014. Available at <http://www.iea.org/publications/freepublications/publication/ESS_Indonesia_2014.pdf>.

———. *Medium-Term Coal Market Report 2014*. Paris: IEA, 2014.

———. *Update on Overseas Investments by China's National Oil Companies: Achievement and Challenges since 2011*. Paris: IEA, 2014.

———. *World Energy Outlook 2014*. Paris: IEA, 2014.

———. *Southeast Asia Energy Outlook 2015*. Paris: IEA, 2015.

International Rivers. "The Myitsone Dam on the Irrawaddy River: A Briefing", 28 September 2011. Available at <http://www.internationalrivers.org/resources/the-myitsone-dam-on-the-irrawaddy-river-a-briefing-3931> (accessed 9 February 2015).

———. "Cheay Areng Dam", 2014. Available at <http://www.internationalrivers.org/campaigns/cheay-areng-dam>.

The Irrawaddy. "Protests Continue Against Letpadaung Copper Mine", 19 January 2015. Available at <http://www.irrawaddy.org/burma/protests-continue-letpadaung-coppermine.html> (accessed 12 March 2015).

Ishee, Jan and Alex Demas. "Going Critical: Being Strategic with Our Mineral Resources", 13 December 2016. Available at <https://www2.usgs.gov/blogs/features/usgs_top_story/going-critical-being-strategic-with-our-mineral-resources/> (accessed 1 June 2016).

Jarvis, Darryl S.L. and Anthony Welch. *ASEAN Industries and the Challenges from China*. Hamshire: Palgrave Macmillan, 2011.

Jiang, Heng. *Out of the Mine Fields and Blind Areas of Overseas Investment Security — Conflict Risk Assessment and Management*. Beijing: China Economic Publishing House, 2013.

Jiang, Julie and Chen Ding. *Update on Overseas Investments by China's National Oil Companies: Achievements and Challenges since 2011*. Paris: OECD/IEA, 2014.

Jiang, Julie and Jonathan Sinton. *Overseas Investments by Chinese National Oil Companies: Assessing the Drivers and Impact*. Paris: OECD/IEA, 2011.

Jiang, Wenran. "Fuelling the Dragon: China's Rise and Its Energy and Resources Extraction in Africa". *The China Quarterly* 199 (2009): 585–609.

Kaspar, Andrew D. "Burma's Extractive Industries Not Digging Deep Enough with Reforms: Report". *The Irrawaddy*, 17 July 2013. Available at <http://www.irrawaddy.org/natural-resources/burmas-extractive-industries-not-digging-deep-enough-with-reforms-report.html> (accessed 3 September 2015).

Kazmin, Amy. "China Boosts Cambodian Relations with $600m Pledge". *Financial Times*, 10 April 2006. Available at <https://www.ft.com/content/127cb9fa-c7fa-11da-a377-0000779e2340>.

K. Chi. "China Attempts to Control Vietnam's Mineral Industries". *VietnamNet Bridge*, 25 January 2014. Available at <http://english.vietnamnet.vn/fms/business/94502/china-attempts-to-control-vietnam-s-mineral-industries.html>.

Keck, Margaret E. and Kathryn Sikkink. *Activists Beyond Borders: Advocacy Networks in International Politics*, vol. 6. Ithaca, NY: Cornell University Press, 1998, pp. 1–38.

Kemp, Melody. "Palawan Tribes Go Cyber to Keep Out Nickel Miner". *Asia Times*, 9 December 2011.

Kennedy, Paul. *The Rise and Fall of the Great Powers*. New York: Vintage, 1968.

Kenney-Lazar, Miles. "Dispossession, Semi-Proletarianization, and Enclosure: Primitive Accumulation and the Land Grab in Laos". Paper presented at the International Conference on Global Land Grabbing, 6–8 April 2011.

———. "Shifting Cultivation in Laos: Transitions in Policy and Perspective". Sector Working Group-Agriculture and Rural Development (SWG-ARD), 2013.

Khine, Nwet Kay. "Foreign-Investment-Induced Conflicts in Myanmar's Mining Sector: The Case of the Monywa Copper Mine". *Perspectives Issue 1: Copper, Coal, and Conflicts: Resources and Resource Extraction in Asia* (June 2013): 48.

Kirsch, Stuart. *Mining Capitalism: The Relationship between Corporations and Their Critics*. Oakland, California: University of California Press, 2014.

Kolstad, Ivar and Arne Wiig. "What Determines Chinese Outward FDI?". Chr. Michelsen Institute, Working Paper No. 2009/3 (2009).

Komesaroff, Michael. "Screwing Up in Foreign Climes". *China Economic Quarterly* (June 2012): 9–11.

Kong, Bo. *China's International Petroleum Policy*. Santa Barbara: Praeger Security International, 2010.

Kotoski, Kali and Sor Chandara. "Royal Group in Talks with China, Indonesia on Proposed Oil Pipeline". *Phnom Penh Post*, 11 April 2016. Available at <http://www.phnompenhpost.com/business/royal-group-talks-china-indonesia-proposed-oil-pipeline> (accessed 1 June 2016).

Kramer, Tom and Kevin Woods. "Financing Dispossession: China's Opium Substitution Programme in Northern Burma". Transnational Institute, Amsterdam, 2012.

Kudo, Toshihiro. "Myanmar's Economic Relations with China: Who Benefits and Who Pays?". In *Dictatorship, Disorder and Decline in Myanmar*, edited by Monique Skidmore and Trevor Wilson. Canberra: ANU E Press, 2008, pp. 87–110.

————. *China's Policy Toward Myanmar: Challenges and Prospects*. Chiba: IDE-JETRO, 2012.

Kurlantzick, Josh. *Charm Offensive: How China's Soft Power is Transforming the World*. New Haven: Yale University Press, 2007.

Kurtz, John and James Van Zorge. "The Myth of Indonesia's Resource Nationalism". *The Wall Street Journal*, 1 October 2013.

Kyaw Yin Hlaing. "Associational Life in Myanmar: Past and Present". In *Myanmar: State, Society and Ethnicity*, edited by Narayanan Ganesan and Kyaw Yin Hlaing. Singapore: Institute of Southeast Asian Studies, 2003, pp. 143–71.

Lai, Hongyi, Sarah O'Hara and Karolina Wysoczanska. "Rationale of Internationalization of China's National Oil Companies: Seeking Strategic Natural Resources, Strategic Assets or Sectoral Specialization?". *Asia Pacific Business Review* 21, no. 1 (2014): 77–95.

Landingin, Roel. "Chinese Foreign Aid Goes Offtrack in the Philippines". *The Reality of Aid, South-South Cooperation: A Challenge to the Aid System* (2010): 87–94.

Laodong Newspaper. "Mối Nguy Từ Các Dự Án Tổng Thầu Epc Rơi Vào Tay Nhà Thầu Trung Quốc" [Dangers from Chinese Investors' Control Over EPC Projects]. Vietnam General Trade Union. Available at <http://laodong.com.vn/kinh-doanh/moi-nguy-tu-cac-du-an-tong-thau-epc-roi-vao-tay-nha-thautrung-quoc-205587.bld>.

Le Hong Hiep. "The Dominance of Chinese Engineering Contractors in Vietnam". *ISEAS Perspective* 2013 #04. Singapore: Institute of Southeast Asian Studies, 17 January 2013.

Lee Ching Kwan. "Raw Encounters: Chinese Managers, African Workers and the Politics of Casualization in Africa's Chinese Enclaves". *The China Quarterly* 199 (2009): 647–66.

————. "se Managers, African Workers and the Politics of Casualization in Africa's Chinese Enclaves". *The China Quarterly* 199 (September 2009): 647–66.

————. "The Spectre of Global China". *New Left Review* 89 (October 2014): 28–65.

Lee Ching Kwan and Yonghong Zhang. "The Power of Instability: Unraveling the Microfoundations of Bargained Authoritarianism in China". *American Journal of Sociology* 118, no. 6 (2013): 1475–1508.

Li, Chenyang. *Annual Report on Myanmar's National Situation (2011–2012)*. China: Social Sciences Academic Press, 2013. [Text in Chinese]

Li, Tania Murray. "Centering Labor in the Land Grab Debate". *The Journal of Peasant Studies* 38, no. 2 (2011): 281–98.

Li Tao. "*Qian xi zhongguo-dongmen de nengyuan hezuo*" [An Analysis of China–ASEAN Energy Cooperation]. *Southeast Asian Studies*, no. 3 (2006).

Li, Xiaoyun, Dan Banik, Lixia Tang, and Jin Wu. "Difference or Indifference: China's Development Assistance Unpacked". *IDS Bulletin* 45, no. 4 (2014): 26.

Liao, Janet Xuanli. "The Chinese Government and the National Oil Companies (NOCs): Who is the Principal?". *Asia Pacific Business Review* 21, no. 1 (2015): 44–50.

Lieberthal, Kenneth and Mikkal Herberg. "China's Search for Energy Security". *NBR Analysis* 17, no. 1 (2006).

Lin Lu, Yuxia Fang, and Xi Wang. "Drug Abuse in China: Past, Present and Future". *Cellular and Molecular Neurobiology* 28, no. 4 (2008): 479–90.

Linh Thư, Xuân Quý, and Minh Thăng. "1 Người Chết Ở Hà Tĩnh" [1 Person Dead in Ha Tinh]. *VietNamNet*, 15 May 2014. Available at <http://vietnamnet.vn/vn/chinh-tri/175609/1-nguoi-chet-o-ha-tinh.html>.

Lintner, Bertil. "Same Game, Different Tactics: China's 'Myanmar Corridor'". *The Irrawaddy*, 13 July 2015. Available at <http://www.irrawaddy.org/magazine/same-game-different-tactics-chinas-myanmar-corridor.html> (accessed 13 July 2015).

Liu, Zhi and Guangsheng Lu. *Report on the Cooperation in the Greater Mekong Sub-region 2012–2013*. China: Social Sciences Academic Press, 2013. [Text in Chinese]

London, Jonathan. *Politics in Contemporary Vietnam: Party, State, and Authority Relations*. Palgrave Macmillan, 2014.

Lu, Jiangyong, Xiaohui Liu, Mike Wright and Igor Filatotchev. "International Experience and FDI Location Choices of Chinese Firms: The Moderating Effects of Home Country Government Support and Host Country Institutions". *Journal of International Business Studies* 45 (2014): 428–49.

Lucarelli, Bart. "The History and Future of Indonesia's Coal Industry: Impact of Politics and Regulatory Framework on Industry Structure and Performance". Program on Energy and Sustainable Development, Freeman Spogli Institute for International Studies, Stanford University, Stanford, California, US. Retrieved 10 May 2010.

Lum, Thomas, Hannah Fischer, Julissa Gomez-Granger and Anne Leland. "China's Foreign Aid Activities in Africa, Latin America, and Southeast Asia". *Congressional Research Service: Report for Congress* (25 February 2009).

Luo, Yadong, Qiuzhi Xue and Binjie Han. "How Emerging Market Governments Promote Outward FDI: Experience from China". *Journal of World Business* 45 (2010): 68–79.

Macaraig, Ayee. "Binay Joint Venture with China Must Be Corruption-Free". Rappler.com, 10 July 2015. Available at <http://www.rappler.com/nation/98940-binayjoint-venture-china-corruption>.

MacLean, Ken. "Unbuilt Anxieties: Infrastructure Projects, Transnational Conflict in the South China/East Sea, and Vietnamese Statehood". *TRaNS: Trans-Regional and -National Studies of Southeast Asia* 4, no. 2 (July 2016): 365–85.

Mahtani, Shiban. "China Rocks Myanmar's Diplomatic Boat". *The Wall Street Journal*, 10 May 2014. Available at <http://www.wsj.com/articles/SB100014240527023046553045795529632384_17846> (accessed 22 July 2015).

Mann, C.C. "Addicted to Rubber". *Science* 325, no. 5940 (2009): 564–66.

Marston, Hunter. "Bauxite Mining in Vietnam's Central Highlands: An Arena for Expanding Civil Society?" *Contemporary Southeast Asia: A Journal of International and Strategic Affairs* 34, no. 2 (2012): 173–96.

Maung Aung Myoe. *In the Name of Pauk-Phaw: Myanmar's China Policy since 1948*. Singapore: Institute of Southeast Asian Studies, 2011.

McAllister, Karen E. "Rubber, Rights and Resistance: The Evolution of Local Struggles Against a Chinese Rubber Concession in Northern Laos". *The Journal of Peasant Studies* 42, nos. 3–4 (2015): 1–21.

McAndrew, John. "Mining Industry Report: The Impact of Corporate Mining on Local Philippine Communities". ARC Publications, 1983.

Ministry of Commerce and Ministry of Foreign Affairs. *Foreign Investment Industrial Guidance Catalogue, Country Directory*. Beijing: August 2004.

Ministry of Commerce et al. *Foreign Investment Industrial Guidance Catalogue, Country Directory*. Beijing: 2007.

———. *Foreign Investment Industrial Guidance Catalogue, Country Directory*. Beijing: 2011.

Ministry of Finance Regulation. Available at <http://www.jdih.kemenkeu.go.id/full Text/2012/75~PMK.011~2012Per.htm> (accessed 17 March 2015).

Monsod, Christian. "Mining a Social Justice Issue". *ABS-CBN News*, 2012. Available at <http://www.abs-cbnnews.com/-depth/03/03/12/christianmonsod-mining-social-justice-issue>.

Moore, Jason W. *Capitalism in the Web of Life: Ecology and the Accumulation of Capital*. London: Verso Books, 2015.

Morck, Randall, Bernard Yeung and Minyuan Zhao. "Perspectives on China's Outward Foreign Direct Investment". *Journal of International Business Studies* 39 (2008): 337–50.

Morris-Jung, Jason. "Reflections on the Oil Rig Crisis: Vietnam's Domestic Opposition Grows". *ISEAS Perspective* 2014 #43. Singapore: Institute of Southeast Asian Studies, 30 July 2014.

———. "The Vietnamese Bauxite Controversy: Towards a More Oppositional Politics". *Journal of Vietnamese Studies* 10, no. 1 (1 February 2015): 63–109.

———. "Vietnam's New Environmental Politics: A Fish Out of Water?" *The Diplomat*, 23 May 2016. Available at <http://thediplomat.com/2016/05/vietnams-new-environmental-politics-a-fish-out-of-water/> (accessed 28 September 2017).

Moyo, Dambisa. *Winner Take All: China's Race for Resources and What It Means for Us*. UK: Penguin, 2012.

Myanmar Centre for Responsible Business (MCRB). *Civil Society Organisations and the Extractives Industries in Myanmar — A Brief Overview*, 2014.

Myint-U, Thant. *Where China Meets India: Burma and the New Crossroads of Asia*. Basingstoke: Macmillan, 2011.

National Development Reform Commission et al. *2006 Catalogue of Industries for Guiding Outward Investment.* Beijing: 2006.

National Statistical Coordination Board. *Poverty Statistics*, 2009.

Nguyen Dinh Liem. "China's FDI in Vietnam: 20 Years in Retrospect". *Journal of Chinese Studies* 8, no. 156 (2014): 31–45.

Nguyễn Huệ Chi, Phạm Toàn, and Nguyễn Thế Hùng. "Kiến Nghị về Quy Hoạch và Các Dự Án Khai Thác Bauxite Ở Việt Nam" [Petition on Bauxite Master Plan and Projects in Vietnam], 12 April 2009. Available at <http://www.boxitvn.net/kien-nghi>.

Nguyễn Hữu Quý. "Trung Quốc đang có âm mưu gì ở Hà Tĩnh và Quảng Trị?" [Does China Have a Secret Plot in Ha Tinh and Quang Tri?]. *Bauxite Vietnam.* Blogger.com, 1 March 2014. Available at <http://boxitvn.blogspot.sg/2014/02/trung-quoc-ang-co-am-muu-gi-o-ha-tinh.html>.

Nguyên, Nam. "Lao động Trung Quốc ở Vũng Áng đủ lập 2 'sư đoàn'" [Chinese Workers in Vung Ang Enough for "Two Military Divisions"]. *Radio Free Asia*, 27 August 2014. Available at <http://www.rfa.org/vietnamese/in_depth/10000-cnese-workers-ll-arrive-ha-tinh-nn-08272014103641.html> (accessed 28 September 2017).

Nguyên Ngọc. "Chương Trình Bauxite Ở Tây Nguyên và Các Vấn Đề Văn Hoá – Xã Hội" [The Central Highlands Bauxite Projects and Their Social and Cultural Issues]. In *Tài Liệu Hội Thảo Khoa Học* [Workshop Proceedings] (Đắk Nông: UBND Dak Nong, Vinacomin and CODE, 2008), pp. 94–97.

———. "Ý Nghĩa Văn Hoá Xã Hội Của Chương Trình Bôxit Tây Nguyên — Diễn Đàn Forum". *Diễn Đàn*, 12 April 2009. Available at <http://www.diendan.org/viet-nam/y-nghia-van-hoa-xa-hoi-cua-chuong-trinh-boxit-taynguyen>.

Nguyen Nha. "ViệT Nam Và Cơ Hội Thoát Trung Lần 4" [Vietnam and the Fourth Opportunity to "Escape" China]. *BBC Vietnamese*, 6 June 2014.

Nguyễn Thanh Sơn. "Đại Kế Hoạch Bô – Xít Ở Tây Nguyên Bị Phản Đối Quyết Liệt" [Fierce Opposition to the Great Bauxite in the Central Highlands]. *Tuan Vietnam*, 24 October 2008. Available at <http://www.tuanvietnam.net/2008-10-24-dai-ke-hoach-bo-xit-o-tay-nguyen-bi-phan-doi-quyet-liet>.

———. "Đại Dự Án Bô – Xít Tây Nguyên: Người Trong Cuộc Đề Xuất Gì?" [The Great Central Highlands Bauxite Project: What Does an Insider Recommend?]. *Tuần Việt Nam*, 26 October 2008. Available at <http://www.tuanvietnam.net/dai-du-an-bo-xit-tay-nguyen-nguoi-trong-cuoc-de-xuat-gi>.

Nguyễn Trọng Vĩnh. "Thư của thiếu tướng đại sứ Nguyễn Trọng Vĩnh" [Letter from Major General Ambassador Nguyễn Trọng Vĩnh]. *Diễn Đàn*, undated. Available at <http://www.diendan.org/viet-nam/thu-cua-thieu-tuong-111ai-su-nguyen-trong-vinh/>.

Nguyen Van Chinh. "Recent Chinese Migration to Vietnam". *Asian and Pacific Migration Journal* 22, no. 1 (2013): 7–30.

————. "China's 'Comrade Money' and Its Social-Political Dimensions in Vietnam". In *Impact of China's Rise on the Mekong Region*. New York: Palgrave Macmillan, 2015, pp. 53–84. Available at <http://www.palgrave.com/br/book/9781137476210>.

Nikkei Asian Review. "Tambang Batubara Bukit Asam: Indonesian Coal Producer Gets $1.2B Loan from China", 28 March 2015. Available at <http://asia.nikkei.com/Business/AC/Indonesian-coal-producer-gets-1.2B-loan-from-China> (accessed 23 November 2015).

Nolan, Peter. *China and the Global Business Revolution*. Basingstoke: Palgrave, 2001.

Nyein Nyein. "NGOs Call For Suspension of Shwe Gas Project". *The Irrawaddy*, 3 October 2012. Available at <http://www.irrawaddy.com/burma/ngos-call-for-suspension-of-shwe-gas-project.html> (accessed 2 May 2015).

Nyiri, Pal. "The Yellow Man's Burden: Chinese Migrants on a Civilizing Mission". *The China Journal* 56 (2006): 84–85.

————. *New Chinese Migration and Capital in Cambodia*. Trends in Southeast Asia 2014 #03. Singapore: Institute of Southeast Asian Studies, 2014.

O'Faircheallaigh, Ciaran. "Negotiating Cultural Heritage? Aboriginal-mining Company Agreements in Australia". *Development and Change* 39, no. 1 (2008): 46.

Oh, Su-Ann and Philip Andrews-Speed. *Chinese Investment and Myanmar's Shifting Political Landscape*. Trends in Southeast Asia 2015 #16. Singapore: Institute of Southeast Asian Studies, 2015.

Okezone Finance. "China Masuk Top Investor Indonesia" [China Became a Top Five Investor in Indonesia], 28 January 2015. Available at <http://economy.okezone.com/read/2015/01/28/20/1098445/china-masuk-top-investor-indonesia> (accessed 27 May 2015).

Oliver, Sarah and Kevin Stahler. "Can Japan Tell Us Where Chinese FDI is Going in ASEAN?". Peterson Institute for International Economics, 3 July 2014. Available at <http://blogs.piie.com/china/?p=3944> (accessed 29 April 2015).

Oxford Institute for Energy Studies. "China's Coal Market — Can Beijing Tame 'King Coal'?", December 2014. Available at <https://www.oxfordenergy.org/wpcms/wp-content/uploads/2014/12/CL-1.pdf> (accessed 29 April 2015).

Pacific Strategies and Assessments (PSA). "Exploitive Chinese Mining in the Philippines". Manila, 2011.

Parameswaran, Prashanth. "Beijing Unveils New Strategy for ASEAN–China Relations". The Jamestown Foundation, *China Brief* XIII, no. 21 (2013): 9–12.

Paul, Sonali. "China's Ramu Nickel Mine in PNG Restarts after Attacks". *Reuters*, 7 August 2014. Available at <http://in.reuters.com/article/2014/08/07/papua-nickel-ramu-idINL4N0QD0GY20140807> (accessed 27 May 2015).

Paulus, Moritz and Johannes Trueby. "Coal Lumps vs. Electrons: How Do Chinese Bulk Energy Transport Decisions Affect the Global Steam Coal Market?". *Energy Economics* 33, no. 6 (November 2011): 1127–37.

Paulus, Moritz, Johannes Trueby, and Christian Growitsch. "Nations as Strategic Players in Global Commodity Markets: Evidence from World Coal Trade". EWI Working Paper. Energiewirtschaftliches Institut an der Universitaet zu Koeln, 2011. Available at <https://ideas.repec.org/p/ris/ewikln/2011_004. html>.

Peluso, Nancy Lee and Peter Vandergeest. "Genealogies of the Political Forest and Customary Rights in Indonesia, Malaysia, and Thailand". *The Journal of Asian Studies* 60, no. 3 (2001): 761–812.

Petrie, Charles and Ashley South. "Mapping of Myanmar Peacebuilding Civil Society", 2013. Available at <https://www.ashleysouth.co.uk/files/EPLO_CSDN_ Myanmar_MappingMyanmarPeacebuildingCivilSociety_CPetrieASouth.pdf> (accessed 2 May 2015).

Pham Minh Ngoc. "'Thoát Trung' Nhưng Cũng Cần Cẩn Trọng" ["Escaping China" But be Cautious]. *The Saigon Times*, 6 June 2014.

Pham Sy Thanh. "Ba Mối Lo Trong Quan Hệ Kinh Tế Việt Nam – Trung Quốc" [Three Mảo Concerns in Vietnam–Sino Economic Relations]. *Finance Review*, 23 June 2014.

Philippine Indigenous Peoples Links. "Farmers, Fisherfolks File Petition for Environmental Protection Order Against Mining", 28 June 2012. Available at <http://www.piplinks.org/farmers,-fisherfolks-file-petition-environmental-protection-order-against-mining.html>.

———. "PH Suspends Blacksand Mining Ops in Region VIII", 29 September 2012. Available at <http://www.piplinks.org/ph-suspends-blacksand-mining-ops-region-viii.html>.

Philippine Statistical Authority. "Provincial Summary: Number of Provinces, Cities, Municipalities and Barangays, by Region". 2015.

Phorn, Bopha. "Defense Minister Says 400 Personnel Will Soon Study in China". *Cambodia Daily*, 12 May 2014. Available at <http://www.cambodiadaily.com/ news/defense-minister-says-400-personnel-will-soonstudy-in-china-58462/>.

Poole, Jennifer. "China to Cut Natural Rubber Import to Increase Local Supply". *Rubber Journal Asia* (2012).

Power, Marcus, Giles Mohan, and May Tan-Mullins. *China's Resource Diplomacy in Africa: Powering Development?* UK: Palgrave Macmillan, 2012.

PT Coalindo Energy Indonesian Coal Index (ICI). Available at <http://coalindoenergy. com/adaros-capacity-of-power-generation-is-potential-to-rise/>.

Purcell, Victor and Alice Li. *The Chinese in Southeast Asia*. London: Oxford University Press, 1965.

Quer, Diego, Enrique Claver, and Laura Rienda. "Political Risk, Cultural Distance, and Outward Foreign Direct Investment: Empirical Evidence from Large

Chinese Firms". *Asia Pacific Journal of Management* 29, no. 4 (13 January 2011): 1089–104.

Radio Free Asia. "Foreign Labourers in Vietnam", 4 August 2009.

Rainforest Rescue. "Philippines: Indigenous Peoples of Palwan Against Mining Operations", 2 June 2011. Available at <https://www.rainforestrescue.org/press-releases/3565/philippines-indigenous-peoples-of-palawanagainst-mining-corporations-1>.

Regis, Emerlina. "Gold Mining Activities as Cause of Poverty of Local Communities in Gold Mining Areas". *Gibón* 4, no. 1 (2004): 81–146.

Reilly, James. "China's Economic Statecraft: Turning Wealth into Power". Lowy Institute for International Policy, University of Sydney, 2013. Available at <http://www.lowyinstitute.org/publications/chinas-economicstatecraft-0> (accessed 3 September 2015).

"Renegosiasi Berhasil, Harga Jual Gas Tangguh Sesuai Harapan" [Renegotiation was Successful, the Price of Tangguh LNG as We Expected]. Available at <http://www.esdm.go.id/berita/migas/40-migas/6862-renegosiasi-berhasil-harga-jual-gas-tangguhsesuai-harapan.html> (accessed 17 March 2015).

Reuters. "China's Coal Use Falling Faster than Expected", 26 March 2015. Available at <http://www.reuters.com/article/2015/03/26/china-coal-idUSL3N0WL32720150326>.

————. "China, Indonesia to Boost Security Ties despite South China Sea Spat", 26 April 2016. Available at <http://www.reuters.com/article/uschina-indonesia-idUSKCN0XN20R>.

Revenue Watch Institute. "The 2013 Resource Governance Index: A Measure of Transparency and Accountability in the Oil, Gas and Mining Sector", 2014. Available at <http://www.resourcegovernance.org/sites/default/files/rgi_2013_Eng.pdf> (accessed 3 September 2015).

RFA's Khmer Service. "Half a Million Cambodians Affected by Land Grabs: Rights Group", 1 April 2014. Available at <http://www.rfa.org/english/news/cambodia/land-04012014170055.html>.

Richardson, Sophie. *China, Cambodia, and the Five Principles of Peaceful Coexistence*. New York: Columbia University Press, 2010.

Risk Asia Consulting. "Fool's Good: The False Economic Promises of the Lafayette Mining Project in Rapu-Rapu". Greenpeace Southeast Asia, 2006.

Risse, Thomas and Kathryn Sikkink. "The Socialization of International Human Rights Norms into Domestic Practices: Introduction". In *The Power of Human Rights: International Norms and Domestic Change*, vol. 66, edited by Thomas Risse, Ropp, Stephen C. Ropp, and Kathryn Sikkink. Cambridge: Cambridge University Press, 1999, pp. 1–38.

Roberts, Dexter. "China's State Enterprises Told to Stop Investing in Vietnam". *Bloomberg Business*, 9 June 2014.

Robinson, Gwen. "Myanmar Cleans House — China's Worst Nightmare?". *Financial Times*, 15 April 2013. Available at <http://blogs.ft.com/beyondbrics/2013/04/15/myanmar-cleans-house-chinas-worst-nightmare/> (accessed 3 May 2015).

Rock, Michael T. "What Can Indonesia Learn From China's Industrial Energy Saving Programs?". *Bulletin of Indonesian Economic Studies* 48, no. 1 (2012).

Romulo, Roberto. "Small Scale Mining: Immeasurable Damage". *The Philippine Star*, 18 October 2013.

Ross, Michael. "Extractive Sectors and the Poor: An Oxfam-America Report", 2001. Available at <http://www.oxfamamerica.org/static/oa3/files/extractive-sectors-and-the-poor.pdf>.

Rutherford, Jef, Kate Lazarus, and Shawn Kelley. *Rethinking Investments in Natural Resources: China's Emerging Role in the Mekong Region*. Phnom Penh: Heinrich Böll Stiftung, WWF and International Institute for Sustainable Development, 2008.

Sack, Robert David. *Human Territoriality: Its Theory and History*. Cambridge: Cambridge University Press, 1986.

Samar News. "Farmers and Fishers of Leyte Cry Freedom From Mining", 12 June 2012. Available at <http://www.samarnews.com/news_clips20/news415.htm>.

Sanderson, Henry and Michael Forsythe. *China's Superbank. Debt, Oil and Influence — How China Development Bank is Rewriting the Rules of Finance*. Singapore: John Wiley, 2013.

Santasombat, Yos. *Impact of China's Rise on the Mekong Region*. New York: Palgrave Macmillan, 2015.

Schertow, John Ahni. "Indigenous Peoples Unite Against Mining in Palawan". *IC Magazine*, 5 June 2010. Available at <https://intercontinentalcry.org/indigenous-peoples-unite-against-mining-in-palawan/>.

Schoenweger, Oliver, Andreas Heinimann, Michael Epprecht, Juliet Lu, and Palikone Thalongsengchanh. *Concessions and Leases in the Lao PDR: Taking Stock of Land Investments*. Bern, Vientiane: Geographica Bernensis, 2012.

Scoones, Ian, Kojo Amanor, Arilson Favareto, and Gubo Qi. "A New Politics of Development Cooperation? Chinese and Brazilian Engagements in African Agriculture". *World Development, China and Brazil in African Agriculture* 81 (May 2016): 1–12.

Scoones, Ian, Lídia Cabral, and Henry Tugendhat. "New Development Encounters: China and Brazil in African Agriculture". *IDS Bulletin* 44, no. 4 (2013): 1–19.

Shambaugh, David. *China Goes Global: The Partial Power*. New York: Oxford University Press, 2013.

Shankleman, Jill. "Going Global: Chinese Oil and Mining Companies and the Governance of Resource Wealth". Wilson Center, July 2011. Available at <https://www.wilsoncenter.org/sites/default/files/Shankleman_Going%20Global.pdf>.

Shi, Weiyi. "Rubber Boom in Luang Namtha: A Transnational Perspective". GTZ RDMA, 2008.

Simpson, Adam. *Energy, Governance and Security in Thailand and Myanmar (Burma): A Critical Approach to Environmental Politics in the South.* UK: Ashgate Publishing Ltd., 2014.

Sine, Richard and Phann Ana. "Cambodia's Largest Land Concession Poses Ominous Threat to Environment". *Cambodia Daily*, 6 April 2002. Available at <https://www.cambodiadaily.com/archives/cambodias-largest-landconcession-poses-ominous-threat-to-environment-30950/>.

Singh, Jewellord T. Nem and Alvin A. Camba. "Neoliberalism, Resource Governance and the Everyday Politics of Protests in the Philippines". In *The Everyday Political Economy of Southeast Asia*, edited by Juanita Elias and Lena Rethel. UK: Cambridge University Press, 2016, pp. 49–71.

Sinomach. "CNEEC Indonesia SUMSEL-5 (2×150 MW) Pithead Coal-fired Power Plant Project Commenced Construction", 25 September 2013. Available at <http://www.sinomach.com.cn/en/MediaCenter/News/201412/t20141209_22074.html>.

Smajgl, Alex, Jianchu Xu, Stephen Egan, Zhuang-Fang Yi, John Ward, and Yufang Su. "Assessing the Effectiveness of Payments for Ecosystem Services for Diversifying Rubber in Yunnan, China". *Environmental Modelling & Software* 69 (2015): 187–95.

Snay Lin. "Burma's 88 Generation Students to Form Political Party". *The Irrawaddy*, 19 March 2013. Available at <http://www.irrawaddy.org/burma/burmas-88-generation-students-to-form-political-party.html> (accessed 22 April 2015).

SourceWatch. "BTN Dumai Power Station". Available at <http://www.sourcewatch.org/index.php/BTN_Dumai_power_station>.

————. "Cilacap Sumber Power Station". Available at <http://www.sourcewatch.org/index.php/Cilacap_Sumber_power_station>.

Stark, Jeffrey, Jennifer Li, and Katsuaki Terasawa. "Environmental Safeguards and Community Benefits in Mining: Recent Lessons from the Philippines". Working Paper No. 1, Foundation for Environmental Security and Sustainability, 2006.

Stern, Tom. "Chinese Investments in the Philippines". *The Journal of Political Risk* 4 (2016). Available at <http://www.jpolrisk.com/chinese-investments-in-the-philippines>.

Stiem, Tylter. "Burma: The War Goes On". *World Policy Blog*, 31 December 2014. Available at <http://www.worldpolicy.org/blog/2014/12/31/burma-war-goes> (accessed 15 September 2015).

Stratfor Global Intelligence. "Indonesia Struggles with an Export Ban". Available at <https://worldview.stratfor.com/article/indonesia-struggles-export-ban> (accessed 26 March 2015).

Strauss, Julia C. and Martha Saavedra, eds. *China and Africa*, vol. 9, *The China Quarterly Special Issues*. Cambridge: Cambridge University Press, 2010.

Sturgeon, Janet C. "Cross-Border Rubber Cultivation between China and Laos: Regionalization by Akha and Tai Rubber Farmers". *Singapore Journal of Tropical Geography* 34, no. 1 (2013): 76.

Sturgeon, Janet C., Nicholas K. Menzies, Yayoi Fujita Lagerqvist, David Thomas, Benchaphun Ekasingh, Louis Lebel, Khamla Phanvilay, and Sithong Thongmanivong. "Enclosing Ethnic Minorities and Forests in the Golden Economic Quadrangle". *Development and Change* 44, no. 1 (2013): 53–79.

Sun, Narin. "Cambodia's Hun Sen Slams U.S. Threats over Aid". *The Wall Street Journal*, 3 August 2013. Available at <http://stream.wsj.com/story/latest-headlines/SS-2-63399/SS-2-293500/>.

Sun, Yingjie. "Sino–Myanmar Oil and Gas Pipelines: An Expensive Lesson for China?". *New York Times*, 11 July 2013. Available at <http://cn.nytimes.com/china/20130711/cc11myanmar/> (accessed 2 March 2015).

Sun, Yun. "China's Tug of War". *The Irrawaddy*, 9 April 2013. Available at <http://www.irrawaddy.org/burma/chinas-tug-of-war.html> (accessed 4 September 2015).

————. "Chinese Investments in Myanmar: What Lies Ahead? Great Powers and the Changing Myanmar". *Stimpson Issue Brief* No. 1 (2013).

Sung, Hsing-Chou. "China's Geoeconomic Strategy: Toward the Riparian States of the Mekong Region". Page unavailable.

Suzuki, Wataru. "Indonesian Coal Producer Gets $1.2B Loan from China". *Nikkei Asian Review*, 28 March 2015. Available at <http://asia.nikkei.com/Business/AC/Indonesian-coal-producer-gets-1.2B-loan-from-China>.

Talk Vietnam. "Vietnamese Told Not to be Too Eager for FDI from China". Available at <http://www.talkvietnam.com/2014/05/vietnamese-told-not-to-be-too-eager-for-fdi-from-china/>.

Tan, Antonio S. *The Chinese in the Philippines, 1898–1935: A Study of Their National Awakening*. Quezon, Philippines: Printed by R.P. Garcia Pub. Co., 1972.

Tan, Danielle. "'Small is Beautiful': Lessons from Laos for the Study of Chinese Overseas". *Journal of Current Chinese Affairs* 41, no. 2 (10 July 2012): 61–94.

Tang-Lee, Diane. "Corporate Social Responsibility (CSR) and Public Engagement for a Chinese State-backed Mining Project in Myanmar: Challenges and Prospects". *Resources Policy* 47 (2016): 28–37.

Taylor, Monique. *The Chinese State, Oil and Energy Security*. Basingstoke: Palgrave Macmillan, 2014.

Thanh Nien Online. "Thousands of Illegal Chinese Labourers in Camau". *Thanh Nien*, 9 August 2011.

Thayer, Carlyle A. "Political Legitimacy of Vietnam's One Party-State: Challenges and Responses". *Journal of Current Southeast Asian Affairs* 28, no. 4 (14 January 2010): 47–70.

Thu Huong Le. "Amid Fish Deaths, Social Media Comes Alive in Vietnam". *The Diplomat*, 19 August 2016. Available at <http://thediplomat.com/2016/05/amid-fish-deaths-social-media-comes-alive-in-vietnam/>.

Tijaja, Julia Puspadewi. "The Proliferation of Global Value Chains: Trade Policy Considerations for Indonesia". *TKN Report*, January 2013. Available at <http://www.iheal.univ-paris3.fr/sites/www.iheal.univ-paris3.fr/files/global_value_chains_indonesia.pdf>.

Titthara, May. "Pheapimex Under Fire Again". *Phnom Penh Post*, 17 January 2012. Available at <http://www.phnompenhpost.com/national/pheapimex-under-fire-again>.

————. "China Reaps Concession Windfalls". *Phnom Penh Post*, 12 April 2012. Available at <http://www.phnompenhpost.com/national/china-reaps-concession-windfalls>.

————. "Fear Accompanies Summons Over Land Disputes". *Phnom Penh Post*, 20 June 2012. Available at <http://www.phnompenhpost.com/national/fear-accompanies-summons-over-land-disputes>.

————. "Kings of Concessions". *Phnom Penh Post*, 25 February 2014. Available at <http://www.phnompenhpost.com/national/kings-concessions>.

Titthara, May and Alice Cuddy. "Areng Valley Dam Activist Summonsed". *Phnom Penh Post*, 27 March 2015. Available at <http://www.phnompenhpost.com/national/areng-valley-dam-activist-summonsed>.

Torode, Greg. "Myanmar Work Will Continue, Vows National Endowment for Democracy". *South China Morning Post*, 1 October 2012. Available at <www.scmp.com/news/asia/article/1050977/myanmar-work-will-continue-vows-national-endowment-democracy> (accessed 9 September 2015).

Transparency International. "Corruption Perception Index 2014: Results", 2014. Available at <https://www.transparency.org/cpi2014/results> (accessed 13 September 2017).

Treanor, Naomi Basik. "China's Hongmu Consumption Boom: Analysis of the Chinese Rosewood Trade and Links to Illegal Activity in Tropical Forested Countries". *Forest Trends*, December 2015. Available at <http://www.forest-trends.org/documents/files/doc_5057.pdf>.

Tu, Jianjun. "Industrial Organization of the Chinese Coal Industry". Stanford University Program on Energy and Sustainable Development, 2011.

Tu, Kevin Jianjun and Sabine Johnson-Reiser. "Understanding China's Rising Coal Imports". Carnegie Endowment for International Peace, 16 February 2012. Available at <http://carnegieendowment.org/files/china_coal.pdf> (accessed 29 April 2015).

Tun, Thiha. "Chinese Investor Assures Transparency If Myanmar Restarts Dam Project". *Radio Free Asia*, 26 December 2013. Available at <http://www.rfa.org/english/news/myanmar/dam-12262013140753.html> (accessed 9 September 2015).

Tung, S. "Formosa Wants to Increase Capital to $28.5 Billion". *VietnamNet Bridge*, 29 May 2013. Available at <http://english.vietnamnet.vn/fms/business/75420/formosa-wants-to-increase-capital-to--28-5-billion.html>.

Tunsjo, Oystein. *Security and Profit in China's Energy Policy: Hedging Against Risk*. New York: Columbia University Press, 2013.

Tuoi Tre News. "Vietnam Metallurgy Association Raises Alarm over Illegal Iron Ore Exports to China", 8 February 2014. Available at <http://tuoitrenews.vn/business/21414/vietnam-metallurgy-association-raises-alarm-over-illegal-ironore-exports-to-china>.

————. "Taiwanese Firm Keeps Demanding More Despite Huge Incentives from Vietnam", 7 September 2014. Available at <http://tuoitrenews.vn/business/20860/taiwan-firm-keeps-asking-for-more-despite-huge-incentivesfrom-vietnam>.

————. "Vietnam Province Pulls Plug on Defense-Sensitive Chinese-Invested Resort Project", 27 November 2014. Available at <http://tuoitrenews.vn/business/24306/vietnam-province-pulls-plug-on-defensesensitive-chineseinvested-resort-project>.

Ueno, Takahiro, Miki Yanagi, and Jane Nakano. "Quantifying Chinese Public Financing for Foreign Coal Power Plants", 2014. Available at <http://www.pp.u-tokyo.ac.jp/research/dp/documents/GraSPP-DP-E-14-003.pdf>.

UK Trade and Investment. *Opportunities for British Companies in Burma's Oil and Gas Sector*, 2015. Available at <https://www.gov.uk/government/publications/opportunities-for-british-companies-in-burmas-oil-and-gas-sector> (accessed 5 September 2015).

Un, Kheang. "State, Society and Democratic Consolidations: The Case of Cambodia". *Pacific Affairs* (2006): 225–45.

United Nations Conference on Trade and Development. *World Investment Report 2015, Annex Tables*. Available at <http://unctad.org/en/Pages/DIAE/World%20Investment%20Report/Annex-Tables.aspx> (accessed 4 April 2015).

Urban, Frauke, Johan Nordensvard, Giuseppina Siciliano, and Bingqin Li. "Chinese Overseas Hydropower Dams and Social Sustainability: The Bui Dam in Ghana and the Kamchay Dam in Cambodia". *Asia & the Pacific Policy Studies* 2, no. 3 (2015): 573–89.

US Geological Survey (USGS). "Appendix C", 13 December 2016. Available at <http://minerals.usgs.gov/minerals/pubs/mcs/2009/mcsapp2009.pdf> (accessed 1 June 2016).

Văn Định. "Tràn Lan Lao Động Trung Quốc Trái Phép". *Tuổi Trẻ Online*, 20 October 2013. Available at <http://tuoitre.vn/chinh-tri-xa-hoi/575548/tran-lan-lao-dong-trung-quoc-traiphep.html#ad-image-0>.

Vandergeest, Peter and Nancy Lee Peluso. "Territorialization and State Power in Thailand". *Theory and Society* 24, no. 3 (1995): 385–426.

Vargas, Anthony. "14 Chinese National Arrested for Illegal Black Sand Mining Activity in Camarines Norte", 7 February 2014.

Vaughan, Martin. "Cambodia's Hun Sen Proves a Feisty Asean Chair". *World Street Journal*, 4 April 2012. Available at <http://blogs.wsj.com/searealtime/2012/04/04/cambodias-hun-sen-proves-a-feisty-asean-chair/>.

Verbrugge, Boris. "The Economic Logic of Persistent Informality: Artisanal and Small-Scale Mining in the Southern Philippines". *Development and Change* 46, no. 5 (2015): 1023–46.

Verbrugge, Boris and Beverly Besmanos. "Undermining the Myths about Small-scale Mining". Bantay Kita Philippines, 2015.

———. "Formalizing Artisanal and Small-scale Mining: Whiter the Workforce?". *Resources Policy* 47, no. 3 (2016): 134–41.

Vietnam Business Forum. "Diversifying Machinery Supply Sources", 23 July 2014. Available at <http://vccinews.com/news_detail.asp?news_id=30784>.

———. "Bauxite Project Still Effective", 21 April 2015. Available at <http://vccinews.com/news_detail.asp?news_id=32009> (accessed 17 May 2015).

Vietnam Economic Forum. "Chinese EPC Contractors Won 90% of the Top Project in Vietnam". *VietnamNet*. Available at <http://community.vef.vn/2010-07-31-trung-quoc-trung-thau-90-cong-trinh-thuong-nguon-cua-viet-nam-?print=1>.

Vietnam Energy. "Nội Địa Hóa Nhà Máy Nhiệt Điện Thấp, Nguyên Nhân Từ Đâu" [Low Local Contents in Thermal Power Projects: What are the Causes?]. Available at <http://nangluongvietnam.vn/news/vn/dien-luc-viet-nam/noi-dia-hoa-nha-may-nhiet-dien-thap-nguyen-nhan-tu-dau.html>.

Vietnam General Statistics Office. "2013 Foreign Direct Investment". Available at <https://gso.gov.vn/default.aspx?tabid=716>.

VietnamNet. "'Phố Trung Quốc' Ở Hà Tĩnh", 8 May 2013. Available at <http://vietnamnet.vn/vn/kinh-te/120201/-pho-trung-quoc-o-ha-tinh.html>.

VietnamNet Bridge. "Thousands of Chinese Workers in Ha Tinh Lack Permits", 10 October 2014. Available at <http://english.vietnamnet.vn/fms/society/113797/thousandsof-chinese-workers-in-ha-tinh-lack-permits.html>.

VnExpress. "Hàng Loạt Dự Án Điện Của Nhà Thầu Trung Quốc Chậm Tiến Độ" [Series of Energy Projects by Chinese Contractors Behind the Schedule]. Available at <http://kinhdoanh.vnexpress.net/tin-tuc/vi-mo/hang-loat-du-an-diencua-nha-thau-trung-quoc-cham-tien-do-2716006.html>.

Võ Nguyên Giáp. "Letter to Prime Minister, Hon. Nguyễn Tấn Dũng", 5 January 2009. Available at <http://www.viet-studies.info/kinhte/Thu_VNGiap_NTDung.pdf>.

———. "Đại Tướng Võ Nguyên Giáp Góp Ý về Dự Án Bô Xít Tây Nguyên". *Tuan Vietnam*, 14 January 2009. Available at <http://community.tuanvietnam.net/2009-01-14-dai-tuong-vo-nguyen-giap-gopy-ve-du-an-bo-xit-tay-nguyen?print=1>.

Vo Thi Thanh and Nguyen Anh Duong. "Revisiting Exports and Foreign Direct Investment in Vietnam". *Asian Economic Policy Review* 6 (2011): 112–31.

Vong, Sokheng and Sebastian Strangio. "Villagers Blockade Kampot Dam Quarry Site Over Airborne Rocks". *Phnom Penh Post*, 11 March 2009. Available at <http://www.phnompenhpost.com/national/villagers-blockade-kampot-dam-quarry-site-over-airborne-rocks>.

Vrieze, Paul. "Civil Society and MPs Draft 'Progressive' Association Registration Law". *The Irrawaddy*, 21 October 2013. Available at <http://www.irrawaddy.org/ burma/csos-mps-draft-progressive-association-registration-law.html> (accessed 5 September 2015).

VTC News. "Cuộc Chơi Sòng Phẳng, Sao Kinh Tế Việt Nam Phải 'Thoát' Trung" [Fair Competition, Why Vietnam Has to "Escape" China?]. Available at <http://vtc.vn/cuoc-choisong-phang-sao-kinh-te-viet-nam-phai-thoat-trung. 1.493780.htm>.

Walz, Julie and Vijaya Ramachandran. "Brave New World: A Literature Review of Emerging Donors and the Changing Nature of Foreign Assistance". *Center for Global Development Working Paper* 273 (2011).

Wang, Chengjin and César Ducruet. "Transport Corridors and Regional Balance in China: The Case of Coal Trade and Logistics". *Journal of Transport Geography*, Changing Landscapes of Transport and Logistics in China, 40 (October 2014): 3–16.

Wang, Mark Yaolin. "The Motivations Behind China's Government-Initiated Industrial Investments Overseas". *Pacific Affairs* 75, no. 2 (2002): 187–206.

Warburton, Eve. "In Whose Interest? Debating Resource Nationalism in Indonesia". *Kyoto Review of Southeast Asia* 15 (March 2014).

Ward, Halina. "Resource Nationalism and Sustainable Development: A Primer and Key Issues". London: IIED, May 2009. Available at <http://pubs.iied. org/G02507/>.

Wells-Dang, Andrew. "The Political Influence of Civil Society in Vietnam". In *Politics in Contemporary Vietnam: Party, State, and Authority Relations*. Palgrave Macmillan, 2014, pp. 162–83.

Wibisono, Makarim. "Indonesia and Global Competitiveness". *Jakarta Post*, 10 October 2011.

Wickberg, Edgar. *The Chinese in Philippine Life, 1850–1898*. Ann Arbor, Michigan: University Microfilms, 1965.

Willis. "Mining Market Review", Spring 2012. Available at <https://www.willis. com/naturalresources/pdf/MiningMarketReview2012.pdf>.

Wilson, Jeffrey D. "Northeast Asian Resource Security Strategies and International Resource Politics in Asia". *Asian Studies Review* 38, no. 1 (2014): 15–35.

———. "Resource Powers? Minerals, Energy and the Rise of the BRICS". *Third World Quarterly* 36, no. 2 (1 February 2015): 223–39.

Wolf, Charles, Xiao Wang, and Eric Warner. "China's Foreign Aid and Government-Sponsored Investment Activities: Scale, Content, Destinations, and Implications". Santa Monica, CA: RAND Corporation, 2013. Available at <http://www.rand. org/pubs/research_reports/RR118.html>.

Womack, Brantly. *China and Vietnam: The Politics of Asymmetry*. Cambridge: Cambridge University Press, 2006.

Wong, Pak Nung, Kathleen Aquino, Kristinne Lara-De Leon and Sylvia Yuen Fun So. "As Wind, Thunder and Lightning: Local Resistance to China's Resource-led Diplomacy in the Christian Philippines". *South East Asia Research* 21, no. 2 (2013): 281–302.

Woods, Kevin. "Ceasefire Capitalism: Military–Private Partnerships, Resource Concessions and Military–State Building in the Burma–China Borderlands". *Journal of Peasant Studies* 38, no. 4 (2011): 747–70.

World Bank. "Investment in Flux". *Indonesia Economic Quarterly* (March 2014): 21.

———. *Seizing the Global Opportunity: Investment Climate Assessment and Reform Strategy for Cambodia*. World Bank Group, 2014.

World Coal. "China Coal Imports from Australia and Indonesia Slip in February", 23 March 2015. Available at <http://www.worldcoal.com/coal/23032015/China-coal-Australia-Indonesia-February-2093/> (accessed 19 April 2015).

World Economic Forum, Asian Development Bank and Accenture. *New Energy Architecture: Myanmar*, 2013. Available at <http://www.adb.org/sites/default/files/publication/30265/new-energy-architecture-mya.pdf> (accessed 5 September 2015).

World Wildlife Fund. "Global Coal: The Acceleration of Market Decline", 2015. Available at <http://d2ouvy59p0dg6k.cloudfront.net/downloads/global_coal__the_acceleration_of_market_decline__report.pdf>.

Wright, Tim. *The Political Economy of the Chinese Coal Industry: Black Gold and Blood-Stained Coal*. Routledge, 2012.

Wu Chongbo. "Forging Closer Sino–Indonesia Economic Relations and Policy Suggestions". *Ritsumeikan International Affairs* 10 (2011): 119–42. Available at <http://r-cube.ritsumei.ac.jp/bitstream/10367/3402/1/asia10_wu.pdf>.

Xinhua. "Cambodia Opens China-funded Bridge for Traffic", 24 January 2011. Available at <http://www.chinadaily.com.cn/china/2011-01/24/content_11907394.htm>.

———. "Chinese Investment in Cambodia up in 2013", 18 January 2014. Available at <http://www.globaltimes.cn/content/838148.shtml>.

———. "China Remains Vietnam's Biggest Trade Partner in 2013", 29 January 2014. Available at <http://www.chinadaily.com.cn/business/chinadata/2014-01/29/content_17264283.htm>.

———. "China Provides Military Trucks and Uniforms to Cambodia". *People Daily*, 7 February 2014. Available at <http://english.peopledaily.com.cn/90786/8528898.html>.

———. "ASEAN–China Trade Expected to Reach 500 bln USD by 2015", 14 November 2014. Available at <http://news.xinhuanet.com/english/china/2014-11/14/c_133788265.htm> (accessed 17 March 2015).

———. "Chinese-built 338 MW Hydropower Dam in Cambodia Begins Operation". *Xinhuanet*, 12 January 2015. Available at <http://news.xinhuanet.com/english/china/2015-01/12/c_133913369.htm>.

Xu, Jianchu. "The Political, Social, and Ecological Transformation of a Landscape: The Case of Rubber in Xishuangbanna, China". *Mountain Research and Development* 26, no. 3 (2006): 254–62.

Xu, Jianchu, Jefferson Fox, John B. Vogler, Zhang Peifang, Fu Yongshou, Yang Lixin, Qian Jie, and Stephen Leisz. "Land-Use and Land-Cover Change and Farmer Vulnerability in Xishuangbanna Prefecture in Southwestern China". *Environmental Management* 36, no. 3 (2005): 404–13.

Yeophantong, Pichamon. "China, Corporate Responsibility and the Contentious Politics of Hydropower Development: Transnational Activism in the Mekong Region". The Global Economic Governance Programme, University of Oxford, 2013.

———. "Cambodia's Environment: Good News in Areng Valley?". *The Diplomat*, 3 November 2014. Available at <http://thediplomat.com/2014/11/cambodias-environment-good-news-in-areng-valley/>.

Yergin, Daniel. "Ensuring Energy Security". *Foreign Affairs* (1 March 2006): 69–77.

Yiu, Daphne W., Chungming Lau and Gary D. Bruton. "International Venturing by Emerging Economy Firms: The Effects of Firm Capabilities, Home Country Networks, and Corporate Entrepreneurship". *Journal of International Business Studies* 38 (2007): 519–40.

Yu, Jincui. "Wooing Old Customers Anew". *Global Times*, 27 November 2012. Available at <http://www.globaltimes.cn/content/746784.shtml> (accessed 2 April 2013).

Zhang, Boting. "Exploring the Issues Related to the Suspension of Myitsone Dam Construction". Sciencenet.cn, 25 April 2012. Available at <http://blog.sciencenet.cn/blog-295826-563583.html> (accessed 6 September 2015). [Text in Chinese]

Zhang, Jin. "China's Oil Industry, International Investment and Developing Countries". In *Handbook on China and Developing Countries*, edited by Carla P. Freeman. Cheltenham, UK: Edward Elgar Publishing, 2015, pp. 287–317.

Zhang, Le. "*Zhuanjia cheng miandian jiaoting misong dianzhan yin dangdi jushi buwen*" ["Expert Says the Suspension of Myitsone Dam Construction is due to Local Instabilities"]. *The Beijing News*, 11 October 2011. Available at <http://dailynews.sina.com/bg/chn/chnpolitics/sinacn/20111011/14162832525.html> (accessed 6 September 2015).

Zhang Qi. "China Collecting New Sources Overseas to Provide Iron Ore". *China Daily*, 7 September 2011. Available at <http://shandong.chinadaily.com.cn/e/2011-09/07/content_13636834.htm> (accessed 1 June 2016).

Zhang, Xiaoxi and Kevin Daly. "The Determinants of China's Outward Investment". *Emerging Markets Review* 12 (2011): 389–98.

Zhao Hong. "China's FDI into Southeast Asia". *ISEAS Perspective* 2013 #08. Singapore: Institute of Southeast Asian Studies, 31 January 2013.

———. "The China–Myanmar Energy Pipelines: Risks and Benefits". *ISEAS Perspective* 2013 #30. Singapore: Institute of Southeast Asian Studies, 15 May 2013.

————. "China–Indonesia Economic Relations: Challenges and Prospects". *ISEAS Perspective* 2013 #42. Singapore: Institute of Southeast Asian Studies, 4 July 2013.

————. *China's Quest for Energy in Southeast Asia: Impact and Implications.* Trends in Southeast Asia 2015 #01. Singapore: Institute of Southeast Asian Studies, 2015.

Zhao Ping. "*Shiyou jingkou zhanglue da tishu*" [Speeding Up Oil Strategy]. *Chinese Foreign Investment*, no. 8 (2005).

Zhao, Suisheng. "The China Model: An Authoritarian State-Led Modernization". In *Handbook on China and Developing Countries*, edited by Carla P. Freeman. Cheltenham, UK: Edward Elgar Publishing, 2015, pp. 21–50.

Zhou Yan. "Indonesia Seeks More Chinese Investment". *China Daily*, 3 May 2011.

Zhu, Charlie, David Lague and Fergus Jensen. "Depleted Oil Field is Window into China's Corruption Crackdown". *Reuters*, 19 December 2014. Available at <http://www.reuters.com/article/2014/12/19/uschina-corruption-indonesia-specialrep-idUSKBN0JX00720141219> (accessed 14 January 2015).

Zin, Min. "Burmese Attitude Toward Chinese: Portrayal of the Chinese in Contemporary Cultural and Media Works". *Journal of Current Southeast Asian Affairs* 31, no. 1 (2012): 115–31.

Zsombor, Peter and Narim Khoun. "Hydro Dam Does Little for Locals, Study Finds". *The Cambodian Daily*, 26 August 2015. Available at <https://www.cambodiadaily.com/news/hydro-dam-does-little-for-locals-studyfinds-92523>.

INDEX

Note: Page numbers followed by "n" denote endnotes.

A

air pollution crisis, 80, 88
All Arakan Students and Youth
 Congress (AASYC), 207
Aluminium Corporation of China
 (Chinalco), 46–48
aluminum ore imports, 69, 70
Alyansa Laban sa Mina (ALAMIN),
 266
American Devon Energy, 110
American domination, 185
Ancestral Land/Domain Watch
 (ALDAW), 267, 268
anti-Chinese sentiments
 in Myanmar, 209–12
 in Philippines, 147, 258, 267
 in resource sector projects,
  235–39
 in Vietnam, 231, 243–48
 violent expression of, 243–48
Aquino, Benigno, 134, 260
Aquino, Corazon, 132, 134
Arakan Oil Watch (AOW), 207
Arroyo, Gloria Macapagal, 132–34
artisanal small-scale mining
 (ASM), 130, 131, 149n7

Chinese, 141, 142–43, 147, 148
 gold mining operations, 140,
  141
 in Philippines, 137, 138
 workers, 132–33
ASEAN–China Framework
 Agreement on Comprehensive
 Economic Cooperation, 232
Asian Development Bank (ADB),
 75
 Greater Mekong Subregion
  Economic Cooperation
  Program, 11
Asian Financial Crisis, 9
Asian Infrastructure Investment
 Bank (AIIB), 1, 67, 68, 74, 76,
 91
"Asia-Pacific Crisis Arch", 72
Asia-Pacific Economic Cooperation
 (APEC), 78n18, 96
 criticism of, 75
 energy mechanism, 75
Asia-Pacific Partnership on Clean
 Development and Climate, 75
Asia World Group, 211
Aung San Suu Kyi, 222

B
Baleh Hydropower Project, 45
Bandung Conference, 67
Ba Ria-Vung Tau province
    (Vietnam), 247
bauxite mining. *See also* Central
    Highlands bauxite mining
    controversy
    Chinese involvement in, 20
    projects, 236, 237
Bauxite Vietnam website, 239
Beijing, 2, 5, 165, 247
    development of western regions,
        158
    international recognition, 156
    principles of non-interference,
        210
biosphere reserves, UNESCO's
    notion of, 275n34
border disputes
    China–India, 67–68
    China–Philippines territorial
        disputes, 21
    China–Vietnam border war, 9,
        243
British Gas Corporation LNG
    project, 110
build-operate-transfer (BOT)
    project, 233
Bureau of Fisheries of Aquatic
    Resources (BFAR), 269

C
Cagayan mining operation, 138
Cambodia. *See also* Sino–Cambodian
    relations
    Chinese investment in, 18–19, 183
    dominant foreign players in, 183
    government, 189
    land and resource governance,
        184
    legal system, 190
    natural resources sector, 190
    sovereignty-in-practice, 185
Cambodian Center for Human
    Rights (CCHR), 191

Cambodian People's Party (CPP),
    186
    political-economic patronage, 189
    political support for, 189
Catholic Bishops Conference of
    the Philippines (CBCP), 268
    National Secretariat for Social
        Action, 269
Central Highlands bauxite mining
    controversy, 240–43
    bauxite-alumina production, 241
    Consultancy on Development,
        240
    social and environmental risks,
        240–41
    Vo Nguyen Giap, 241–42
Cheay Areng dam project, 194, 196
China–ASEAN Investment
    Cooperation Fund, 11
China Banking Regulatory
    Commission, 199
China Development Bank, 44, 47, 91
China Exim Bank, 44, 47, 91
China Foundation for Poverty
    Alleviation (CFPA), 220, 222
China Huaneng Group, 96
China–Indonesia coal relationship,
    15–16
    coastal imports, 84
    domestic market obligation, 89
    duopolistic entanglement, era of,
        82–87
    duopolistic relationship, 88–90
    economy, 82–83
    electricity industry, 85
    export-oriented coal industry,
        86
    foreign investment goals, 92
    global resource governance, 80
    import–export, 80, 90
    investment, 90–96
    labour-intensive coal mining, 81
    mining industry, 82, 83
    Mining Law (2009), 89
    National Energy Policy, 90
    overview, 80–82

production, import, and export
    (2000–12), 86
steam and metallurgical coal
    market, 87
transport system, 85
"two markets, two resources"
    policy, 85
vagaries in, 97
China–Indonesia energy sector,
    105, 109–14, 119
Chinalco. *See* Aluminium
    Corporation of China
China National Machinery and
    Equipment Group (CNMEG),
    133–34
China National Offshore Oil
    Corporation (CNOOC), 29, 42,
    43, 110
China National Petroleum
    Corporation (CNPC), 29, 42,
    43, 110
China Petrochemical Corporation
    (Sinopec), 29, 42
China Pipeline Bureau, 74
China Power Investment
    Corporation (CPI), 216–17
China–Southeast Asia relationship,
    80
China Southern Power Grid
    (CSG), 197
China's peripheries and resource
    imports
    aluminum ore imports, 69, 70
    "Asia-Pacific Crisis Arch", 72
    border disputes, 67–69
    challenges for, 63–64
    China's consumption and
        demand, 59–62
    collaborative development, 73–76
    "first island chain" strategy, 72
    India, 66
    Indonesia, 67
    influence of powers, 72–73
    Kazakhstan, 65, 66
    Mongolia, 66–67
    nickel imports, 69, 70

reserves and reserve base,
    notion of, 65
Russia, 65–66
tin imports, 71
unbalanced development in
    peripheral regions, 64–67
youhao countries, 65
Chinese Academy of Engineering,
    58
Chinese Academy of International
    Trade and Economic
    Cooperation (CAITEC), 217
Chinese Academy of Social
    Sciences, 216
Chinese interlocutors, 205, 212,
    215, 218, 220
Chinese investments in Southeast
    Asian resources, 28
    corporate motivations, 37–38
    Dunning's framework, 32–34
    *Foreign Investment Industrial
        Guidance Catalogue* (2004),
        41
    "Going Out" policy, 35, 38
    host government motivations, 39
    hydropower sector, 44–46
    internationalization of Chinese
        companies, 35–36
    mining sector, 46–49
    motivations of Chinese
        government, 38
    nature of engagement, 39–40
    oil and gas sector, 42–44
    overseas investment, objectives
        for, 32, 33
    overview of, 29–32
    Southeast Asia's energy
        demand, 30, 31
Chinese laws and regulations, 198
Chinese mestizos, 144, 145, 147,
    148
Chinese mining companies, 46–49
    opposition to, 266–71
    in Philippines, 135, 151n21
Chinese Ministry of Foreign
    Affairs, 221

"Chinese model", 4
Chinese National Offshore Oil
    Corporation (CNOOC), 247
Chinese Sinohydro Corporation,
    193
Chinese State Farms Corporation,
    192, 194, 195
Chinese state-owned enterprises
    (SOEs), 206–7, 215
civil society organizations (CSOs),
    212, 214–15, 220, 223n1
coal production
    in China, 86
    in Indonesia, 109
Cojuancos, 145
Comprehensive Strategic
    Partnership Agreement (2010),
    183
concession agreements, rubber
    sector, 171, 174
Conference on Interaction and
    Confidence-Building Measures
    in Asia (CICA) Summit, 74
Consultancy on Development
    (CODE), 240
contract farming, 171, 174
corporate social responsibility
    (CSR) projects, 221, 223,
    228n65
Council on Foreign Relations,
    99n7, 99n15
crisis arch. *See* "Asia-Pacific Crisis
    Arch"
cross border trade, 166, 208
Cua Khem Cape, 238

D
Dak Nong province, bauxite
    mining projects in, 236, 237
"Declaration on Asia-Pacific
    Partnership on Clean
    Development and Climate,
    The", 75
Department of Environment and
    Natural Resources (DENR),
    257, 269

Development Assistance
    Committee (OECD), 158
development cooperation, 5–6,
    176, 177
    Chinese, 158–61
    land acquisitions as, 160–61
    through rubber plantations in
        Northern Laos, 167–75
domestic market obligation (DMO)
    policy, 89, 109
domestic oil production, 89
Dunning's framework, of overseas
    investment
    "eclectic paradigm", 34, 51n15
    objectives for, 32, 33
    Ownership-Location-
        Internalization framework,
        34, 35

E
EarthRights International (ERI),
    207, 218
Economically Progressive
    Ecosystem Development
    (EcoDev), 207, 221, 226n42
economic land concessions, 190,
    193–95
effective sovereignty, concept of,
    184
88 Generation, 216, 226n43
electricity sector, 85, 89
Embassy in Myanmar, 222
energy demand, in Indonesia, 89
Energy Information
    Administration, 99n5
Engineering, Procurement and
    Construction (EPC) contracts,
    11, 234, 237
Environmental and Social Impact
    Assessments (ESIA), 209, 215,
    222
Environmental Impact Assessment
    (EIA), 193
Environmental Impact Assessment
    Review Committee (EIARC),
    273n12

Environmental Impact Statement
  (EIS), 273n12
Executive Order No. 79
  (Philippines), 262–63
Exim Bank of China, 91, 192
external dependence, 59, 61

F
Filipino–China relations, 131
Filipino elites, 145
Financial Technical Assistance
  Agreement (FTAA), 132, 134
financing infrastructure projects,
  187
"first island chain" strategy, 72
Five Principles of Peaceful
  Coexistence, 185
foreign direct investment (FDI)
  China, 12, 13, 112, 158
  Dunning's model for, 32–34
  Indonesia, 105, 113
  in Indonesian and Philippine
    mining, 136
  stock in ASEAN states (2015),
    137
*Foreign Investment Industrial
  Guidance Catalogue* (2004), 41
Formosa Ha Tinh steel factory,
  243–48
  foreign investors, 245
  local economy and jobs, 244–45
  oil rig crisis, 247–48, 254n55
  riots in, 243, 247–48
Formosa Steel Complex, 237
Fukkien migrants, 144

G
General Department of Geology
  and Minerals of Vietnam,
  235
geopolitics, 72, 229–55
Global Environmental Institute,
  220–21
global financial crisis (2008), 15,
  20, 82, 105
global resource governance, 80

Global South, 219
  Chinese investments in, 131
"Going Global/Going Out" policy,
  35, 38, 91, 183, 184
"Golden Triangle" countries, 165
government-organized NGOs
  (GONGOs), 217, 222
government-to-government
  relations, 190
Greenpeace International, 99n11

H
Habibie, Bacharuddin Jusuf, 117
Hagyi dam, 45
Hai Phong Thermal Power Plant, 237
Haiyang Shiyou 981, 247
Hakka migrants, 144
Ho Chi Minh City, 247
Hu Jintao, 119
"Human Security and Land Rights
  in Cambodia", 195
Hun Sen, 187, 195
hydropower sector, 44–46

I
"implicit security guarantee", 188
import concentration, 61
Indian Ocean lane, China's
  dependence on, 72
Indonesia
  bilateral energy cooperation, 110
  British Gas Corporation LNG
    project, 110
  coal economy, 88
  coal-fired power plants in, 80,
    92, 96
  coal production and
    consumption, 109
  distillation products exports in,
    114
  economic growth, 105, 106
  energy exploration and
    development, 110
  energy sectors, 105–9
  export market, 106, 107, 112,
    114, 120, 121

export-oriented coal industry,
    86
foreign direct investment, 105,
    113
gross domestic product, 105–7
Huadian subsidiary in, 94
largest export markets, 109
long-term strategic energy
    investment, 105
metallurgical coal, 87
mineral resource nationalism,
    106, 115–19
national economic security, 121
National Energy Policy, 90
oil and gas production in,
    109–11
Presidential Elections (2009),
    117
production sharing contract,
    110
resource and mineral industries,
    105
state revenue, 106, 108
steam coal, 87
Indonesia–China relations, 67, 69
Indonesia Medium Term
    Development Plan (2015–19),
    123
Indonesian government, 89, 96, 97,
    116, 118
Indonesia's crude oil price (ICP),
    106
informal mining, 149–50n7
International Energy Agency
    (IEA), 98n1, 99n4, 101n50, 109,
    126n10
internationalization of Chinese
    companies, 35–36
International Law of the Sea, 247
International Non-Governmental
    Organization (INGO)
    networks, 218–19, 222
International Organizations (IOs),
    218
Intex Resources Corporation (IRC),
    266

J
Jiang Zemin, 158
Jinghong Tropical Crops Research
    Institute, 162
Jinrun Rubber Company, 166
Jokowi (Joko Widodo), 92, 94

K
Kachin Independence Army (KIA),
    210
Kachin Independence Organization
    (KIO), 210
Kalimantan
    coal projects in, 96
    export-oriented developments
        in, 97
Kamchay dam project, 193, 196
Kampong Chhnang province
    (Cambodia), 192
Kampot province (Cambodia), 193
Kazakhstan–China relations, 65, 66
Koh Kong province (Cambodia),
    Cambodia 192, 196
Ky Anh, Ha Tinh province
    (Vietnam), 243, 244

L
laissez-faire approaches, 158
land acquisitions, 160
land concessions, 190, 195
land grabbing, 191–93, 198
Lao Meng Khin, 187, 192
Laos–China relations, 178
    development cooperation, 166
    economic cooperation, 157
    labour supply, 172–73
    Lao opium cultivation, 168
    resource imports, 61
    rubber investments, 18, 160, 176
    tax holidays for rubber
        concessions, 170
    "turning land into capital",
        160–61
large-scale mining (LSM), 130,
    132, 137
    investments in, 146

locations of, 139
multinational, 141, 147
operations, 142
latex processing activities, 175
Law on Concessions, 190
Law Relating to Formation
     of Associations and
     Organizations (Law No. 6/88)
     (Myanmar), 213
Law Relating to Peaceful
     Assembly and Peaceful
     Procession (Myanmar), 213
Law Relating to the Registration
     of Organizations (Myanmar),
     213–14
Law, Stephen, 211
Letpadaung copper mining
     project, 2, 19, 205–7, 209, 222
liquefied natural gas (LNG), 106, 109
Lo Hsing Han, 211
Lopezes, 145
Luang Namtha (Laosso), 168, 170, 175
Ly Yong Phat, 187

M
"Malacca Dilemma", 208
Malacca Straits, 15
market economy, China's
     transition to, 9, 159
metallurgical coal, 87, 125n2
Metallurgical Corporation of
     China (MCC), 48–49
military power/support, 188
mine-mouth plants, 96
Mineral Production Sharing
     Agreement (MPSA), 132, 134
mineral resources in China, 14–15
Mines and Geosciences Bureau
     (MGB) (Philippines), 257, 260
Mining Act (1995) (Philippines),
     261, 262
Mining Law 2009 (Indonesia), 89,
     115, 117
Ministry of Industry, Mines and
     Energy (MIME) (Cambodia),
     197

Ministry of Labour, Invalids
     and Social Affairs (MOLISA)
     (Vietnam), 254n48
Mondulkiri province (Cambodia),
     192
Mongolia–China relations, 66–67
multinational mining investments,
     130, 142
Myanmar Centre for Responsible
     Business (MCRB), 226n39
Myanmar–China Pipeline Watch
     Committee (MCPWC), 207,
     214
Myanmar–China relations, 221–23
     anti-Chinese sentiment, 209–12
     Chinese investments, 12–13,
          19–20
     dismissive attitudes towards
          civil society, 214–15
     economic interests, 208
     energy and resource investment,
          206–9
     geopolitical importance, 208
     hydropower projects, 44–46
     lack of legitimacy, 215–18
     land grabs, 209–10
     legal status due to state
          repression, 212–14
     mining projects, 48
     mutual mistrust and suspicion,
          212
     oil and gas pipelines, 19, 43–44,
          206–9, 214, 222
     protests, 205
     radical approach, 218–21
     tin imports, 71
Myanmar civil society, 19, 212,
     224n21, 228n64
     activism, 218–19, 222
     Chinese interlocutors and, 220,
          221
     dismissive attitudes towards,
          214–15
     lack of legal status, 212–13
     registration of organizations,
          213–14

repressive law, 213
state-backed projects, 220
Myanmar Development Resource
Institute (MDRI), 221
Myanmar Extractives Industry
Transparency Initiative
(MEITI), 212
Myanmar Oil & Gas Enterprise
(MOGE), 210–11
Myitsone dam, 2, 19, 45, 46, 222
anti-Myitsone dam
organizations, 217
Chinese-backed, 2, 205, 206
construction, 210
impacts of, 209, 216
opposition to, 208
suspension of, 216, 220

N
Nam Dinh province (Vietnam),
239
National Commission on
Indigenous Peoples (NCIP)
(Philippines), 268
National Development and Reform
Commission (China), 85
National League of Democracy
(NLD) (Myanmar), 222
national oil companies (NOCs),
105, 110
in China, 29, 42–43
in Southeast Asia, 43
natural resource extraction
projects
in Cambodia, 190
CCHR land reform project, 191
Chinese investments in, 197–98
human rights violations, 194–97
investment projects in, 193
land grabbing, 191–93
popular protest, 194–97
regulatory capture, 190–91
social and environmental
impacts, 193–94
natural resources, global
investments in, 8

New Development Bank (China),
91
New Silk Road Economic Zone, 73
Nhan Co bauxite-alumina project,
237
nickel imports, peripheral sources
of, 69, 70
Nicua Mining Corporation, 268
non-governmental organization
(NGO), 212, 215, 248
activists, 215–16
government-organized, 217, 222
registration, 213
northern Laos, 156, 169, 176
Opium Replacement Program in,
166–67
rubber plantations in, 167–75
North–South development
cooperation, 158

O
OECD's Development Assistance
Committee, 158
oil and gas sector, 42–44
oil rig crisis, 247–48, 254n55
One Belt One Road (OBOR)
initiative, 1, 11, 63, 65, 76
Opium Replacement Program
(ORP), 17, 155, 156, 161, 164,
176
aims, 165
companies, 169, 172, 173–75
implementation, 165, 171
latex processing activities, 175
in northern Laos, 166–67
policy, Chinese exceptionalism
in, 169
rubber projects, 168
Yunnan Province, 165, 166
outward direct investment (ODI),
China, 35
overseas investment
Dunning's framework, 32–34
objectives for, 32, 33
Ownership-Location-Internalization
(OLI) framework, 34, 35

P

Pacific Strategies and Assessments
    (PSA), report of 2011, 263
Pak Lay Dam, 45
Papua New Guinea–China
    relations, 48–49
Paung Ku, 207
Pearl River Delta, 83–84
People's Liberation Army (China),
    163
People's Mining Act (1991)
    (Philippines), 132, 260, 262
People's Republic of China (PRC),
    130, 147, 166
Perusahaan Listrik Negara (PLN),
    92
PetroChina, 42
    acquisition of SPC, 43
PetroChina Daqing Oilfield
    Company, 38
Pew Research Global Attitudes
    Survey, 122
Pheapimex-Fuchan Ltd., 196
Philippine Board of Investments,
    131
Philippine Investment Promotion
    Agencies, 131
Philippine Mining Act (1995), 132
Philippine mining, Chinese
    investments in, 130–37
    Chinese diaspora networks and
        political relations, 143–47
    differ from multinational
        mining, 148
    economy and mining sector,
        132–37
    foreign mining companies, 132–34
    investment co-ownership, 146
    migration from China to
        Philippines, 144–45
    minerals, smuggling of, 143
    mining companies, 135, 151n21
    multinational companies, 146
    overseas developmental
        assistance, 133
    small-scale mining, 137–43

territorial disputes in South
    China Sea, 134–35
Philippine mining industry, 257
    anti-Chinese sentiments, 258, 267
    Chinese involvement in, 257–58
    Executive Order No. 79, 262–63
    large-scale productions, 260–61
    laws and regulations, 261–63
    local economic development,
        264–66
    local opposition to Chinese
        mining, 266–71
    Mining Act (1995), 261, 262
    mining conflict, 20–21, 261
    Pacific Strategies and
        Assessments report (2011),
        263
    People's Small-scale Mining Act
        (1991), 260, 262
    positive impacts, 265
    poverty-alleviation, 264–65
    pro-mining groups, 272n3
    Rapu-Rapu Polymetallic Project,
        262, 265
    Regalian Doctrine, 261
    small-scale productions, 260–62
    statistics, 259
Philippines–China territorial
    disputes, 21
Philippine Statistical Authority,
    131
Physicians for Human Rights,
    218
pipeline projects, Myanmar–China,
    43–44
poverty-alleviation, 264–65
Presidential Election, in Indonesia,
    117
production sharing contract (PSC),
    110

Q

Quang Binh province (Vietnam),
    245
Quang Ninh Thermal Power Plant,
    236

R

Ramos, Fidel, 132
Rang Dong industrial park, 239
Rapu-Rapu Polymetallic Project,
    262, 265
Regalian Doctrine, 261
Report on China Mineral
    Resources (2015), 77n4
resource conflict, 7–8
resource nationalism, in Indonesia,
    115–19
resource production, 2, 6–7
resource sector investments, 2–3,
    5, 14
Revenue Watch Institute, 209,
    225n25
Royal Cambodian Armed Force
    (RCAF), 188, 192
rubber industry, 17–18, 161, 170
    in China, 161–62
    contract farming, 171
    cultivation techniques, 162
    cup lump, 173
    drop in prices, 156, 167, 173–75
    investments, 170
    labour, 162–63, 172–73
    plantations, potential for, 172–73
    production, 163–64
    promotion of, 161
    in Xishuangbanna. See
        Xishuangbanna, rubber in
Russia–China relations, 65–66
    border tensions, 69
    promoting regional cooperation,
        74

S

Sambor Hydropower Project, 45
Scarborough Shoal, 238, 258, 269
Shanghai Cooperation Organization
    (SCO), 66
Shwe gas project, 211
Sihanouk, Norodom, 185
Singapore Petroleum Corporation
    (SPC), 43
Sino–Cambodian relations, 184

countering donor leverage,
    188–89
developmental state policy, 187
economic patronage and
    political stability, 186–87
Five Principles of Peaceful
    Coexistence, 185
pipeline projects, 74
security cooperation, 188
sovereign power and social
    control, 187–88
Sinochem, 29, 42
Sino–Indian relations, 66
    boundary disputes, 68
    Indian Ocean lane, China's
        dependence on, 72
    iron ores imports, 68
Sino–Indonesian energy
    cooperation, 105, 118, 122–24
Sinopec. See China Petrochemical
    Corporation
small-scale mining, Chinese
    tendencies for, 17, 137–43
Social Development Mineral Plan,
    142
South China Sea, 2, 147, 238
    territorial disputes in, 134, 136
    triangle on, 245, 246
Southeast Asia
    China's Cold War-era influence
        in, 183
    China's investments in, 3, 9
    China's return to, 8–12
    energy demand, 30, 31
    fossil fuel production and trade,
        30, 31
    geopolitical tensions on, 12
    resource investments in, 12–22
    two-way trade, Chinese, 10
South–South cooperation, 158
Southwest China, injection drug
    use in, 165
sovereignty, 18, 71, 96, 160,
    182–99, 269, 277
State Grid Corporation of China,
    74

State Law and Order Restoration
    Council (SLORC), 213
state-owned enterprises (SOEs)
    categories of, 46
    in China, 29, 35–36
steam coal, 87
strategic minerals
    China's demand for, 59–62
    consumption of, 59, 60
    definition of, 57–58
    external dependence, 59, 61
    import concentration, 61
    mineral import statistics, 61, 62
    from periphery regions. *See*
        China's peripheries and
        resource imports
    scientific–historical viewpoint,
        63
"string of pearls" strategy, 64
Sumatra, coal projects in, 96
Susilo Bambang Yudhoyono, 119
"Sydney Declaration", 75

T
Taipans, 145
Taiwanese steel factory, 239
Tan, Lucio, 267
Tan Rai bauxite-alumina project,
    237
Tasang Dam, 45
Temporary Environmental
    Protection Order (TEPO), 257
10th Five-Year Plan, China's, 183
territoriality, 7, 277
territorial trap, 6, 7
territory, 7, 18, 184, 186, 278
Thazin Development Foundation,
    214
thermal coal, 125n2
transshipment smuggling, 138
Tropical Crops Research Institute,
    162
"turning land into capital"
    strategy, 160
21st Century Maritime Silk Road,
    73

U
Unahin Lagi Natin Ang Diyos-Bito
    Lake Fisherfolks Association
    (UNLAD-BLFA), 268–69
UNESCO's notion of biosphere
    reserves, 275n34
Union of Myanmar Economic
    Holdings Ltd. (UMEHL),
    211
United Nations Office of Drugs
    and Crime (UNODC), 168,
    180n44
United States, 123, 136
    blocked rubber import, 161
    use of coal, 81
Unlawful Associations Act, 213
UN Security Council, 211
US Embassy's small grants
    programme, 217
US Geological Survey (USGS), 57

V
Vietnam, anti-Chinese protest in,
    20, 229, 249
    bauxite mining, 230, 236, 237
    Central Highlands bauxite
        mining controversy, 240–43
    China domination over Vietnam,
        238
    Chinese investment in, 231–34
    customs records, 251n17
    economy, 249
    EPC contractors, 234
    foreign direct investment, 232,
        233
    Formosa Ha Tinh steel factory,
        243–48
    illegal export problem, 235–36
    import of Chinese labour force,
        236–37
    project quality, failure to ensure,
        236
    resource and energy sector
        projects, 235–39
    trade deficit with China, 237

Vietnam Association of Mechanical Industry (VAMI), 237
Vietnam–China relations, 48
border war, 9, 243
Vietnam Coal and Minerals Corporation (Vinacomin), 240, 241
Vietnam Energy Association, 236, 255n25
Vietnam Foundry and Metallurgy Science and Technology Association, 235
Vietnam's Exclusive Economic Zone, 247
Vietnam Steel Association, 245
Vinh Tan 1 power plant, 232
Vo Nguyen Giap, 241–42
Vung Ang Economic Zone, 246

W
Wanbao Mining, 48
Western Development Strategy, 158
World Coal Association, 125
World Shine Hong Kong Co. Ltd., 238
World Trade Organization (WTO), 163, 232
World Wildlife Fund, 99n6, 199

X
Xi Jinping, 76, 92, 122
Xishuangbanna, rubber in, 156–57, 161, 169, 176
economic transformation, 164
labour, 162–63
production, 163–64
research and development efforts, 162

Y
Yarlung Zangbo River, 68
Yeywa dam, 44
youhao countries, 65
Yunnan Academy of Social Sciences, 215, 216
Yunnan Alternative Development Association, 155
Yunnan Natural Rubber, 175
Yunnan Province (China), 177
ORP implementation, 165, 166
rubber in, 161, 162
Yunnan State Farms, 162, 163, 169, 172

Z
Zhang Boting, 216, 226n47
Zhongxing Semiconductor Co. (ZTE) scandal, 134